Single-Polymer Composites

Single-Polymer Composites

Samrat Mukhopadhyay
Bapan Adak

CRC Press
Taylor & Francis Group
Boca Raton London New York

CRC Press is an imprint of the
Taylor & Francis Group, an Informa business

CRC Press
Taylor & Francis Group
6000 Broken Sound Parkway NW, Suite 300
Boca Raton, FL 33487-2742

Printed on acid-free paper

International Standard Book Number-13: 978-1-138-57532-5 (Hardback)

Library of Congress Cataloging-in-Publication Data

Names: Mukhopadhyay, Samrat, author. | Adak, Bapan, author.
Title: Single-polymer composites / Samrat Mukhopadhyay, Bapan Adak.
Description: Boca Raton: CRC Press, 2019. | Includes bibliographical references and index.
Identifiers: LCCN 2018018590 | ISBN 9781138575325 (hardback; alk. paper) | ISBN 9781351272247 (ebook)
Subjects: LCSH: Polymeric composites.
Classification: LCC TA418.9.C6 M839 2019 | DDC 620.1/92—dc23
LC record available at https://lccn.loc.gov/2018018590

Visit the Taylor & Francis Web site at
http://www.taylorandfrancis.com

and the CRC Press Web site at
http://www.crcpress.com

Samrat Mukhopadhyay dedicates this book to his childhood teacher Dr. Ashish Chattopadhyay, who has been his perennial inspiration in the quest to appreciate science

and

Ms. Haimanti Mukhopadhyay for her sustained motivation towards meaningful endeavors.

Bapan Adak wants to dedicate this book to his parents for their endless love, support, and encouragement.

Contents

Preface xiii
Authors xv

1. Single-Polymer Composites: General Considerations 1
1.1 Introduction 1
1.2 Initial Research 2
1.3 General Considerations for Single-Polymer Composites 3
1.3.1 Elevation of Melting Point 3
1.3.2 Initial Morphology 6
1.3.3 Structural Changes with Temperature 6
1.3.4 Thermal Mismatch 7
1.3.5 Compaction Pressure 9
1.3.6 Transcrystallinity 10
1.3.7 Crystallization Behavior and Cooling History 11
1.4 Conclusion 12
References 13

2. Transcrystallinity in Single-Polymer Composites 19
2.1 Introduction 19
2.2 Causes of Transcrystallinity 19
2.3 Importance of Fiber Introduction Temperature on Transcrystallinity 21
2.4 Transcrystalline Growth as a Function of Initial Temperature and Degree of Undercooling 22
2.5 Effect of Surface Change on Transcrystallinity 24
2.6 Matrix Morphology 27
2.6.1 Effect of Transcrystallinity 28
2.7 Conclusion 29
References 29

3. Single-Polymer Composites from Polyolefins 33
3.1 Introduction 33
3.2 Single-Polymer Composites with Varying Starting Materials Based on PE 34
3.2.1 TENFOR 34
3.2.1.1 Gel-Spun Fibers 35
3.2.2 Oriented Fibers and Tapes 36
3.2.3 Use of a Combination of Different Grades of Polyethylene 38
3.2.3.1 HDPE and LDPE 39

3.2.3.2 UHMWPE/HDPE 39
3.2.3.3 UHMPE/LDPE 40
3.3 Single-Polymer Composites from Polypropylene 40
3.3.1 Different Approaches to Single-Polymer Composites from Polypropylene 40
3.3.1.1 Microcellular Injection Molding 40
3.3.1.2 Undercooling Melt Film Stacking Method 44
3.3.1.3 Hot Compaction of Woven Materials 46
3.3.1.4 Film-Stacking Method 49
3.3.2 Importance of Starting Material 50
3.3.2.1 α- and β- Polymorphs of Isotactic PP Homopolymer and Random Copolymer 50
3.3.2.2 PP Yarns and Materials α and β Crystal Forms of Isotactic PP Homopolymer 50
3.3.2.3 PP Tape 51
3.3.3 Advances in Testing Methods 53
3.4 Conclusion 54
References 58

4. Single-Polymer Composites from Polyamides 63
4.1 Introduction 63
4.2 Single-Polymer Composites from Nylons Based on Routes of Manufacturing 66
4.2.1 Resin Transfer Molding 66
4.2.2 Film-Stacking Technique 67
4.2.3 Film-Casting Technique 69
4.2.4 Microencapsulation 70
4.2.5 In situ Polymerization 72
4.3 Comparisons and Concluding Remarks 74
References 77

5. Single-Polymer Composites from Polyesters 81
5.1 Introduction 81
5.2 Single-Polymer Composites from Different Starting Materials 82
5.2.1 Fibers 82
5.2.2 Tapes 85
5.2.3 Double-Covered Uncommingled Yarn 86
5.2.4 Bicomponent Multifilament Yarns 90
5.2.5 Woven Sheets 91
5.3 Comparative Study and Conclusions 93
References 94

6. PLA-Based Single-Polymer Composites 97
6.1 PLA Self-Reinforced Composites Based on Composite Manufacturing 97

6.1.1 Importance of Temperature 103
6.1.2 Importance of Time 105
6.2 Application of PLA Self-Reinforced Composites 106
6.3 Comparative Analyses and Concluding Remarks 108
References 112

7. All-Cellulose Composites: Concepts, Raw Materials, Synthesis, Phase Characterization, and Structure Analysis 115
7.1 Introduction 115
7.2 Cellulose: Chemistry and Overview 116
7.2.1 Solid-State Structures of Native Cellulose 116
7.2.2 Polymorphism of Cellulose 117
7.2.3 Physical and Chemical Properties of Cellulose 118
7.3 Sources of Cellulose 119
7.4 Pros and Cons of Cellulosic Materials for Making Bio-composites 120
7.5 Basic Concepts of All-Cellulose Composites 122
7.6 Classification of All-Cellulose Composites 123
7.6.1 ACCs Based on Type of Matrix Phase 123
7.6.2 ACCs Based on Type of Reinforcement 124
7.6.3 ACCs Based on Alignment of Reinforcements 124
7.7 Different Forms of Cellulosic Materials for the Preparation of ACCs 124
7.8 Manufacturing of Non-Derivatized All-Cellulose Composites 126
7.8.1 Cellulose Dissolution 126
7.8.1.1 Cellulose–Solvent Systems for Manufacturing Non-Derivatized ACCs 126
7.8.1.2 Mechanisms of Cellulose Dissolution 129
7.8.2 Cellulose Regeneration 132
7.8.3 Drying 133
7.9 Synthesis of ACCs and Different Processing Routes 133
7.9.1 Impregnation Technique 133
7.9.2 Partial Dissolution Technique 135
7.9.3 Other Approaches 137
7.9.3.1 Derivatized ACCs 138
7.9.3.2 Non-Derivatized and Non-Solvent Approach 138
7.10 Phase Characterization of Cellulose in All-Cellulose Composites 139
7.10.1 Wide-Angle X-ray Diffraction Analysis 140
7.10.2 CP/MAS^{13}C NMR Spectra Analysis 143
7.10.3 FTIR Spectra Analysis 145
7.10.4 Raman Spectra 147
7.11 Microstructural Analysis of Different ACCs 148

7.12 Conclusion ... 151
References ... 151

8. Properties of Non-Derivatized All-Cellulose Composites ... 159
8.1 Introduction ... 159
8.2 Mechanical Properties ... 160
8.2.1 Factors Affecting Mechanical Properties of ACCs ... 160
8.2.2 Mechanics of ACCs ... 160
8.2.3 Rule of Mixtures: General and Modified Equations ... 161
8.2.4 Tensile Properties of Unidirectional ACCs ... 161
8.2.5 Tensile Properties of Isotropic ACCs ... 167
8.2.6 Flexural Properties of ACCs ... 176
8.2.7 Impact Properties of ACCs ... 178
8.2.8 Peel Strength of ACC Laminates ... 179
8.2.9 Fracture Behavior of ACCs ... 179
8.3 Viscoelastic and Thermal Properties of ACCs ... 184
8.3.1 Dynamic Mechanical Analysis ... 184
8.3.2 Thermogravimetric Analysis ... 187
8.3.3 Thermal Expansion Coefficient ... 188
8.4 Optical Transparency of ACCs ... 189
8.5 Other Miscellaneous Properties of ACCs ... 192
8.5.1 Density ... 192
8.5.2 Thickness ... 193
8.5.3 Fluid Permeability and Barrier Property ... 193
8.5.4 Hydrophobicity ... 195
8.5.5 Swelling and Re-swelling of ACC-gel ... 196
8.5.6 Drug-Release Property ... 197
8.6 Biodegradability of ACCs ... 197
8.7 Conclusion ... 199
References ... 200

9. Derivatized All-Cellulose Composites ... 207
9.1 Introduction ... 207
9.2 Derivatizing Solvents ... 208
9.3 Philosophy of Making DACCs ... 208
9.4 Different Types of DACC ... 209
9.5 Benzylated Cellulose-Based DACCs ... 209
9.5.1 Synthesis ... 209
9.5.2 Structure and Properties ... 210
9.6 Esterified Cellulose-Based DACCs ... 211
9.6.1 Synthesis ... 211
9.6.2 Structure and Properties ... 213
9.7 Oxypropylated Cellulose-Based DACCs ... 213
9.7.1 Synthesis ... 213
9.7.2 Structure and Properties ... 215

9.8 Carbamated Cellulose-Based DACCs 216
9.8.1 Synthesis 216
9.8.2 Structure and Properties 216
9.9 Synthesis and Properties of DACC Produced by TEMPO-Mediated Oxidation 219
9.10 Comparison of Mechanical Properties: Non-Derivatized ACC vs DACC 220
9.11 Synthesis and Properties of ACC Fibers or Nanofibers 220
9.12 Thermoset DACC 221
9.12.1 Synthesis 221
9.12.2 Structure and Properties 222
9.13 Conclusion 222
References 223

10. Applications, Current Difficulties, and Future Scope of Single-Polymer Composites 227
10.1 Introduction 227
10.2 Probable Applications of SPCs 227
10.2.1 Synthetic Polymer–Based SPCs 227
10.2.2 Companies Using Synthetic Polymer–Based SPCs in Commercial Applications 228
10.2.2.1 Curv 228
10.2.2.2 Armordon 229
10.2.2.3 PURE 230
10.2.3 All-Cellulose Composites 231
10.3 Current Difficulties, Major Challenges, and Future Scope of SRCs 234
10.3.1 Synthetic Polymer–Based SPCs 234
10.3.1.1 Current Difficulties and Major Challenges 234
10.3.1.2 Future Scope 235
10.3.2 All-cellulose Composites 236
10.4 Conclusion 239
References 239

Index 245

Preface

Composites have been used in myriad applications—industrial, domestic, medical, defense, aerospace, and construction. The increased usage has generated a need for informed professionals as well as dedicated literature toward advancing the theory and design of composite materials. A rich body of literature now exists in texts and handbooks on composites. Theories on manufacturing, modeling, analysis, and design of composite structures have been understood in fairly good detail.

A lot of research has been directed toward improvements of the interface and subsequently the interphase, since failures in composite materials often have been a result of weak boundary regions. Interface has been described as the boundary between two layers of different microstructure. However, such boundaries frequently have chances of chemical interaction. An interphase has been defined as "the volume of material affected by the interaction at the interface." Interphase has been widely used in understanding the phenomenon of adhesion and to specify the presence of a chemically/mechanically altered zone between adjacent phases.

To have a successful composite, the interphase must be strong and adequate toward maximum compatibility. A good interphase is essential for composite materials to endure stress. Incompatibility between fiber and matrix often leads to a weak interphase. Thus, there has been continuous research toward modification of either the reinforcing materials or matrix phase. A third component like compatibilizer/modifier has been used to improve interphase and enhance stress transfer. However, these modifiers are often considered as "foreign" material in systems with requirements of high purity.

The quest of green materials to be used in medical applications, to avoid fiber–matrix compatibility issues of conventional polymeric composites as described in the earlier paragraphs and also the urge toward development of materials with potential of being reused, necessitated the concept of single-polymer composites (SPCs). Under this class of composites, both the phases are either of the same polymer or polymeric systems. These materials have evolved as a class of composite materials with mechanical properties comparable to heterogeneous composites. Being fully recyclable they have specific economic and environmental benefits.

The authors felt that there is no single reference book dedicated to SPCs, in spite of the fact that there has been substantial research in the last four decades using polymers like polyolefins, polyamides, polyesters, and cellulose. This book can be used as reference material for researchers or as a textbook in colleges, where the subject of composites is being seriously taught. This book would be equally useful for composite industries and research labs with direction toward environmental-friendly commitments. This book would be cherished by entrepreneurs who are looking for new opportunities in eco-friendly product design and recyclable materials. The authors hope that this book would serve as an invaluable resource for practical composite design, since mechanical properties of SPCs have been discussed in minute detail. Though the authors have tried to explain things in reasonable detail, a basic book of composites would be required for someone fresh to the field, to complement understanding. Every chapter features potential of research using SPCs, which would be beneficial for practicing engineers and research fraternity.

Great care has been exercised in organizing the material in this book. Nonetheless, the authors would be thankful to readers for pointing out errors, if any are found. The readers are also welcome to suggest any areas concerning SPCs which they feel would have enhanced the scope and usefulness of this book.

The deliberated class of composites are lightweight, recyclable, and with acceptable mechanical (tensile, flexural, and impact) properties for many useful applications. We hope these efforts would further bolster the research community in exploring more environmentally conscious, responsible product designs and process developments in the field of reinforced materials.

Samrat Mukhopadhyay
Bapan Adak
IIT, New Delhi
February 23, 2018

Authors

Dr. Samrat Mukhopadhyay is an Associate Professor at the Department of Textile Technology, Indian Institute of Technology (IIT), Delhi. A gold medalist from the University of Kolkata, he subsequently completed his master's and PhD from IIT Delhi. He has worked with Arvind Mills, Ahmedabad, and has experience of teaching in various colleges in India. Dr. Mukhopadhyay was with the Fibrous Materials Research Group, University of Minho, Portugal as a Postdoctoral Scientist with the prestigious FCT grant. He has been working with synthetic and natural fibers, fiber-reinforced composites and concrete systems, non-destructive testing of composite systems, sustainable approaches in textile chemical processing and technological interventions in the handloom sector. Presently he is part of research groups working on clothing for extreme weather protection (DRDO), multi-functional clothings (Design Innovation Grants, MHRD) and industrial application oriented projects. Dr. Mukhopadhyay has been working with single-polymer composites (SPCs) for the last two decades. Part of his PhD work involved the use of polypropylene filaments in SPCs. He was the Principal Investigator of a project sponsored by GAIL, Government of India, on development of high-impact structures from SPCs from HDPE. The project was successfully executed and has been selected among few projects to be funded for second stage of execution. He has also worked on all-cellulose composites systems jointly with the coauthor.

Email: samrat@textile.iitd.ac.in

Bapan Adak obtained his B.Tech degree in 2010 from Govt. College of Engineering and Textile Technology, Serampore (West Bengal, India) in Textile Engineering. During his B.Tech, he worked on "Dyeing of silk with natural coloring matters with or without mordants." In 2010, he joined Arvind Ltd, Gujarat, India, and continued his work for 3 years as a manufacturing manager. He received his M.Tech in Fiber Science and Technology from the Department of Textile Technology, Indian Institute of Technology (IIT), Delhi, in 2015. During M.Tech, he worked on cellulose-based SPCs popularly termed as "All-cellulose composites." Currently he is pursuing a PhD from the Department of Textile Technology, Indian Institute of Technology (IIT),

Delhi, India. His PhD topic, "Studies on High Gas Barrier and Weather Resistant Polyurethane Nanocomposite Films and Laminates," is part of an ongoing project sponsored by Aerial Delivery Research & Development Establishment (ADRDE, Agra), Defence Research and Development Organization (DRDO, India).

Email: bapan.adak1588@gmail.com

1

Single-Polymer Composites: General Considerations

1.1 Introduction

Majority of composites are manufactured with two different polymers. With such composites there is the issue of recyclability, and there is also a problem in developing a good bond between different polymers that comprise the fiber and the matrix phases. There is, thus, an increasing interest in the area of single-polymer composites, as they offer the opportunity to bridge the performance gap between isotropic unfilled polymers and glass-reinforced polymers, while offering possibilities for improved lightweight and recyclability over traditional composites. This is particularly the case when the fiber and the matrix are both thermoplastic in nature.

Single-polymer composites (SPC) are at odds with conventional definition of composite materials. A fiber-reinforced composite generally consists of three components: (i) the fibers as the discontinuous or dispersed phase, (ii) the matrix as the continuous phase, and (iii) the fine interphase region, also known as the interface. In general, composites refer to a group of chemically different materials in which one component serves as reinforcement (load-bearing component), while the other serves the purpose of matrix (for embedding, protection, and stress transfer). However, SPCs have some similarities with traditional composites. The mechanical characteristics (stiffness, strength) of the reinforcing component even in the single polymer system differ from those of the matrix. An additional resemblance is that reinforcements in SPCs are generally fibers, tapes, and related woven or nonwoven textile architectures.

SPCs are also referred to as single polymer, single phase, self-reinforced, mono-material, homogeneous, composites. However, in a broad way, SPCs have been approached from two philosophies: (i) SPCs manufactured from the same polymer and (ii) SPCs manufactured from the same polymer type [1]. Both these classes of composites will be discussed in the subsequent chapters.

There are several advantages of using SPCs.

a. *Better interphase*: Similar to traditional composites, the stress transfer occurs via an interface/interphase in SPCs. In traditional composites, weak van der Waals forces are known to act across the interfacial region. This does not result in satisfactory bonding. Hence, either the reinforcement or the matrix needs to be surface treated (sizing, coating, etc.) or it necessitates the use of coupling agents. However, molecular entanglements and even H-bonding may help in enhanced adhesion and for better stress transfer between the matrix and the reinforcement via the interphase of SPCs.
b. *Lightweight structures*: Single-polymer composites starting from lightweight polymers like polyolefins are lightweight. They are thermoformable and exhibit improved impact resistance and high specific stiffness and tensile strength. Products based on SPCs not only offer the possibility to dramatically reduce part weight but also serve to minimize the processing waste.
c. *Retention of properties at extreme temperatures*: Many of the SPCs maintain high impact resistance even at subzero temperatures, offering outstanding opportunities for cold weather and refrigeration applications.
d. *Recyclability*: SPCs may match with traditional composites in various application fields based on their favored recycling and beneficial performance/cost balance. The recycling characteristics are superior to many conventional materials such as glass-reinforced composites.

1.2 Initial Research

Single-polymer composites are novel structures where both the reinforcement and matrix phases are derived from identical polymeric material. In one of the most explored routes, an array of oriented polymer fibers or tapes is heated to a temperature where a thin skin of material on the surface of each fiber or tape is melted and then subjected to compaction. On cooling, this material recrystallizes to form the matrix of a self-reinforced composite. This method of producing the "matrix" phase of the SPC, by selectively melting the surface of the oriented reinforcing phase, is an important aspect of the hot compaction process.

Initial research at Leeds University [2,3] produced hot-compacted oriented fibers and tapes, by choosing suitable conditions of temperature and pressure such that a thin skin of each fiber or tape is "selectively" melted. On cooling, this molten material recrystallizes to bind the whole structure together. The resulting hot-compacted material is therefore composed of a single polymeric material and, by virtue of molecular continuity between the phases, is found to have excellent fiber/matrix adhesion. Also, as a result of melting the skin of each fiber or tape, there are no matrix wetting problems.

Reinforcing a common polyethylene with polyethylene fibers leads to a strong and stiff SPC [4]. Work on such a self-reinforced high-density polyethylene (HDPE) was first published by Capiati and Porter [5]. Highly oriented polyethylene film strips, which are the predecessors of today's commercially available ultra-high-molecular-weight (UHMW) polyethylene fibers, were used as reinforcement. In a study by Loos et al. [6], the composite of isotactic polypropylene (iPP) was prepared by compression molding of unoriented iPP thin films. They used films of 80-μm thickness, together with highly oriented iPP fibers that were fixed on a glass slide, at 170°C for 5 min. The procedure resulted in the melting of the iPP matrix material, while the constrained fibers remained unaffected, thus forming the single polymer composite of iPP. Morphological studies on the prepared homogeneity of composites via optical and scanning electron microscopy illustrated the formation of iPP transcrystalline layers in the vicinity of iPP fibers.

There has been a lot of research regarding the nature of reinforcement in this class of composites, which will be dealt with in detail in the subsequent chapters. An excellent classification has been done by Kmetty et al. [7]. Compared to the matrix, the reinforcing structure has either different (a) crystalline and/or (b) supermolecular structures or is given by a preform, a prefabricate that entails different textile architectures with higher crystallinity. Related technologies, practiced, for example, in fiber-spinning operations, are grouped into multistep productions of single-component self-reinforced polymeric materials (Figure 1.1).

1.3 General Considerations for Single-Polymer Composites

1.3.1 Elevation of Melting Point

The basis for manufacturing some of the SPCs, which use the same class of polymer, was the elevation in the melting point of filaments in a constrained state. The Gibbs free energy equation is given as

$$\Delta G = \Delta H - T\Delta S,$$

where ΔG is the free energy, ΔH is the change in enthalpy, ΔS is the change in entropy, and T is the melting point. The free energy ΔG being zero at melting, the melting point (T) is the ratio of ΔH to ΔS. Partial retention of orientation till melting due to constraint implies lower entropy due to fewer available

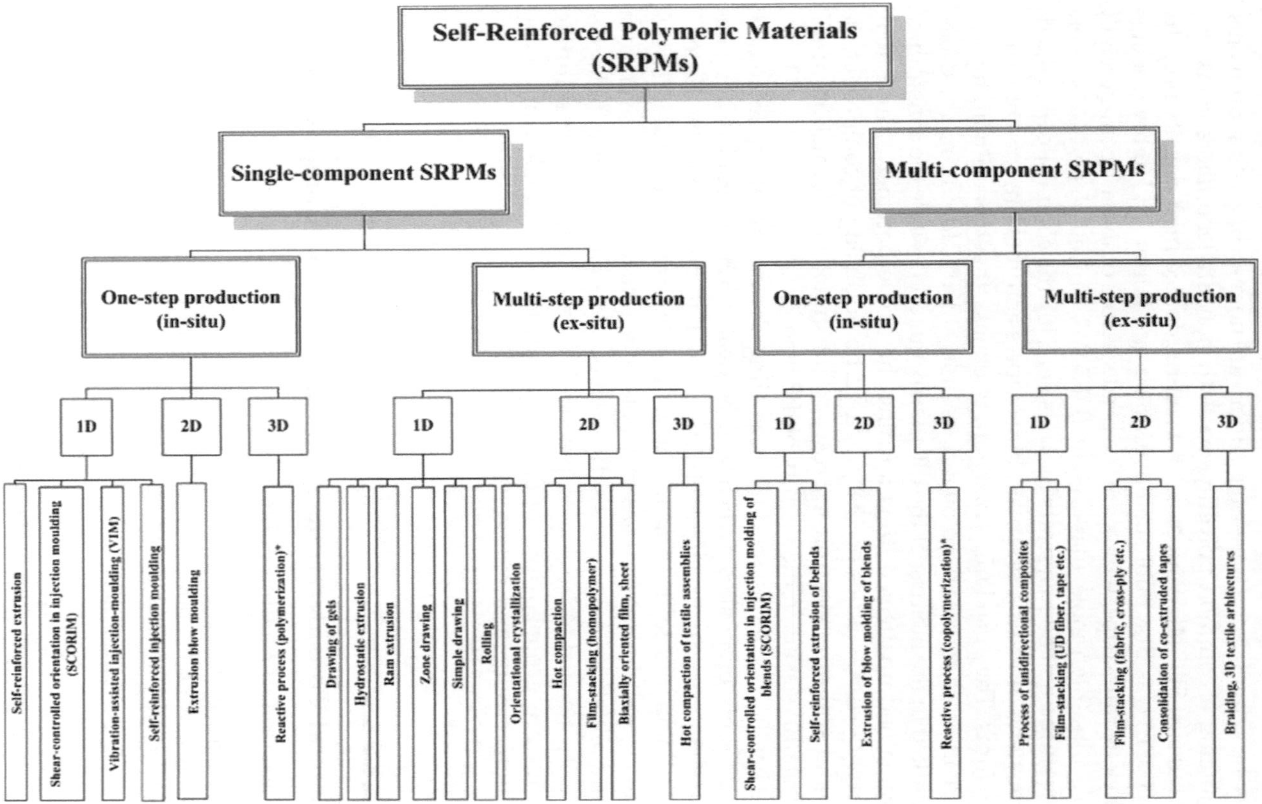

FIGURE 1.1
Classification of self-reinforced polymeric materials. (After Kmetty, A. et al., *Prog. Polym. Sci.*, 35, 1288, 2010. With permission.)

conformations for molecules in the amorphous zones. If the enthalpy change is assumed to be constant during the melting process, a reduced entropy results in an elevation of the melting point. The rise of melting point under constrained state has been experienced by many scientists for polyethylene [8–10] and polypropylene [11].

Similar observations are made by Lacroix et al. [9], where constrained HDPE filaments melted at a higher temperature allow a larger processing window (Figure 1.2).

The authors explained triple peaks of HDPE filaments due to melting of a part of the orthorhombic phase, a lattice transition from orthorhombic to hexagonal and melting of the hexagonal phase, as also argued by van Mele et al. [12]. If shrinkage of the fibers during heating is prevented, the first peak, and with this the melting of the orthorhombic phase, is reduced.

In the work by Mukhopadhyay et al. [13], there is a shift in the melting peak for filaments in a constrained state, as observed by Differential Scanning Calorimetry (DSC). The filament, manufactured through a special drawing process, exhibits a sharp endotherm and provides tolerance to manufacture the SPC. A double peak is observed for the constrained filament. The double peak in this case may be caused due to the difference in heat flow to the filaments wound around the aluminum pan in the DSC crucible. X-ray measurements for polypropylene filaments annealed at 155°C through the gradient drawing process [14] indicated that it was an oriented monoclinic structure.

Polypropylene is generally not stable at elevated temperatures and can often show significant shrinkage, and hence changes in morphology, due to a large frozen-in stress developed during manufacture. This is an important

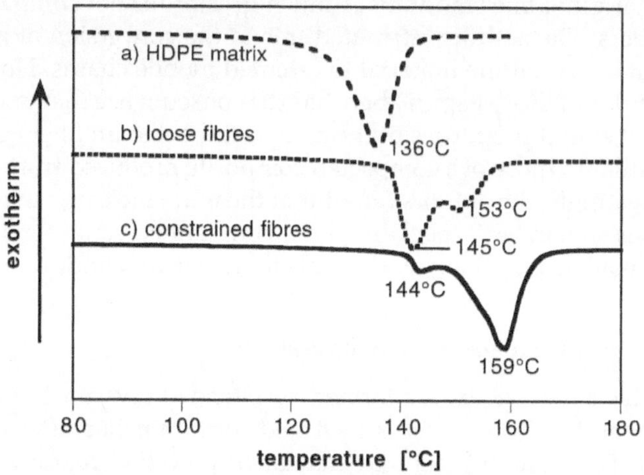

FIGURE 1.2
DSC scans of (a) HDPE matrix, (b) UHMWPE fiber lying loose in the pan, and (c) UHMWPE fiber fixed in the pan. (From Lacroix, F. V. et al., *Polymer*, 40, 843, 1999. With permission from ScienceDirect, www.sciencedirect.com/science/article/pii/S0032386198003097#FIG1.)

aspect to be considered in the manufacture of SPCs because the properties of the final compacted sheet will depend not only on the initial properties of the component material but also on how the structure changes during the compaction process.

1.3.2 Initial Morphology

Polypropylene in general does not have such a refined crystalline structure as polyethylene, so it is more sensitive to elevated temperatures close to the melt (i.e., the compaction temperature) and also shows a greater superheating effect. On the contrary, due to high crystallinity, polyethylene is stable to temperatures only a few degrees below the melting range, and the peak melting point does not change significantly.

In polyethylene, the oriented phase is characterized by a very high modulus (40–90 GPa) and a relatively low strain to failure (2%–4%). The matrix shows a very low modulus (1 GPa) and is ductile. Consequently, the properties of the compacted sheet are dominated by the oriented phase and the matrix properties can almost be neglected.

In polypropylene, the oriented phase has a modulus between 10 and 15 GPa, and the matrix around 1.5 GPa, so the matrix can make a significant contribution to the properties of the sheet. Additionally, the oriented polypropylene fibers or tapes have a high failure strain (15%), making it crucial that the matrix remain coherent until the oriented phase fails.

Yan et al. [15], who used gel-spun polyethylene fibers for preparing SPCs, compared the results with hot compaction of melt-spun fiber. SPECTRA 1000 gel-spun polyethylene fibers were successfully compacted by the group. The loss in original fiber modulus is much greater for hot compaction with gel-spun fibers. The authors attributed this to the conversion of a considerable amount of crystalline material to oriented mobile chains. However, the higher modulus of the gel-spun fibers has the consequence that under certain conditions, the final properties of the compacted gel-spun fibers composites are very similar to those of a compacted composite produced from melt-spun fibers. Interestingly, the authors found that the main melting range with gel-spun fibers is so narrow, especially under constrained conditions, that samples can be made controllably within this temperature window.

1.3.3 Structural Changes with Temperature

UHMW polyethylene fibers are formed as extended crystals, but during the heat treatment, the fiber surfaces melt and form lamellae after crystallization [16–18]. The same nucleation behavior of polyethylene on a polyethylene substrate is described for melted and crystallized skin of hot-compacted melt-spun polyethylene fibers, for the shish–kebab crystallization in injection-molded polyethylene [19], and for the epitaxial crystallization of linear low-density polyethylene on high-density polyethylene [20].

For samples melted incompletely, there was a significant effect of the compaction temperature on the melting point, and although the shapes of the melting endotherms changed, there was relatively little effect on the degree of crystallinity, for compaction temperatures between 249°C and 261°C. The melting point of the fibers compacted at 249°C was higher than that of the original fiber, and the melting point increased as the compaction temperature was increased to 262°C. For compaction temperatures above the melting point of the fibers, a large decrease was observed in both melting point and the degree of crystallinity. These results are in agreement with those reported by Rasburn et al. for PET fiber compaction [21].

According to Mukhopadhyay et al. [11], for polyethylene–polyethylene SPCs, temperatures around 138°C were inadequate for any sort of adhesion. A temperature of 140°C resulted in reasonably good adhesion, while another degree higher showed optimum properties. Every degree higher than 141°C resulted in the deterioration of composite mechanical properties, while at 144°C the composite structure became brittle. The experiments confirmed the importance of temperature in the formation of SPCs. In SPCs where processing temperatures approach the melting regime of polymers, temperature is the single most crucial parameter affecting the ultimate properties of the composites.

In a study by Mukhopadhyay et al. [11], it was found that the character of polypropylene films changes when compacted with a system of filaments (Figure 1.3). The unoriented smectic structure of the matrix gets transformed into an α-monoclinic structure characterized by sharp peaks. The percentage X-ray crystallinity also increases from 55% to 78% and supports the change in melting behavior. Pressure also plays an important role in consolidation.

At temperatures around 154°C, there is considerable filament wetting, and the nature of fracture is shown in Figure 1.4.

At a temperature of 156°C, the matrix ductility reduces, and there is a transverse fracture in the form of matrix cracking with the majority of the fibers still intact. Such a sample is taken out in between a tensile test (at elongation of 8%) and seen through an optical microscope. The same sample could take up 80% of the breaking load on subsequent testing. At temperatures above 165°C, the composite as a whole becomes too brittle and tends to break along the boundary of the fibers even without the application of any force.

1.3.4 Thermal Mismatch

For any material, a low thermal expansion is ideally required, and if different materials are to be used, the thermal expansions should be matched, or at least comparable. The situation becomes more complicated when the material in question shows anisotropic thermal expansion behavior. In particular, highly oriented polymers can show an extreme thermal expansion

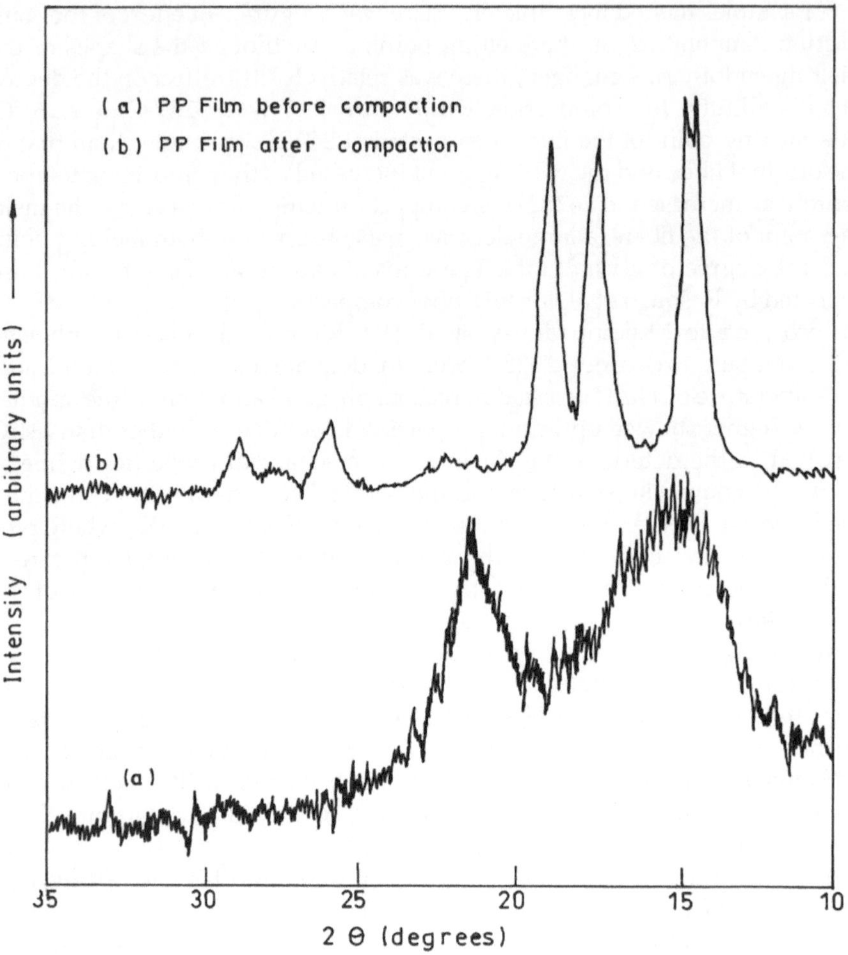

FIGURE 1.3
Wide angle x-ray diffraction profiles of polypropylene films before (a) and after (b) compaction. (From Mead, W. T. and Porter, R. S., *J. Appli. Polym. Sci.*, 22, 3249, 1978. With permission.)

anisotropy, with a low, possibly negative, thermal expansion in the orientation direction, and a large positive value perpendicular to that direction.

Single-fiber composites can get rid of the difference of thermal coefficients of the matrix and the reinforcement that causes large internal stresses to be induced in the composite during thermal cycling, often resulting in permanent dimensional changes and a pronounced strain hysteresis [22–24]. Besides recyclability and thermal mismatch, the interest in the concept of SPCs is based upon the idea that interfacial bonding should improve if the matrix and reinforcement are made from the same polymer [25,26].

Bozec et al. [27] also showed that thermal expansion of a hot-compacted composite is controlled by a number of factors, including the molecular

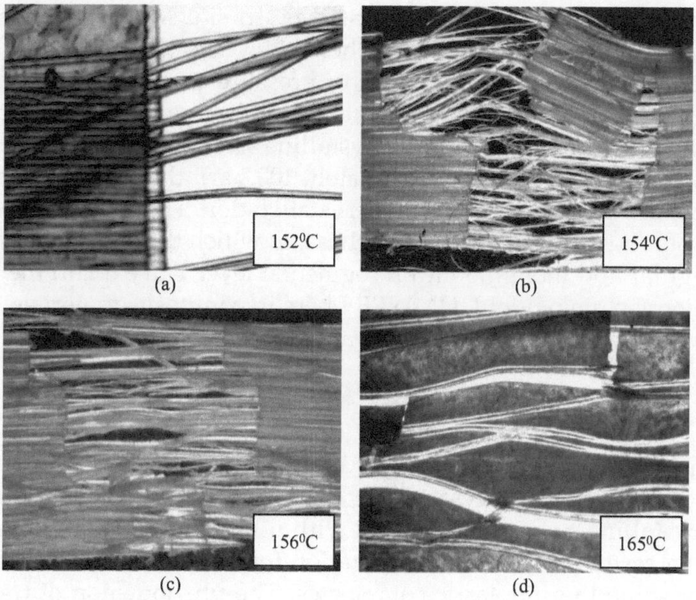

FIGURE 1.4
Optical micrographs of fractured samples consolidated at different temperatures (x 12.5). (From Mead, W. T. and Porter, R. S., *J. Appli. Polym. Sci.*, 22, 3249, 1978. With permission.): (a) partial consolidation at 152°C; (b) better consolidation at 154°C and matrix break; (c) superior consolidation at 156°C as both fibers and matrix share the load; (d) more than required temperature of 165°C results in a brittle composite structure.

structure of the individual component tapes/fibers, the woven architecture of the composite, and the elastic anisotropy on the tapes/fibers. Shrinkage and modulus measurement have shown that, for polypropylene in particular, the structure of the component elements is dramatically affected by the compaction process, so that the thermal expansion behavior of the final compacted woven sheet is dependent on the final molecular structure after compaction, rather than the starting structure.

1.3.5 Compaction Pressure

Compaction pressure has an important role to play, and generally its effect is threefold: first aiding heat transfer to the fiber assembly, second, ensuring consolidation during compaction, and third, helping restrain shrinkage of the fibers or tapes. The third aspect is particularly important for polypropylene where high shrinkage is often seen as the melting point is approached. A general discussion of the effect of compaction on composite morphology in fiber-reinforced composites can be further accessed from the findings of research groups [28–31].

Recently, an X-ray diffraction study was performed by a research group [32]. The effect of pressure on the phase transformations observed upon the heating of oriented fibers of fully extended-chain ultra-high-modulus polyethylene (UHMWPE) was evaluated by in situ X-ray diffraction measurements under pressure with the help of synchrotron radiation. In the developed fiber structures, the effect of pressure on phase transitions was significantly stronger than that predicted and observed previously in chain-folded and also in extended-chain crystalline structures. Additionally, even at relatively low pressure (approximately 100 bar), the hexagonal phase is observed during both melting and recrystallization. The existence and metastability of mesomorphic hexagonal phase, in which the chain conformation is less regular and more mobile on segmental level, can explain the basis of compaction technology of UHMWPE fibers in composite materials. Similar findings were made by few other research groups [33–36].

1.3.6 Transcrystallinity

Wu et al. [37] and Billon et al. [38,39] defined transcrystallization as a nucleation-controlled process taking place under quiescent conditions in a semicrystalline polymer in contact with other materials, e.g., fibers. It is basically a case of oriented crystallization found at the fiber/matrix interphase for some thermoplastic composites. The phenomenon of transcrystallinity has been investigated in great detail by different scientists, who have come up with their observations. Since it is an interfacial phenomenon, the different causes of transcrystallinity, their kinetics, procedures to detect transcrystallinity, and effects on composite properties are reviewed in the next section.

For transcrystallinity to occur, it is necessary that the nucleation rate in the fiber exceeds that in the matrix bulk, the lateral growth on the fiber is less favored, and the columnar growth is ensured [40]. The phenomenon may be related to enhanced stress-transfer capability along the fiber–matrix interface [41,42]. A possible reason for transcrystallization is heterogeneous nucleation on fiber surfaces. Transcrystallization has also been reported in SPCs.

In SPCs, because of the different morphologies of the filaments/tapes and films, a difference in the melting temperatures was observed. Microscopic investigations of these samples showed a third, transcrystalline, morphology between the film strips and the matrix. Crystalline lamellae in this region had grown perpendicular to the fiber direction. Using commercial gel-spun UHMW polyethylene fibers as reinforcement for HDPE, transcrystalline morphologies were also found [43,44]. Possible reasons for the transcrystallization are heterogeneous and homogeneous nucleation on fiber surfaces. Because of the good lattice match between the fiber and the matrix and the highly favorable energetics, a homogeneous, epitaxial nucleation seems to be the more probable alternative. The morphologies of transcrystalline layers vary with the temperature cycle used to manufacture the composites.

Stern et al. [45] observed four different transcrystalline morphologies, depending on the cooling history, with the same fiber–matrix combination.

Three of these were obtained by cooling samples at different cooling rates from a temperature between the melting temperatures of the fiber and matrix. The composite with the fourth morphology was crystallized under isothermal conditions at a temperature only a few degrees below the melting temperature of the matrix.

Capiati et al. [46] showed the growth of transcrystalline regions in the melt matrix at the polyethylene–polyethylene interface and a partial melting between the fiber and the matrix. In these cases, the outer layer of the fiber could potentially co-crystallize with the matrix. Wang et al. [47] showed that HDPE and UHMPE can co-crystallize, which is confirmed by the single endotherm observed by them. In a study by Mead et al. [48], the optimum temperature in the polyethylene–polyethylene composite has been assessed by growth of transcrystalline regions from the melt matrix without a considerable modulus reduction of the annealed ultra-oriented and reinforcement fiber or film.

The presence of a transcrystalline interphase has been reported to improve the mechanical properties of composite materials [49,50], leading to a more uniform distribution of strain along the fiber, increased resistance to compressive and radial stresses (arising from the differential shrinkage on cooling of matrix and reinforcement), and enhanced interfacial bond strength [51].

It is, however, to be noted that transcrystallinity plays an important role only in aligned filament reinforcement. Stern et al. [52], in their study on chopped fiber reinforcement, showed that there was no effect on transcrystallinity.

1.3.7 Crystallization Behavior and Cooling History

The crystallization behavior and morphology of polyethylene-based SPCs have been investigated by Lacroix [53], using light microscopy and low-voltage scanning electron microscopy techniques. The surface crystals on UHMW polyethylene fibers act as nucleation centers for the high-density polyethylene matrix, which may result from epitaxial crystallization. After crystallization from the melt and independent of air-cooled or isothermal crystallization conditions, a transcrystalline layer was found having lamellar crystals grown perpendicular to the fiber axis.

Li et al. [54] showed that single polymer fiber/matrix composites of iPP were prepared by introducing the high-performance fibers into its supercooled melts. With this procedure, the morphologies of the resulting composites as a function of fiber introduction temperature (T_{in}) were studied in detail by means of optical microscopy. Such composites of iPP have been successfully prepared by utilizing the fact that polymers crystallize only under a state of supercooling. The induced supermolecular structures of iPP by its homogeneity fibers as a function of fiber introduction temperature were studied by means of optical microscopy. The results showed that the induced supermolecular structures of iPP at the fiber/matrix interfacial layer can be ascribed to the transcrystallization of iPP triggered by strong heterogeneous nucleation rather than the shearing produced by fiber introduction. As fiber introduction temperature increased, an increasing content of iPP surrounding the iPP fibers was seen. Transcrystallization layers have been observed when T_{in} was set at 173°C. This is related to the melting, or at least surface partial melting, of the iPP fibers.

Results of Hine et al. [55] with woven compacted polypropylene tapes have proved valuable from a commercial standpoint, because the hot-compacted sheets showed an interesting portfolio of properties including lightweight, good stiffness and strength, and outstanding impact performance even at low temperatures. Equally important, the hot-compacted polypropylene sheets were found to be thermoformable [56].

1.4 Conclusion

Various parameters affecting the production of SPCs are discussed. It is seen that though temperature plays the most important role in the process, various other parameters such as initial morphology, pressure, and method of crystallization play a significant role in the end properties of composites. Hot-compacted SPCs have found their use in interesting practical applications such as automotive undertray, helicopter dome, and loudspeaker applications. The role of morphology affecting the properties of such composites very close to their melting zones needs further systematic investigation. In Chapters 2–6, research on SPC systems, starting from different polymers, will be explored.

Interesting alternatives to SPCs from synthetic materials are self-reinforced cellulosic composites [57–65]. Here, natural fibers are treated to plasticize their exteriors, such as wood pulp fibers and sisal, which become thermoplastic. These fibers now have a skin/core structure with the stiffness of a fibrous core, with a deformable thermoplastic coating. By applying a processing route similar to hot compaction, this thermoplastic coating is melted into a matrix that produces sisal/sisal composites possessing moduli. Because these composites are totally biodegradable, disposal is easy, although this low stability may limit their applications. Selective dissolution has been one popular approach [66–70] investigated by some research groups. Several other routes have been investigated [71], and all-cellulose nanocomposites [72–84] have also been explored. The advancement in characterization of this class of composites has been explored [85,86]. Chapters 7–9 are dedicated to these classes of materials. Chapter 10 is an overall summary of SPCs, their present application, and future research directions.

References

1. Karger-Kocsis, J., and T. Bárány. Single-polymer composites (SPCs): Status and future trends. *Composites Science and Technology* 92(2014): 77–94.
2. Hine, P. J., I. M. Ward, R. H. Olley, and D. C. Bassett. The hot compaction of high modulus melt-spun polyethylene fibres. *Journal of Materials Science* 28, no. 2 (1993): 316–324.
3. Olley, R. H., D. C. Bassett, P. J. Hine, and I. M. Ward. Morphology of compacted polyethylene fibres. *Journal of Materials Science* 28, no. 4 (1993): 1107–1112.
4. Lacroix, F. V., M. Werwer, and K. Schulte. Solution impregnation of polyethylene fibre/polyethylene matrix composites. *Composites Part A: Applied Science and Manufacturing* 29, no. 4 (1998): 371–376.
5. Capiati, N. J., and R. S. Porter. The concept of one polymer composites modelled with high density polyethylene. *Journal of Materials Science* 10, no. 10 (1975): 1671–1677.
6. Loos, J., T. Schimanski, J. Hofman, T. Peijs, and P. J. Lemstra. Morphological investigations of polypropylene single-fibre reinforced polypropylene model composites. *Polymer* 42, no. 8 (2001): 3827–3834.
7. Kmetty, Á., T. Bárány, and J. Karger-Kocsis. Self-reinforced polymeric materials: A review. *Progress in Polymer Science* 35, no. 10 (2010): 1288–1310.
8. Capiati, N. J., and R. S. Porter. The concept of one polymer composites modelled with high density polyethylene. *Journal of Materials Science* 10, no. 10 (1975): 1671–1677.
9. Mead, W. T., and R. S. Porter. The preparation and tensile properties of polyethylene composites. *Journal of Applied Polymer Science* 22, no. 11 (1978): 3249–3265.
10. Lacroix, F. V., J. Loos, and K. Schulte. Morphological investigations of polyethylene fibre reinforced polyethylene. *Polymer* 40, no. 4 (1999): 843–847.
11. Mukhopadhyay S., B. L. Deopura, and R. Alagirusamy. *Studies on High Modulus High Tenacity Polypropylene Fibers and Single Polymer Composites*. PhD thesis, IIT Delhi, February 2004. http://eprint.iitd.ac.in/bitstream/2074/3889/1/TH-3107.pdf.
12. Van Mele, B., and E. Verdonck. Physico-chemical characterization of the fibre/matrix interaction in polyethylene fibre/epoxy matrix composites. *Composite Interfaces* 3, no. 2 (1995): 83–100.
13. Mukhopadhyay, S., B. L. Deopura, and R. Alagirusamy. Drawing of polypropylene filaments on a gradient heater. *Journal of the Textile Institute* 96, no. 5 (2005): 349–354.
14. Mukhopadhyay, S., B. L. Deopura, and R. Alagirusamy. Studies on production of polypropylene filaments with increased temperature stability. *Journal of Applied Polymer Science* 101, no. 2 (2006): 838–842.
15. Yan, R. J., P. J. Hine, I. M. Ward, R. H. Olley, and D. C. Bassett. The hot compaction of SPECTRA gel-spun polyethylene fibre. *Journal of Materials Science* 32, no. 18 (1997): 4821–4832.
16. Wasiak, A., and P. Sajkiewicz. Orientation distributions and melting behaviour of extended and folded-chain crystals in gel-drawn, ultra-high-molecular-weight polyethylene. *Journal of Materials Science* 28, no. 23 (1993): 6409–6417.

17. Kabeel, M. A., D. C. Bassett, R. H. Olley, P. J. Hine, and I. M. Ward. Compaction of high-modulus melt-spun polyethylene fibres at temperatures above and below the optimum. *Journal of Materials Science* 29, no. 18 (1994): 4694–4699.
18. Olley, R. H., D. C. Bassett, P. J. Hine, and I. M. Ward. Morphology of compacted polyethylene fibres. *Journal of Materials Science* 28, no. 4 (1993): 1107–1112.
19. Martinez-Salazar, J., J. V. Garcia Ramos, and J. Petermann. On the fine structure of shish-kebabs in injection moulded polyethylene. *International Journal of Polymeric Materials* 21, no. 3–4 (1993): 111–121.
20. Loos, J., F. Katzenberg, and J. Petermann. Epitaxial crystallization of linear low-density polyethylene on high-density polyethylene. *Journal of Materials Science* 32, no. 6 (1997): 1551–1554.
21. Rasburn, J., P. J. Hine, I. M. Ward, R. H. Olley, D. C. Bassett, and M. A. Kabeel. The hot compaction of polyethylene terephthalate. *Journal of Materials Science* 30, no. 3 (1995): 615–622.
22. Wakashima, K. L., M. Otsuka, and S. Umekawa. Thermal expansions of heterogeneous solids containing aligned ellipsoidal inclusions. *Journal of Composite Materials* 8, no. 4 (1974): 391–404.
23. Kural, M. H., and B. K. Min. The effects of matrix plasticity on the thermal deformation of continuous fiber graphite/metal composites. *Journal of Composite Materials* 18, no. 6 (1984): 519–535.
24. Mitra, S., I. Dutta, and R. C. Hansen. Thermal cycling studies of a cross-plied P100 graphite fibre-reinforced 6061 aluminium composite laminate. *Journal of Materials Science* 26, no. 22 (1991): 6223–6230.
25. Capiati, N. J., and R. S. Porter. The concept of one polymer composites modelled with high density polyethylene. *Journal of Materials Science* 10, no. 10 (1975): 1671–1677.
26. Mead, W. T., and R. S. Porter. The preparation and tensile properties of polyethylene composites. *Journal of Applied Polymer Science* 22, no. 11 (1978): 3249–3265.
27. Le Bozec, Y., S. Kaang, P. J. Hine, and I. M. Ward. The thermal-expansion behaviour of hot-compacted polypropylene and polyethylene composites. *Composites Science and Technology* 60, no. 3 (2000): 333–344.
28. Robitaille, F., and R. Gauvin. Compaction of textile reinforcements for composites manufacturing. I: Review of experimental results. *Polymer Composites* 19, no. 2 (1998): 198–216.
29. Chen, B., and T. W. Chou. Compaction of woven-fabric preforms in liquid composite molding processes: Single-layer deformation. *Composites Science and Technology* 59, no. 10 (1999): 1519–1526.
30. Chen, B., E. J. Lang, and T. W. Chou. Experimental and theoretical studies of fabric compaction behavior in resin transfer molding1. *Materials Science and Engineering: A* 317, no. 1–2 (2001): 188–196.
31. Saunders, R. A., C. Lekakou, and M. G. Bader. Compression in the processing of polymer composites 1. A mechanical and microstructural study for different glass fabrics and resins. *Composites Science and Technology* 59, no. 7 (1999): 983–993.
32. Rein, D. M., L. Shavit, R. L. Khalfin, Y. Cohen, A. Terry, and S. Rastogi. Phase transitions in ultraoriented polyethylene fibers under moderate pressures: A synchrotron X-ray study. *Journal of Polymer Science Part B: Polymer Physics* 42, no. 1 (2004): 53–59.
33. Shavit-Hadar, L., D. M. Rein, R. Khalfin, Y. Cohen, A. E. Terry, and S. Rastogi. The path to "single-component composites": An in situ X-ray study of melting

and crystallization of extended-chain polyethylene fibers under pressure. *Nuclear Instruments and Methods in Physics Research Section B: Beam Interactions with Materials and Atoms* 238, no. 1–4 (2005): 39–42.
34. Rastogi, S., and J. A. Odell. Stress stabilization of the orthorhombic and hexagonal phases of UHM PE gel-spun fibres. *Polymer* 34, no. 7 (1993): 1523–1527.
35. Ratner, S., A. Weinberg, E. Wachtel, P. Moret, and G. Marom. Phase transitions in UHMWPE fiber compacts studied by in situ synchrotron microbeam WAXS. *Macromolecular Rapid Communications* 25, no. 12 (2004): 1150–1154.
36. Rein, D. M., R. L. Khalfin, and Y. Cohen. Analysis of melting transitions in extended-chain polymer crystals. *Journal of Polymer Science Part B: Polymer Physics* 42, no. 12 (2004): 2238–2244.
37. Wu, C. M., M. Chen, and J. Karger-Kocsis. Transcrystallization in syndiotactic polypropylene induced by high-modulus carbon fibers. *Polymer Bulletin* 41, no. 2 (1998): 239–245.
38. Billon, N., and J. M. Haudin. Influence of transcrystallinity on DSC analysis of polymers. *Journal of Thermal Analysis* 42, no. 4 (1994): 679–696.
39. Billon, N., C. Magnet, J. M. Haudin, and D. Lefebvre. Transcrystallinity effects in thin polymer films. Experimental and theoretical approach. *Colloid and Polymer Science* 272, no. 6 (1994): 633–654.
40. Folkes, M. J., and S. T. Hardwick. Direct study of the structure and properties of transcrystalline layers. *Journal of Materials Science Letters* 6, no. 6 (1987): 656–658.
41. Bessell, T., and J. B. Shortall. The crystallization and interfacial bond strength of nylon 6 at carbon and glass fibre surfaces. *Journal of Materials Science* 10, no. 12 (1975): 2035–2043.
42. Campbell, D., and M. M. Qayyum. Enhanced fracture strain of polypropylene by incorporation of thermoplastic fibres. *Journal of Materials Science* 12, no. 12 (1977): 2427–2434.
43. Ishida, H., and P. Bussi. Surface induced crystallization in ultrahigh-modulus polyethylene fiber-reinforced polyethylene composites. *Macromolecules* 24, no. 12 (1991): 3569–3577.
44. Stern, T., E. Wachtel, and G. Marom. Epitaxy and lamellar twisting in transcrystalline polyethylene. *Journal of Polymer Science Part B: Polymer Physics* 35, no. 15 (1997): 2429–2433.
45. Stern, T., G. Marom, and E. Wachtel. Origin, morphology and crystallography of transcrystallinity in polyethylene-based single-polymer composites. *Composites Part A: Applied Science and Manufacturing* 28, no. 5 (1997): 437–444.
46. Capiati, N. J., and R. S. Porter. The concept of one polymer composites modelled with high density polyethylene. *Journal of Materials Science* 10, no. 10 (1975): 1671–1677.
47. Wang, X. Y., and R. Salovey. Melting of ultrahigh molecular weight polyethylene. *Journal of Applied Polymer Science* 34, no. 2 (1987): 593–599.
48. Mead, W. T., and R. S. Porter. The preparation and tensile properties of polyethylene composites. *Journal of Applied Polymer Science* 22, no. 11 (1978): 3249–3265.
49. Burton, R. H., and M. J. Folkes. Interfacial morphology in short fibre reinforced thermoplastics. *Plastics and Rubber Processing and Applications* 3(1983): 129–35.
50. Peacock, J. A., B. Fife, E. Nield, and C. Y. Barlow. Influence of interface on macroscopic properties. In *Composite Interfaces*, p. 129, eds H. Ishida and J. L. Koenig. Elsevier, New York, 1986.

51. Capiati, N. J., and R. S. Porter. The concept of one polymer composites modelled with high density polyethylene. *Journal of Materials Science* 10, no. 10 (1975): 1671–1677.
52. Stern, T., A. Teishev, and G. Marom. Composites of polyethylene reinforced with chopped polyethylene fibers: Effect of transcrystalline interphase. *Composites Science and Technology* 57, no. 8 (1997): 1009–1015.
53. Lacroix, F. V., J. Loos, and K. Schulte. Morphological investigations of polyethylene fibre reinforced polyethylene. *Polymer* 40, no. 4 (1999): 843–847.
54. Li, H., S. Jiang, J. Wang, D. Wang, and S. Yan. Optical microscopic study on the morphologies of isotactic polypropylene induced by its homogeneity fibers. *Macromolecules* 36, no. 8 (2003): 2802–2807.
55. Hine, P. J., M. Ward, and J. Teckoe. The hot compaction of woven polypropylene tapes. *Journal of Materials Science* 33, no. 11 (1998): 2725–2733.
56. Prosser, W., P. J. Hine, and I. M. Ward. Investigation into thermoformability of hot compacted polypropylene sheet. *Plastics, Rubber and Composites* 29, no. 8 (2000): 401–410.
57. Huber, T., J. Müssig, O. Curnow, S. Pang, S. Bickerton, and M. P. Staiger. A critical review of all-cellulose composites. *Journal of Materials Science* 47, no. 3 (2012): 1171–1186.
58. Adak, B., and S. Mukhopadhyay. All-cellulose composite laminates with low moisture and water sensitivity. *Polymer* 141(2018): 79–85.
59. Bapan, A., and M. Samrat. Jute based all-cellulose composite laminates. *Indian Journal of Fibre & Textile Research (IJFTR)* 41, no. 4 (2016): 380–384.
60. Gandini, A., A. A. da Silva Curvelo, D. Pasquini, and A. J. de Menezes. Direct transformation of cellulose fibres into self-reinforced composites by partial oxypropylation. *Polymer* 46, no. 24 (2005): 10611–10613.
61. Nishino, T., I. Matsuda, and K. Hirao. All-cellulose composite. *Macromolecules* 37, no. 20 (2004): 7683–7687.
62. Soykeabkaew, N., N. Arimoto, T. Nishino, and T. Peijs. All-cellulose composites by surface selective dissolution of aligned ligno-cellulosic fibres. *Composites Science and Technology* 68, no. 10–11 (2008): 2201–2207.
63. Adak, B., and S. Mukhopadhyay. Effect of pressure on structure and properties of lyocell fabric-based all-cellulose composite laminates. *Journal of the Textile Institute* 108, no. 6 (2017): 1010–1017.
64. Adak, B., and S. Mukhopadhyay. Effect of the dissolution time on the structure and properties of lyocell-fabric-based all-cellulose composite laminates. *Journal of Applied Polymer Science* 133, no. 19 (2016).
65. Adak, B., and S. Mukhopadhyay. A comparative study on lyocell-fabric based all-cellulose composite laminates produced by different processes. *Cellulose* 24, no. 2 (2017): 835–849.
66. Nishino, T., and N. Arimoto. All-cellulose composite prepared by selective dissolving of fiber surface. *Biomacromolecules* 8, no. 9 (2007): 2712–2716.
67. Soykeabkaew, N., C. Sian, S. Gea, T. Nishino, and T. Peijs. All-cellulose nanocomposites by surface selective dissolution of bacterial cellulose. *Cellulose* 16, no. 3 (2009): 435–444.
68. Gindl, W., T. Schöberl, and J. Keckes. Structure and properties of a pulp fibre-reinforced composite with regenerated cellulose matrix. *Applied Physics A* 83, no. 1 (2006): 19–22.

69. Han, D., and L. Yan. Preparation of all-cellulose composite by selective dissolving of cellulose surface in PEG/NaOH aqueous solution. *Carbohydrate Polymers* 79, no. 3 (2010): 614–619.
70. Duchemin, B. J. C., A. P. Mathew, and K. Oksman. All-cellulose composites by partial dissolution in the ionic liquid 1-butyl-3-methylimidazolium chloride. *Composites Part A: Applied Science and Manufacturing* 40, no. 12 (2009): 2031–2037.
71. Schuermann, H., T. Huber, and M. P. Staiger. Prepreg style fabrication of all cellulose composites. In *Proceedings of the 19th International Conference on Composite Materials*, Montreal, Canada, pp. 5626–5634. 2013.
72. Oksman, K., Y. Aitomäki, A. P. Mathew, G. Siqueira, Q. Zhou, S. Butylina, S. Tanpichai, X. Zhou, and S. Hooshmand. Review of the recent developments in cellulose nanocomposite processing. *Composites Part A: Applied Science and Manufacturing* 83 (2016): 2–18.
73. Gindl, W., and J. Keckes. All-cellulose nanocomposite. *Polymer* 46, no. 23 (2005): 10221–10225.
74. Zhang, J., N. Luo, X. Zhang, L. Xu, J. Wu, J. Yu, J. He, and J. Zhang. All-cellulose nanocomposites reinforced with in situ retained cellulose nanocrystals during selective dissolution of cellulose in an ionic liquid. *ACS Sustainable Chemistry and Engineering* 4, no. 8 (2016): 4417–4423.
75. Bondeson, D., P. Syre, and K. O. Niska. All cellulose nanocomposites produced by extrusion. *Journal of Biobased Materials and Bioenergy* 1, no. 3 (2007): 367–371.
76. Qi, H., J. Cai, L. Zhang, and S. Kuga. Properties of films composed of cellulose nanowhiskers and a cellulose matrix regenerated from alkali/urea solution. *Biomacromolecules* 10, no. 6 (2009): 1597–1602.
77. Pullawan, T., A. N. Wilkinson, and S. J. Eichhorn. Influence of magnetic field alignment of cellulose whiskers on the mechanics of all-cellulose nanocomposites. *Biomacromolecules* 13, no. 8 (2012): 2528–2536.
78. Pullawan, T., A. N. Wilkinson, L. N. Zhang, and S. J. Eichhorn. Deformation micromechanics of all-cellulose nanocomposites: Comparing matrix and reinforcing components. *Carbohydrate Polymers* 100(2014): 31–39.
79. Yousefi, H., M. Faezipour, T. Nishino, A. Shakeri, and G. Ebrahimi. All-cellulose composite and nanocomposite made from partially dissolved micro-and nanofibers of canola straw. *Polymer Journal* 43, no. 6 (2011): 559–564.
80. Yousefi, H., T. Nishino, M. Faezipour, G. Ebrahimi, and A. Shakeri. Direct fabrication of all-cellulose nanocomposite from cellulose microfibers using ionic liquid-based nanowelding. *Biomacromolecules* 12, no. 11 (2011): 4080–4085.
81. Larsson, P. A., L. A. Berglund, and L. Wågberg. Ductile all-cellulose nanocomposite films fabricated from core–shell structured cellulose nanofibrils. *Biomacromolecules* 15, no. 6 (2014): 2218–2223.
82. Pullawan, T., A. N. Wilkinson, and S. J. Eichhorn. Orientation and deformation of wet-stretched all-cellulose nanocomposites. *Journal of Materials Science* 48, no. 22 (2013): 7847–7855.
83. Hooshmand, S., Y. Aitomäki, M. Skrifvars, A. P. Mathew, and K. Oksman. All-cellulose nanocomposite fibers produced by melt spinning cellulose acetate butyrate and cellulose nanocrystals. *Cellulose* 21, no. 4 (2014): 2665–2678.
84. Zhang, J., N. Luo, X. Zhang, L. Xu, J. Wu, J. Yu, J. He, and J. Zhang. All-cellulose nanocomposites reinforced with in situ retained cellulose nanocrystals during

selective dissolution of cellulose in an ionic liquid. *ACS Sustainable Chemistry and Engineering* 4, no. 8 (2016): 4417–4423.
85. Gindl, W., K. J. Martinschitz, P. Boesecke, and J. Keckes. Structural changes during tensile testing of an all-cellulose composite by in situ synchrotron X-ray diffraction. *Composites Science and Technology* 66, no. 15 (2006): 2639–2647.
86. Benoît, J. C. D., R. H. Newman, and M. P. Staiger. Phase transformations in microcrystalline cellulose due to partial dissolution. *Cellulose* 14, no. 4 (2007): 311–320.

2

Transcrystallinity in Single-Polymer Composites

2.1 Introduction

A review of literature shows that semicrystalline polymers have usually been reinforced with various reinforcements like fibers to form composites with enhanced mechanical properties. These reinforcements can result in variations in morphology and crystallinity of the interphase regions. Various research groups have observed that fibers may act as heterogeneous nucleating agents and nucleate crystallization along the interface with adequately high concentration of nuclei. These nuclei that are formed hinder the lateral extension and start initiating growth in one direction. This is perpendicular to fiber surfaces and results in a columnar crystalline layer, known as transcrystallinity (TC) or transcrystalline layers (TCL), with less thickness [1,2] (Figure 2.1). Jenckel et al. [3] described TC for the first time in 1952.

2.2 Causes of Transcrystallinity

There are different factors that affect the phenomenon of transcrystallinity (Figure 2.2) [4]. As observed from various experiments, transcrystallinity has been seen to be a function of:

a. epitaxy between the fiber and the matrix [5],
b. mismatch of thermal coefficients between the fiber and the matrix [6,7],
c. topography of the fiber [8],
d. chemical composition of the fiber surface [9],
e. thermal conductivity of the fiber [10],

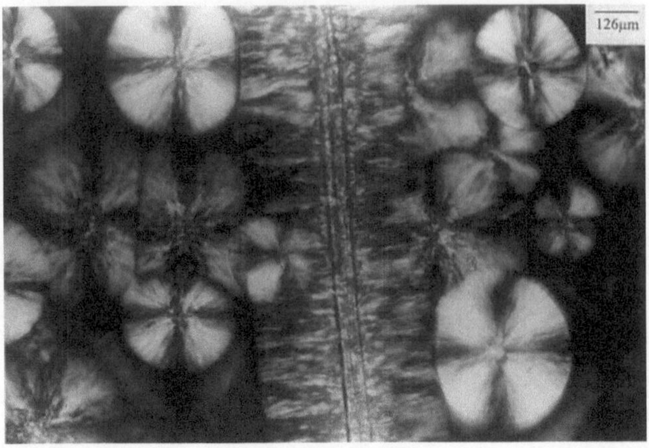

FIGURE 2.1
Optical micrograph of the transcrystallinity layer in dew-retted flax/iPP2 melting. (From Zafeiropoulos, N. E. et al., *Composi. A*, 32, 525, 2001. With permission from ScienceDirect, https://ars.els-cdn.com/content/image/1-s2.0-S1359835X00000580-gr22.jpg; www.sciencedirect.com/science/article/pii/S1359835X00000580, Figure 22; After Ref www.sciencedirect.com/science/article/pii/S1359835X00000580.)

FIGURE 2.2
Transmission optical micrograph showing the morphology of a polypropylene single-fiber model composite isothermally crystallized at 145°C for 3 days—showing the growth of transcrystalline layers. (From Loos, J. et al., *Polymer*, 42, 3827, 2001. With permission from ScienceDirect, www.sciencedirect.com/science/article/pii/S0032386100006601#FIG1, Figure 4 from original reference.)

f. surface energy of the fiber [11],
g. crystallinity of substrate and processing conditions—the important ones being cooling rate and temperature.

Often transcrystallinity may be a result of the combination of the given factors.

2.3 Importance of Fiber Introduction Temperature on Transcrystallinity

Zhang et al. [12] demonstrated that single-polymer composites (SPCs) were prepared from isotactic polypropylene (iPP). They were manufactured by introducing iPP fibers into the molten or supercooled homogeneous iPP matrix. Polarized optical microscopy (POM) and a universal tensile test machine were used to investigate the influences of fiber introduction temperature (Ti) on the resultant morphology of transcrystallinity (transcrystallinity) and mechanical properties of SPCs. The authors also studied the effects of interfacial crystallization on mechanical properties. It was observed that tensile strength of SPCs increased initially to reach a maximum value at Ti = 160°C, and then decreased with further increase in the Ti. Wide-angle X-ray diffraction (WAXD), scanning electron microscopy (SEM), and POM were employed to understand the mechanical enhancement mechanism. It is interesting to note that the improved tensile strength of SPCs was found to be dependent on the synergistic effects of transcrystallinity and high orientation degree of iPP fibers. Good adhesion between the iPP fiber and the matrix also played a role.

In the research work done by the group, a change in the trend of tensile strength was observed. With the increase of the Ti, the tensile strength of SPCs was noted to first rise and reach a maximum value at Ti = 160°C and then decrease to the minimum value at Ti = 175°C. Thus, the group decided to select SPC-145, SPC-160, and SPC-175 as characteristic samples to examine relationship between interfacial features and mechanical property. The SEM pictures are presented in Figure 2.3. A groove-like interspace was found to have existed for SPC-145, 160. The figures, according to the researchers, indicate that interfacial adhesion between iPP fiber and matrix in case of SPC-160 was higher than that in SPC-145. The research group proposed that this could be considered a main reason for the increased tensile strength for SPC-160 by 20.6% compared to that of SPC-145. However, in the case of SPC-175, the boundary lines between the iPP fiber and the matrix were found to have completely disappeared. This was an indication of an excellent interface. Additionally, the iPP fiber was hardly distinguished from the matrix, as evident from Figure 2.3. SPC-175 shows best interfacial adhesion between the iPP fiber and the iPP matrix. The research group found that superior interfacial adhesion happened at the expense of orientation relaxation due to the melting of iPP fiber. This was confirmed by the FTIR results.

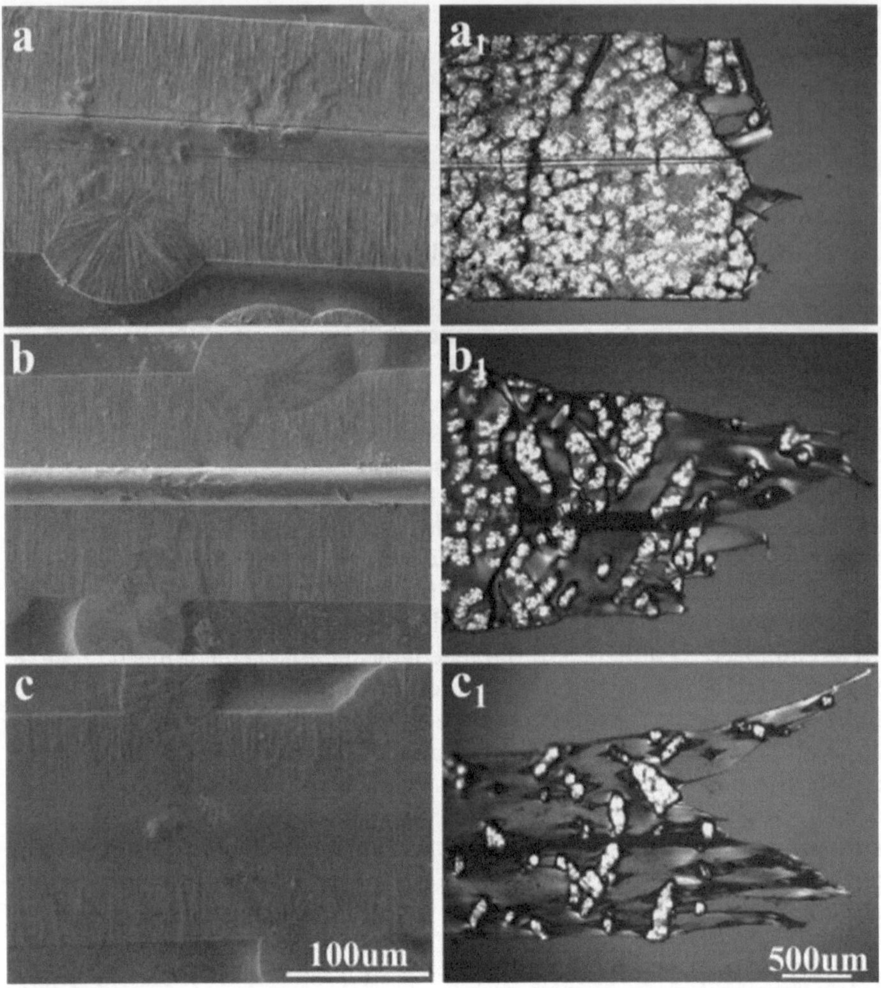

FIGURE 2.3
SEM images of interfacial features of SPCs, and (a1-c1) POM micrographs of the fractured zone of the SPCs. (a) and (a1): SPC-145; (b) and (b1): SPC-160; (c) and (c1): SPC-175. (From Zhang, L. et al., *Polymer*, 90, 18, 2016. With permission from ScienceDirect, www.sciencedirect.com/science/article/pii/S003238611630132X, Figure 8a–c.)

2.4 Transcrystalline Growth as a Function of Initial Temperature and Degree of Undercooling

The group of Li et al. [13] studied supermolecular structures of iPP fiber/matrix composites. The formation of composites were reviewed as a function of crystallization temperature. The authors observed partial melting of the iPP fibers, which could have facilitated the initiation of β-iPP crystal growth. However, it was observed that interfacial morphology of iPP SPCs was encouraged by its own fiber and was dependent on the crystallization temperature. It was found that transcrystalline structures of negative radial β_{III}-iPP or banded β_{IV}-iPP could be produced within the crystallization temperature range of 105–137°C, whereas transcrystallization zone of pure negative radial α_{II}-iPP crystals was observed at higher crystallization temperatures.

Transcrystalline growth was observed as a function of temperature and as a function of the degree of undercooling in their research. Figure 2.4a shows an optical micrograph of the iPP fiber/matrix composite. This was prepared by initiating iPP fibers into their molten matrix at 173°C. This was subsequently isothermally crystallized at 105°C for 30 min. The researchers for the sake of direct comparison showed an optical micrograph of the iPP fiber/matrix SPC. This composite was prepared by the introduction of iPP fibers into supercooled matrix at 160°C, which was followed by isothermal crystallization at 105°C for 30 min, as shown in Figure 2.4b. It was observed by the researchers that the different initial conditions of iPP fibers during

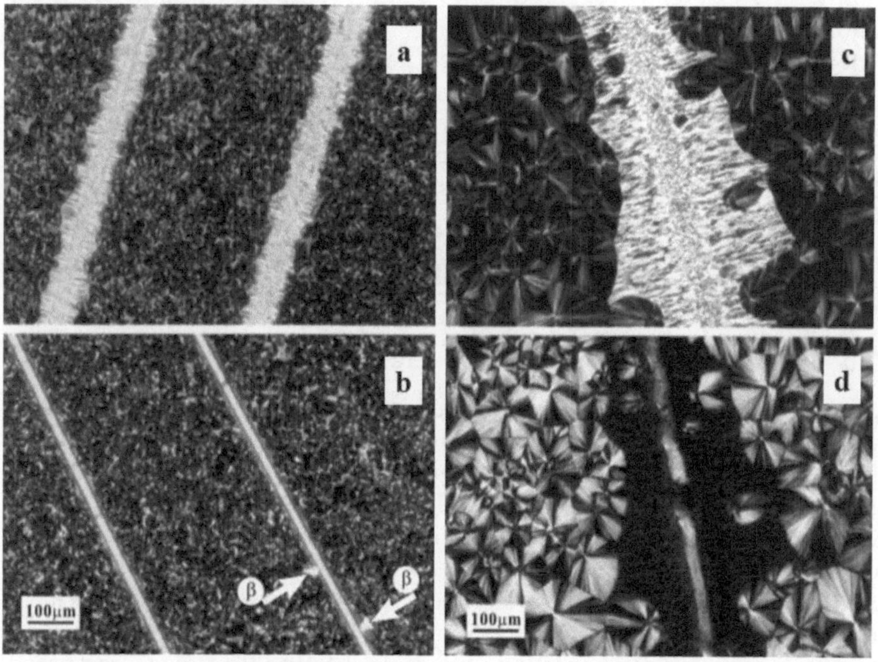

FIGURE 2.4
Polarized optical micrographs of iPP SPCs crystallized isothermally at 105°C for 30 min. The temperatures of fiber introduction were (a) 173°C and (b) 160°C. The arrows in Figure 2.4b indicate the fan-shaped β-iPP crystals. (From Li, H. et al., *Polymer*, 45, 8059, 2004. With permission from ScienceDirect, www.sciencedirect.com/science/article/pii/S0032386104009048; www.sciencedirect.com/science/article/pii/S0032386104009048, Figures 1 and 2.)

crystallization led to the formation of different interfacial morphologies of investigated composites. The width of transcrystalline structures also varied as a function of the degree of undercooling.

A difference in spherulite and transcrystalline formation as a function of crystallization temperature is shown in Figure 2.4b. It was observed by the research group that an increase in crystallization temperature led to a decrease in average nucleus density of the iPP bulk. This resulted in an increment of average spherulite size. Lateral width of the TCL surrounding the iPP fiber was also found to increase. According to the previous research on TCL, such zones should be composed mainly of radial β-iPP crystals. However, in this research, some leaf-shaped α-iPP inclusions randomly dispersed in the β-form transcrystalline iPP were clearly visible. This, according to the researchers, was caused by nucleation and slower crystal growth of the α-iPP with respect to its β-counterpart. The researchers carried out the subsequent melting test, thus confirming that column structure surrounding the iPP fiber was composed of β-iPP crystals.

Polarized optical micrographs of an iPP fiber/matrix composite were crystallized isothermally at 116°C for 30min. The temperature of fiber introduction was 173°C, (c) as prepared sample and (d) after melting at 158°C for 5min.

It is interesting to note that the optical nature of the transcrystallinity layers changed at a function of crystallization temperature.

Figure 2.5a and b shows no regular Maltese Cross, which can be clearly observed in Figure 2.5c with outstanding contrast. This suggests the existence of different supermolecular constructs of these spherulites.

2.5 Effect of Surface Change on Transcrystallinity

According to the study by Vaisman and his group [14], ultra-high-molecular-weight polyethylene (UHMWPE) fibers were treated by photochemical bromination. The analysis of the fibers by XPS and attenuated total reflection Fourier-transform infrared (ATR-FTIR) showed that this process led to the introduction of C–Br and C–OH moieties. Composites were fabricated using either treated or untreated fibers and high-density polyethylene (HDPE) for the matrix. WAXD analysis showed that the treated fibers, by offering a higher concentration of crystallization nuclei, generated a denser TCL with higher specific radial orientation with respect to the fiber axis—compared with the untreated fiber. Furthermore, the introduction of polarity onto the fiber surface enabled analysis of the complex relaxation behavior of PE/PE composites by dielectric spectroscopy. It showed the typical α-, β-, and γ-relaxation processes of polyethylene, combined with the effect of the TCL, generating—amongst other changes—a strong β-transition.

FIGURE 2.5
Polarized optical micrographs of iPP SPCs prepared by introducing the iPP fibers into the molten iPP matrices at 173°C and subsequently crystallized isothermally at (a) 126°C, (b) 130°C, and (c) 133°C. (From Li, H. et al., *Polymer*, 45, 8059, 2004. With permission from ScienceDirect, www.sciencedirect.com/science/article/pii/S0032386104009048; www.sciencedirect.com/science/article/pii/S0032386104009048#Fig3.)

Figure 2.6 a and b gives an idea of the interfacial region of composites prepared using untreated and 20-min brominated fibers, respectively. The researchers treated both the samples identically with permanganic solution. It was observed that both fibers generated a thick TCL (of the order of the fiber width) that had seemingly nucleated inherently with the fiber skin and had grown to form the fiber and transcrystalline growth. Radially oriented lamellae relative to the fiber axis were observed, with some indication of lamella twisting. The TCL around the brominated sample was also found to be slightly denser and more compact.

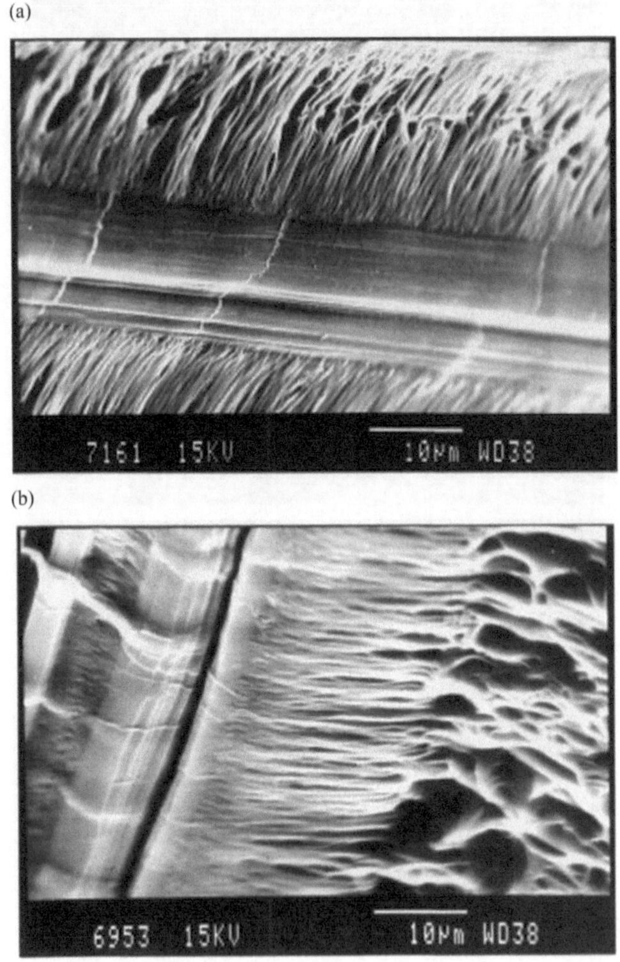

FIGURE 2.6
SEM pictures of the transcrystalline interfacial region in composites based on (a) untreated fibers and (b) surface-modified fibers. (From Vaisman, L. et al., *Polymer*, 44, 1229, 2003. With permission from ScienceDirect, www.sciencedirect.com/science/article/pii/S0032386102008480; www.sciencedirect.com/science/article/pii/S0032386102008480#FIG3.)

Transcrystallinity of HDPE had the same crystalline morphology of the bulk spherulites, depending on the thermal history. Epitaxial nucleation at the fiber surface occurred so that the lamella c-axis (the chain axis) was parallel to the fiber axis. In all the cases the TCL exhibited an oriented crystalline structure in which the orthorhombic crystalline lamellae grew radially into the matrix bulk. The growth proceeded by lamella twisting, wherein the a–c plane rotated around the b-axis (growth direction).

2.6 Matrix Morphology

Polyethylene-based single polymer microcomposites were prepared by Stern et al. [15]. A TCL was developed on the surface of the fiber using different processing conditions. Liquid nitrogen or ice-water quenching, air cooling, and isothermal crystallization were the three conditions used for transcrystalline growth. The morphology of the TCL was found to be controlled mainly by crystalline morphology of the matrix. The morphology was governed by the processing conditions of composite. A smooth and banded transcrystalline morphology matured under ice-water quenching and air cooling conditions, respectively, corresponding to smooth and banded spherulites in the matrix was observed by the researchers. Under isothermal conditions, an apparent rodlike morphology was found to develop in the matrix. Accordingly, the TCL obtained under the same conditions exhibited a similar morphology. X-ray diffraction studies have shown that the crystal structure of all the observed morphologies was orthorhombic. Characterization through X-ray diffraction disclosed an oriented crystalline structure in the TCL. This was in contrast with the isotropic bulk matrix. The predominant orientation perceived was such that the a-axis was at an angle of approximately 39° with the fiber axis, while the b-axis was focused radially outwards from the fiber surface.

The researchers observed that the incidence of a TCL and its magnitudes in the composite was a function of many factors mainly a) type of fibers, b) matrix used, and c) processing conditions. However, the morphology of the TCL observed in a given composite has been found to be governed mainly by the crystalline morphology of the matrix, which in turn was dictated by processing conditions. It is interesting to note that the research group found four different transcrystalline morphologies. Amongst these, three were obtained by avoiding isothermal crystallization, namely, ice-water quenching, liquid nitrogen quenching, or air cooling, while the fourth was obtained under isothermal conditions. According to their inference, there were probably numerous intermediate morphologies that could be possible according to cooling conditions. However, in the TCL obtained under isothermal conditions, the morphology did not change with isothermal time or temperature.

The inherent morphologies of the transcrystalline zones, however, remained the same throughout. All these morphologies were recognized as orthorhombic. This was reflected in measured d-spacings, a different (rodlike) habit apparently formed under isothermal conditions.

2.6.1 Effect of Transcrystallinity

The effect of transcrystallinity has been widely debated by the research groups. However, it has been generally agreed that due to its anisotropy, the formation of a TCL has critical effects on performances of fiber/polymer interfaces. The major effect was in terms of mechanical properties of composites. There are many groups who reported that transcrystallinity resulted in the improvement of properties of some fiber/polymer interfaces and composites. The approach by which a TCL occurred was not fully understood and there are no rules to foresee the appearance of TCL in a particular fiber/matrix system; the effects of TCL on the interfaces and properties of composites remains debatable. Several authors have reported that transcrystallinity can improve shear transfer at the interface, and, consequently, the mechanical properties of composites. According to this group of researchers, a) the fiber material type and its topology and surface coating, b) the matrix type and c) thermal history have mainly affected TCL formation in these composites.

The microstructure of polyethylene (PE)/PE composites, consisting of the HDPE matrix and UHMWPE fibers were investigated by Teishev et al. [16]. Single-fiber composites were prepared and analyzed in a hot-stage crystallization unit attached to a polarizing microscope. The main aim of the research was to observe the effect of the conditions of crystallization on transcrystalline (Tc) growth at the fiber–matrix interface. It was found that a uniform TCL was established on the UHMWPE fiber from the HDPE melt under isothermal conditions. However, it was seen that rapid cooling from the melt disallowed generation of the TCL. The research group measured mechanical properties of unidirectional composite under two conditions—with or without the formation of transcrystalline zone. A comparison of the transverse strength predicted by theoretical models with the experimental values revealed good interfacial adhesion in the PE/PE system. It was shown that the Tc growth had a negligible effect on the composite mechanical properties in the longitudinal direction, whereas it resulted in a 50% decrease of the transverse tensile strength and strain to failure.

It was further observed that transverse tensile strength in single polymer PE composites was significantly higher than that in the PE fiber–reinforced epoxy with untreated fibers and approaches the value reported for plasma-treated fibers. The contrast of transverse strength predicted by the theoretical models by the researchers with values from experiments indicated a good bonding in the PE/PE system. While the longitudinal mechanical properties of the composites are not affected by the absence/presence of the transcrystalline growth, it results in a sharp reduction in the transverse tensile strength and strain to failure. In view of the SEM investigations, it was concluded that the effect was due to a premature brittle failure occurring in a zone of weakness formed adjacent to the fiber surface. It was thought that this zone appears where growing transcrystalline layers meet.

2.7 Conclusion

Transcrystallinity has been extensively studied in fiber-reinforced composite systems [17–26], and there are few papers that have studied the effect of transcrystallinity on mechanical properties of composite systems in much detail [27–30]. This chapter aims to cover the phenomenon of transcrystallinity in SPC systems. Although transcrystallinity has been shown to better properties of some fiber/polymer interfaces and composites, the mechanism by which this happens is not fully understood. There have been no general rules to predict the appearance of transcrystallinity in a combination of particular fiber/matrix system. The effects of transcrystalline growth on interfaces and properties of composites have been debated. Several authors have reported that transcrystallinity can improve shear transfer at the interface, and, consequently, the mechanical properties of composites, whereas others claimed that it has no, or even a negative, effect on these properties. The effects of introduction of fibers, initial temperature and degree of undercooling, change of surface have been discussed in detail in this chapter. In one of the interesting studies, it has been shown that while transcrystallinity hardly improved longitudinal mechanical properties—it, in fact, deteriorated the transverse properties. In general it has been noticed that fiber topography, surface coating, nature of the matrix and processing conditions of composites have influenced transcrystalline growth. The phenomenon of transcrystalline growth has also been found highly specific to the fiber–matrix combination. Few issues like effect of substrate surface energy on transcrystalline growth, surface-induced crystallization, thermal conductivity of fiber can be examined in more detail for single polymer systems.

References

1. Quan, H., Z. M. Li, M. B. Yang, and R. Huang. On transcrystallinity in semicrystalline polymer composites. *Composites Science and Technology* 65, no. 7–8 (2005): 999–1021.
2. Zafeiropoulos, N. E., C. A. Baillie, and F. L. Matthews. A study of transcrystallinity and its effect on the interface in flax fibre reinforced composite materials. *Composites Part A: Applied Science and Manufacturing* 32, no. 3–4 (2001): 525–543.

3. Jenckel, E., E. Teege, and W. Hinrichs. Transkristallisation in hochmolekularen Stoffen. *Kolloid-Zeitschrift* 129, no. 1 (1952): 19–24.
4. Mukhopadhyay, S., B. L. Deopura, and R. Alagiruswamy. Interface behavior in polypropylene composites. *Journal of Thermoplastic Composite Materials* 16, no. 6 (2003): 479–495.
5. Loos, J., T. Schimanski, J. Hofman, T. Peijs, and P. J. Lemstra. Morphological investigations of polypropylene single-fibre reinforced polypropylene model composites. *Polymer* 42, no. 8 (2001): 3827–3834.
6. Thomason, J. L., and A. A. Van Rooyen. Transcrystallized interphase in thermoplastic composites. *Journal of Materials Science* 27, no. 4 (1992): 897–907.
7. Sukhanova, T. E., F. Lednický, J. Urban, Y. G. Baklagina, G. M. Mikhailov, and V. V. Kudryavtsev. Morphology of melt crystallized polypropylene in the presence of polyimide fibres. *Journal of Materials Science* 30, no. 9 (1995): 2201–2214.
8. Gray, D. G., and J. E. Guillet. Open tubular columns for studies on polymer stationary phases by gas chromatography. *Journal of Polymer Science Part C: Polymer Letters* 12, no. 4 (1974): 231–235.
9. Chen, E. J. H., and B. S. Hsiao. The effects of transcrystalline interphase in advanced polymer composites. *Polymer Engineering & Science* 32, no. 4 (1992): 280–286.
10. Cai, Y., J. Petermann, and H. Wittich. Transcrystallization in fiber-reinforced isotactic polypropylene composites in a temperature gradient. *Journal of Applied Polymer Science* 65, no. 1 (1997): 67–75.
11. Schonhorn, H., and F. W. Ryan. Effect of morphology in the surface region of polymers on adhesion and adhesive joint strength. *Journal of Polymer Science Part B: Polymer Physics* 6, no. 1 (1968): 231–240.
12. Zhang, L., Y. Qin, G. Zheng, K. Dai, C. Liu, X. Yan, J. Guo, C. Shen, and Z. Guo. Interfacial crystallization and mechanical property of isotactic polypropylene based single-polymer composites. *Polymer* 90(2016): 18–25.
13. Li, H., X. Zhang, Y. Duan, D. Wang, L. Li, and S. Yan. Influence of crystallization temperature on the morphologies of isotactic polypropylene single-polymer composite. *Polymer* 45, no. 23 (2004): 8059–8065.
14. Vaisman, L., M. F. González, and G. Marom. Transcrystallinity in brominated UHMWPE fiber reinforced HDPE composites: morphology and dielectric properties. *Polymer* 44, no. 4 (2003): 1229–1235.
15. Stern, T., G. Marom, and E. Wachtel. Origin, morphology and crystallography of transcrystallinity in polyethylene-based single-polymer composites. *Composites Part A: Applied Science and Manufacturing* 28, no. 5 (1997): 437–444.
16. Teishev, A., and G. Marom. The effect of transcrystallinity on the transverse mechanical properties of single-polymer polyethylene composites. *Journal of Applied Polymer Science* 56, no. 8 (1995): 959–966.
17. Chen, E. J. H., and B. S. Hsiao. The effects of transcrystalline interphase in advanced polymer composites. *Polymer Engineering & Science* 32, no. 4 (1992): 280–286.
18. Klein, N., G. Marom, A. B. A. U. D. T. Pegoretti, and C. B. A. U. D. T. Migliaresi. Determining the role of interfacial transcrystallinity in composite materials by dynamic mechanical thermal analysis. *Composites* 26, no. 10 (1995): 707–712.
19. Pompe, G., and E. Mäder. Experimental detection of a transcrystalline interphase in glass-fibre/polypropylene composites. *Composites Science and Technology* 60, no. 11 (2000): 2159–2167.

20. Felix, J. M., and P. Gatenholm. Effect of transcrystalline morphology on interfacial adhesion in cellulose/polypropylene composites. *Journal of Materials Science* 29, no. 11 (1994): 3043–3049.
21. Joseph, P. V., K. Joseph, S. Thomas, C. K. S. Pillai, V. S. Prasad, G. Groeninckx, and M. Sarkissova. The thermal and crystallisation studies of short sisal fibre reinforced polypropylene composites. *Composites Part A: Applied Science and Manufacturing* 34, no. 3 (2003): 253–266.
22. Nuriel, H., N. Klein, and G. Marom. The effect of the transcrystalline layer on the mechanical properties of composite materials in the fibre direction. *Composites Science and Technology* 59, no. 11 (1999): 1685–1690.
23. Sanadi, A. R., and D. F. Caulfield. Transcrystalline interphases in natural fiber-PP composites: effect of coupling agent. *Composite Interfaces* 7, no. 1 (2000): 31–43.
24. Klein, N., G. Marom, and E. Wachtel. Microstructure of nylon 66 transcrystalline layers in carbon and aramid fibre reinforced composites. *Polymer* 37, no. 24 (1996): 5493–5498.
25. Dasari, A., Z. Z. Yu, and Y. W. Mai. Transcrystalline regions in the vicinity of nanofillers in polyamide-6. *Macromolecules* 40, no. 1 (2007): 123–130.
26. Assouline, E., E. Wachtel, S. Grigull, A. Lustiger, H. D. Wagner, and G. Marom. Lamellar orientation in transcrystalline γ isotactic polypropylene nucleated on aramid fibers. *Macromolecules* 35, no. 2 (2002): 403–409.
27. Teishev, A., S. Incardona, C. Migliaresi, and G. Marom. Polyethylene fibers-polyethylene matrix composites: Preparation and physical properties. *Journal of Applied Polymer Science* 50, no. 3 (1993): 503–512.
28. Stern, T., A. Teishev, and G. Marom. Composites of polyethylene reinforced with chopped polyethylene fibers: effect of transcrystalline interphase. *Composites Science and Technology* 57, no. 8 (1997): 1009–1015.
29. Marais, C., and P. Feillard. Manufacturing and mechanical characterization of unidirectional polyethylene-fibre/polyethylene-matrix composites. *Composites Science and Technology* 45, no. 3 (1992): 247–255.
30. Borysiak, S. Fundamental studies on lignocellulose/polypropylene composites: effects of wood treatment on the transcrystalline morphology and mechanical properties. *Journal of Applied Polymer Science* 127, no. 2 (2013): 1309–1322.

3

Single-Polymer Composites from Polyolefins

3.1 Introduction

Polyolefins, as defined by *Encyclopaedia Britannica* [1], are "a class of synthetic resins prepared by the polymerization of olefins." Olefins are hydrocarbons whose molecules contain a pair of carbon atoms linked together by a double bond. The polyolefins can be classified based on their monomeric unit and chain structures as ethylene-based polyolefins, propylene-based polyolefins, higher polyolefins, and polyolefin elastomers. Generally, ethylene-based polyolefins are created under low conditions of pressure using transition metal catalysts, resulting in predominantly linear chain structure. The alternative route is to apply high-pressure conditions using oxygen or peroxide initiators. This process results in predominantly branched chain structures of various crystallinity levels and densities. There has been a lot of research regarding conditions of polymerization resulting in control of branching and crystallinity of such polymers. Propylene-based polyolefins are, however, produced with transition metal catalysts resulting in linear chain structures with stereospecific arrangement of the propylene units or special stereoblock structures from a single-site catalyst.

The first reference to single-polymer composites (SPCs) dates back to 1975. Capiati et al. [2] showed that SPC materials were made using a variance in melting points between the components. Transcrystalline growth was observed in the melt matrix and at the interface. Partial melting between fiber and matrix resulted in a strong and intimate interfacial bond, with a gradient in morphologies of high-density polyethylene (PE). The interfacial strength was evaluated with pull-out test. The researchers commented that the interfacial strength in the PE composites was mainly due to the unique epitaxial bonding rather than the radial forces from compressive shrinkage.

Marais et al. [3], in their subsequent research, indicated that SPCs were manufactured using high-modulus PE fibers in a polyethylene matrix (PE/PE). The initial discussion was what matrix characteristics would be essential to obtain a good composite with fiber volume fraction about 0.7. It has been observed that in the fiber direction, strength and modulus values were in-between those of carbon/epoxy and aramid/epoxy composites. As with the other class of composites, the usual weakness was observed for off-axis properties. Material irradiation had a negligible effect on both fracture and time-dependent properties. Physical aspects like crystallinity and electromagnetic properties of PE/PE were also discussed.

A major issue in these class of composites was compatibility between the fiber and matrix, which although a single polymer type, were usually of diverse molecular structure. Furthermore, wetting out of the fiber by viscous thermoplastic "matrix" can be a problem, unless a solvent route was used.

The selective surface melting phenomenon was, however, investigated intensely in some of the studies. Production of solid-section highly oriented PE by compaction of melt-spun PE fibers has been explained in one of these studies [4]. Various characterizing techniques like differential scanning calorimetry (DSC), X-ray diffraction, and electron microscopy have been used to determine the structure of the polymer. Selective surface melting of the fibers was found to be essential to form a polyethylene/polyethylene composite of very high integrity. The challenge is to maintain a very high share of the strength and stiffness of the fibers, which often gets disturbed when polymers are subjected to high temperatures close to their melting points. I M Ward [5] did an extensive review of these series of attempts in one of the Advances in Polymer Science series and the reader may refer to the whole chapter.

3.2 Single-Polymer Composites with Varying Starting Materials Based on PE

3.2.1 TENFOR

Ward's research group [6] used TENFOR as the starting material for manufacturing SPCs. TENFOR is the trade name given to high-modulus PE fibers produced by the melt-spinning/hot-drawing route. It was realized that under the right conditions of temperature and pressure, it is possible to melt a very small proportion of each fiber. On cooling, this molten material recrystallizes to bind the structure together and fills all the interstitial voids in the sample, leading to a substantial retention of the original fiber properties. For a hexagonal close-packed array of cylinders, only 10% of melted material was needed for this purpose. With low compaction temperature, there was insufficient melt to fill the interstices. It was observed that the fibers distorted into polygonal shapes. It was also noticed that the transverse strength developed was insufficient. Above the optimum temperature, the amount of polymer melt increased, causing reduced stiffness of the composite. It was found that melt recrystallized and nucleated on the oriented fibers, giving similarly oriented growth. The researchers also observed that in regions of high melting and cooling, sufficiently rapid expansion away from the nucleus was accompanied by a supportive rotation in chain orientation. This is similar to banding in spherulites.

3.2.1.1 Gel-Spun Fibers

The general concept of gel spinning [7] has been given in Figure 3.1 and discussed in detail in several references [8,9]. Gel-spun fibers were also experimented by Ward's group [10]. The compaction of gel-spun high-molecular-weight PE fiber has been examined for a variety of compaction temperatures between 142°C and 155°C.

The gel-spun PE fibers, with the commercial name SPECTRA 1000, were unidirectionally wound around a frame. This was then arranged in a parallel configuration in a matched metal mold, which was placed in a heated compression press. The required compaction temperature was maintained. The temperature of the fiber assembly was monitored by a thermocouple in the mold. A pressure of 2.8 MPa (400 psi) was applied and the sample was left for 10 minutes.

The temperature dependence of the shrinkage of the SPECTRA 1000 fiber has been found to be very interesting. With increasing temperature, the shrinkage of the fiber increases very slowly, being lower than 10% until the temperature reaches 140°C. This is very important if the tension in the experimental system has to be maintained. As soon as the temperature passes 140°C, the shrinkage surges rapidly and reaches its maximum value fast. It has been observed that this pattern of shrinkage behavior is typical of both

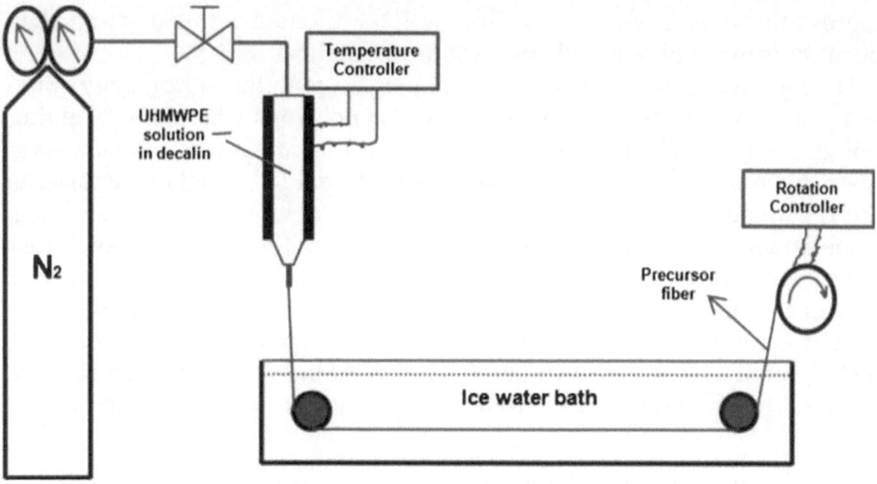

FIGURE 3.1
A schematic of gel spinning of PE fibers. (From Zhang, W. et al., *Composi.* B, 51, 276, 2013. With permission from ScienceDirect, www.sciencedirect.com/science/article/pii/S1359836813001169#f0005, Figure 1.)

gel-spun and melt-spun–drawn PE fibers. The researchers found that crystalline morphology effectively constrains the structure so that further shrinkage is observed when the melting range is reached and there is adequate flexibility for the molecular chains to slide within the crystalline domains.

Differential scanning calorimetry, scanning electron microscopy (SEM), and broad-line nuclear magnetic resonance (NMR) techniques were used to study the structure of the compacted materials and to determine mechanisms of compaction. The researchers found that if the fibers were compacted in the range of 142°C–154°C, there was no evidence of significant surface melting. Therefore, all the ensuing compacted composites had similar morphologies and mechanical properties. In addition, the peak melting temperatures of the compacted materials were very similar and close to that of the partially constrained fiber. However, when the compaction temperature was increased to 155°C, there was evidence that fibers had melted to some extent and had recrystallized from the continuous fiber crystals into a new form of lamellar crystals during the process.

The importance of the temperature window was highlighted in this research. Single-polymer composites often have their best properties in a very narrow process window. DSC and SEM studies revealed that no obvious surface melting and recrystallization occurred during the process of hot compaction in the temperature range of 144°C–154°C. However, the rigid crystalline fraction measured by NMR for all compacted materials was significantly lower than that of the original fiber. Significant improvement in transverse strength also was found at lower compaction temperatures. Structural investigations show how the fibers deform so as to interlock, and localized welding occurs, so as to bond each fiber to its neighbor. The importance of the process window was very well highlighted in this study and has been confirmed subsequently by various researchers.

The researchers found a very interesting feature in the morphology of the compacted materials. Well-packed irregular polygons with some "welding points" were observed. Compared with the hot compaction of melt-spun fiber, the loss in original fiber modulus was found to be higher, because, as the researchers opined, a considerable amount of crystalline material gets transformed to oriented mobile chains. However, it is also to be noted that the higher modulus of the gel-spun fiber had the consequence that under certain conditions the final properties of the compacted gel-spun fiber composites are very similar to those of a compacted composite produced from melt-spun fiber. This research gave a very good base for understanding the comparative analyses of SPCs from melt-spun versus gel-spun material.

3.2.2 Oriented Fibers and Tapes

There has been research on SPCs starting with oriented fibers and tapes. Assemblies of oriented fibers and tapes were produced over a range of processing temperatures. Figure 3.2 shows a woven cloth made from

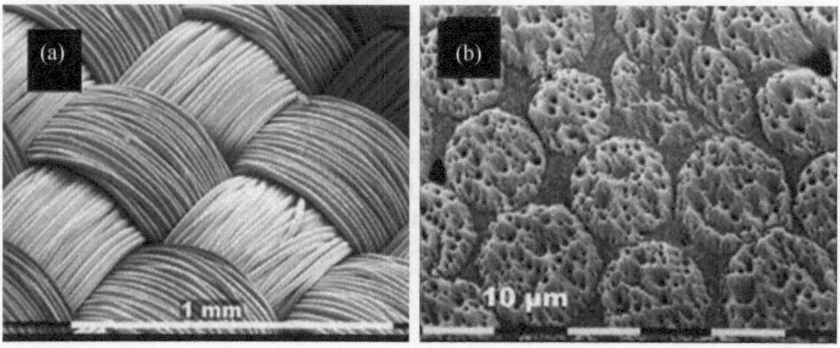

FIGURE 3.2
Typical PE-based SRC: (a) a woven cloth made from multifilament bundles; (b) permanganic etched image of compacted woven melt-spun multifilaments. (From Schneeberger, C. et al., *Composi. A*, 103, 69, 2017. With permission from ScienceDirect, www.sciencedirect.com/science/article/pii/S0079670011001237, Figure 1.)

multifilament bundles, and a permanganic etched image of compacted woven melt-spun multifilaments from a PE-based single polymer reinforced composites. The next challenge was to choose the right temperature. It was chosen so that only the interleaved film melted, and in other cases, both the film and the fiber surfaces melted. This permitted a traditional film-stacking process. In most cases, the optimum compaction temperature was about 1°C below the point at which substantial crystalline melting occurred. At this optimum temperature, DSC melting studies showed that about 30% of the original oriented phase had been lost to bonding the structure together. The researchers observed a growth of transcrystalline regions in the melt matrix at the interface. Additionally, partial melting was observed between fiber and matrix. These were indications of a strong and intimate interfacial bond with a gradient in morphologies for high-density PE. The pull-out test is a simple and adequate method for evaluating the interfacial shear strength of one polymer composites. However, the interfacial strength in the PE composites is due mainly to the unique epitaxial bonding rather than the radial forces from compressive shrinkage.

Deng and Shalaby [11] found that tensile modulus, strength, and resistance to creep were significantly increased when ultra-high-molecular-weight polyethylene (UHMWPE) fibers were used in conjunction with a UHMWPE matrix. It was observed that tensile strength of the resultant SRCs increased with fiber content, according to the law of mixtures. Interestingly, transverse strength remained unchanged for fiber contents of <7%. While the double notch impact strength of the composites was higher than plain UHMWPE, there was no alteration in the wear properties between the composites and the plain UHMWPE. Cross-sectional and tensile fracture surfaces of the composites were examined using SEM, and the overall results indicated that UHMWPE SRCs have good potential in load-bearing biomedical applications. Thus, overall findings indicated that the tensile and impact strength improved for SPCs.

Guan et al. [12] described the effect of mold temperature on mechanical performance and microstructure of high-density polyethylene (HDPE) SRCs prepared by melt deformation in an oscillating stress field. The research led to the findings that Young's modulus and yield strength of the HDPE increased from 1 to 3.5 GPa and from 23 to 87 MPa, respectively. Interestingly, DSC curves of portions taken from the SRCs exhibited double-peak melting endotherms, depending on the distance from the sample surface. The low-temperature peak was attributed to the presence of spherulites, while the high-temperature peak was attributed to the melting of the shish–kebab crystals that made up 20% of the crystalline phase. In addition, wide-angle X-ray diffraction (WAXD) measurements revealed a preferred orientation of the molecular chains, which, combined with the production of shish–kebab crystals, resulted in significantly improved mechanical properties.

3.2.3 Use of a Combination of Different Grades of Polyethylene

Some of the most common grades of polyethylene include low-density polyethylene (LDPE), linear low-density polyethylene (LLDPE), HDPE, and ultra-high-molecular-weight polypropylene (UHMPE) (Figure 3.3). HDPE is chemically the closest in structure to pure PE. It consists principally of unbranched molecules with very few flaws to disturb its linearity [13]. LDPE contains extensive concentrations of branches that obstruct the crystallization process, resulting in relatively low densities. The branches primarily consist of ethyl and butyl groups, together with some long-chain branches. LLDPE resins consist of molecules with linear PE backbones. Short alkyl groups are attached to these backbones at random intervals. These materials are generally produced by the copolymerization of ethylene with 1-alkenes. Other variants include medium-density polyethylene, ultra-low-molecular-weight

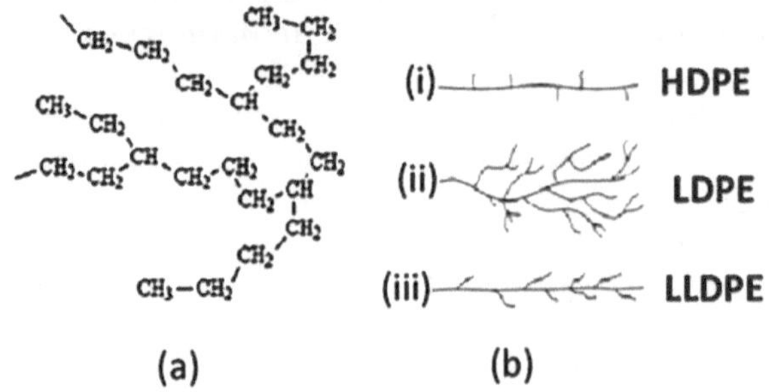

FIGURE 3.3
Structure of polyethylenes. (From Gao, C. et al., *Prog. Polym. Sci.* 37, 767, 2012. With permission from ScienceDirect, www.sciencedirect.com/science/article/pii/S2213343715000056#fig0005, Figure 1.)

polyethylene (ULMWPE or PE-WAX), high-molecular-weight polyethylene (HMWPE), high-density cross-linked polyethylene, cross-linked polyethylene (PEX or XLPE), very-low-density polyethylene (VLDPE), and chlorinated polyethylene.

Various combinations have been used by researchers for the manufacturing of self-reinforced composites to investigate structure and properties.

3.2.3.1 HDPE and LDPE

In an article by Zhang et al. [14], a significant improvement in toughness of HDPE by LDPE in injection molding was reported. When the researchers compared the final properties of the composites with self-reinforced pure HDPE, the tensile strength of HDPE/LDPE (80/20 wt%) was found to be maintained at the same level, and toughness increased. Multilayer structure on the fracture surface of self-reinforced HDPE/LDPE specimens was observed by the scanning electron microscope. The central layer of the surface fractured in a ductile manner, whereas the shear layer breakage was somewhat brittle. An increase of strength and modulus was observed by the researchers. They opined that it was a combined effect of high orientation of macromolecules along the flow, crystallization, and a structure consisting of shish–kebab crystals. Co-crystallization between HDPE and LDPE was also confirmed from DSC and WAXD.

3.2.3.2 UHMWPE/HDPE

The second combination in PE was tried by Zhuang et al. [15]. They investigated the damage mechanisms in self-reinforced PE composite laminates (UHMWPE/HDPE) under tensile loading. They used the acoustic emission (AE) technique for fracture analyses. Fracture surfaces were examined using a scanning electron microscope. The research group used model specimens showing a dominant failure mechanism. They then established correlations between observed damage growth mechanisms and AE results in terms of the events amplitude. The researchers commented that these correlations could be used to monitor the damage growth process in the UHMWPE/HDPE composite laminates exhibiting multiple modes of damage. Results from this study revealed that the AE technique is a viable and effective tool for identifying damage mechanisms.

In composite fracture science, damage growth process is an accumulation of basic damage mechanisms, such as fiber–matrix interfacial debonding, matrix cracking, fiber breakage, and delamination. Results revealed that fiber–matrix interfacial debonding, matrix plastic deformation and cracking, fiber pull-out, fiber breakage, and interlaminar delamination were associated with AE events with specific amplitude range. These correlations can be used to monitor the damage growth process in the UHMWPE/HDPE composite laminates exhibiting multiple modes of damage.

3.2.3.3 UHMPE/LDPE

Another combination was used by Lacroix et al. [16], a novel processing route for ultra-high-modulus PE fiber and low-density PE matrix composites. The processing route was different from the previous procedures. It involved an impregnation step in a saturated matrix solution. The growth of a matrix layer on the fiber surfaces in such a solution has been shown by SEM micrographs. Further processing steps are the preparation of prepregs and hot compacting to form PE/PE composites. In this study, unidirectional composites were manufactured at three different temperatures. Mechanical testing showed a high modulus and tensile strength. The composite compressive strength is comparatively low as a result of the low compressive strength of the PE fiber.

3.3 Single-Polymer Composites from Polypropylene

Polypropylene (PP) (Figure 3.4) has the physical characteristics of low specific gravity, rigidity, heat resistance, and superior workability. In addition, since it is reasonably low in cost, it is used in a range of applications, such as films, furniture, and industrial components for automobiles. It has been more than 50 years since 1954, when G. Natta et al. of Italy were successful in synthesizing high-molecular-weight, highly crystalline PP. Polypropylene has higher strength than other polyolefins, low density, and acceptable thermal resistance and is available in many grades differing in molar mass and copolymer type and distribution. It has two main limitations: PP is brittle at lower temperatures, has moderate strength and poor long term mechanical properties like creep. Polypropylene was first manufactured by Natta in 1955 using organometallic catalysts. A wide range of polypropylene homo and copolymers were developed, and the polymer is amongst the most widely used in the automobile industry.

3.3.1 Different Approaches to Single-Polymer Composites from Polypropylene

3.3.1.1 Microcellular Injection Molding

Wang et al. [17] performed an insert-microcellular injection-molding [18] (Figure 3.5) process on an injection-molding machine equipped with a

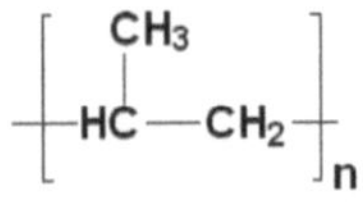

FIGURE 3.4
General structure of polypropylene.

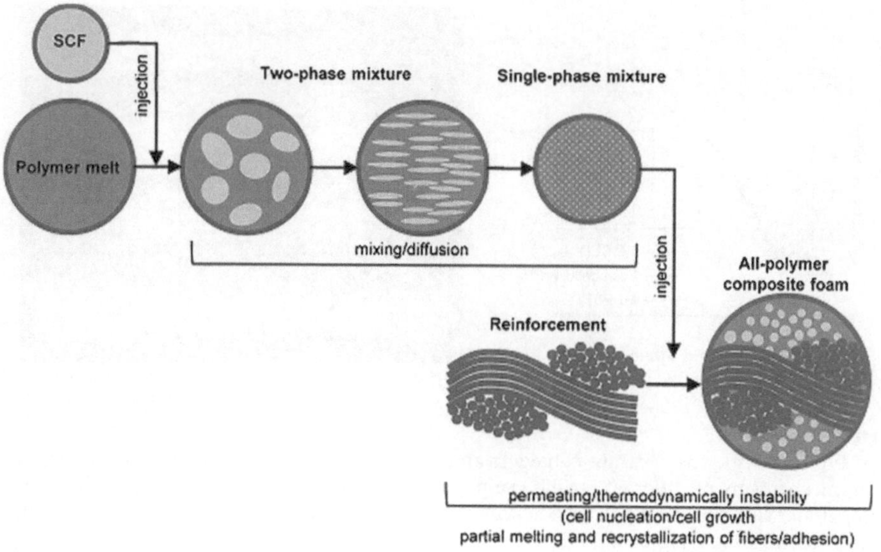

FIGURE 3.5
Schematic of microcellular injection molding. (From Cantero, G. et al., *Compos. Sci. Technol.*, 63, 1247, 2003. With permission from ScienceDirect, www.sciencedirect.com/science/article/pii/S0263822317319189#f0005, Figure 1.)

supercritical fluid (SCF) system. Microcellular injection molding is one of the recent physical foaming processes that use SCF such as nitrogen and carbon dioxide. It is interesting to note that microcellular foam has huge numbers of cells less than 100 µm. Microcellular injection molding, used in the industry, brings a lot of benefit such as weight reduction and improved dimensional stability.

This methodology has been used by a few research groups for SPCs. Microcellular PP SPCs unite advantages of SPCs with those of the microcellular plastics. These new class of materials hold the promise for reduced weight, enhanced recyclability, and improved fiber–matrix interface. In comparison with the solid PP, weight reductions of the tensile and impact microcellular PP SPCs (MPPSPCs) could be up to 12.9% and 3.3%, respectively. There is a significant improvement in the tensile and impact properties. The tensile and impact strengths of the MPPSPCs were improved by 59% and 1799%, respectively. On the basis of the tensile properties, the injection temperature of 220°C and injection speed of 70 mm/s were the optimum processing conditions for the MPPSPC samples. The typical morphology of the MPPSPC sample includes five different layers: sandwiched fabric layer, transition layer between fabric and core, center core layer, transition layer between skin and center core, and skin layer.

The typical tensile stress–strain behaviors of the different samples are all demonstrated in Figure 3.6a. It is interesting to note that microcellular PP exhibited reduction of both strength and strain at break, compared

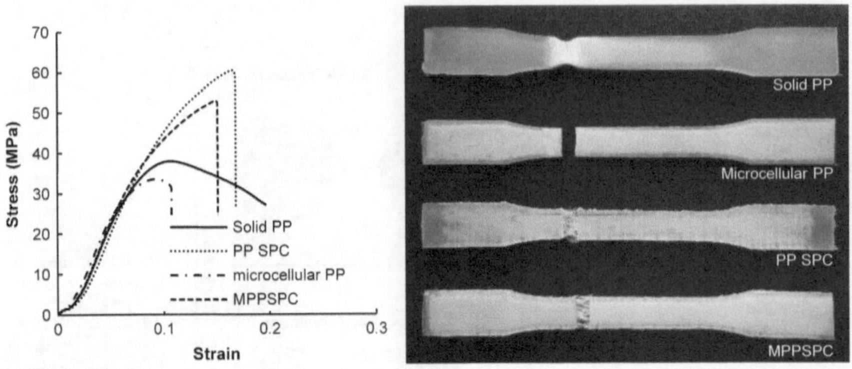

(a) (b)

FIGURE 3.6
(a) Typical tensile stress–strain behaviors and (b) an illustrative comparison of different failure mechanisms of different tensile samples. (From Wang, J. and D. Chen., *Compos., A*, 90, 567, 2016. With permission from ScienceDirect, www.sciencedirect.com/science/article/pii/S1359835X1630272X, Figure 4.)

to the other samples. Both MPPSPCs and PP SPCs showed a significant increase in strength, and PP SPC showed slightly higher strength and strain at break than the MPPSPC. The mechanical properties of MPPSPC foams were influenced by both the presence of the reinforcements and the cell morphology; the effect of reinforcements mainly depends on the interfacial bonding. The research group commented on the toughness of these different classes of polypropylene materials. In comparison with the microcellular PP, the MPPSPC was found to be not only stiffer and stronger, but also had higher toughness. Figure 3.6b shows an illustrative comparison of different failure mechanisms of different tensile samples. In the case of solid PP, it exhibited high extensibility and necking, indicating elastic deformation in the failure section due to its high ductility. Whereas, the PP SPC, microcellular PP, and MPPSPC all presented plastic deformation. Plastic deformation is a process in which permanent deformation is triggered by an adequate load. It produces a permanent alteration in the shape or size of a solid body, resulting from the application of sustained stress beyond the elastic limit. Very clear fracture surfaces indicated that the embrittlement occurred on microcellular PP, which may be due to the microcellular structures. From the plastic deformation in the failure section of the PP SPC and MPPSPC, it can be deduced that interfacial bonding was achieved.

The mechanical properties of the samples prepared by various manufacturing processes were compared. The tensile test results of solid PP, PP SPC, microcellular PP, and MPPSPC under the same process conditions are shown in Figure 3.7. It demonstrates that the PP SPC had much higher tensile strength, followed by the MPPSPC, and the microcellular PP had the lowest

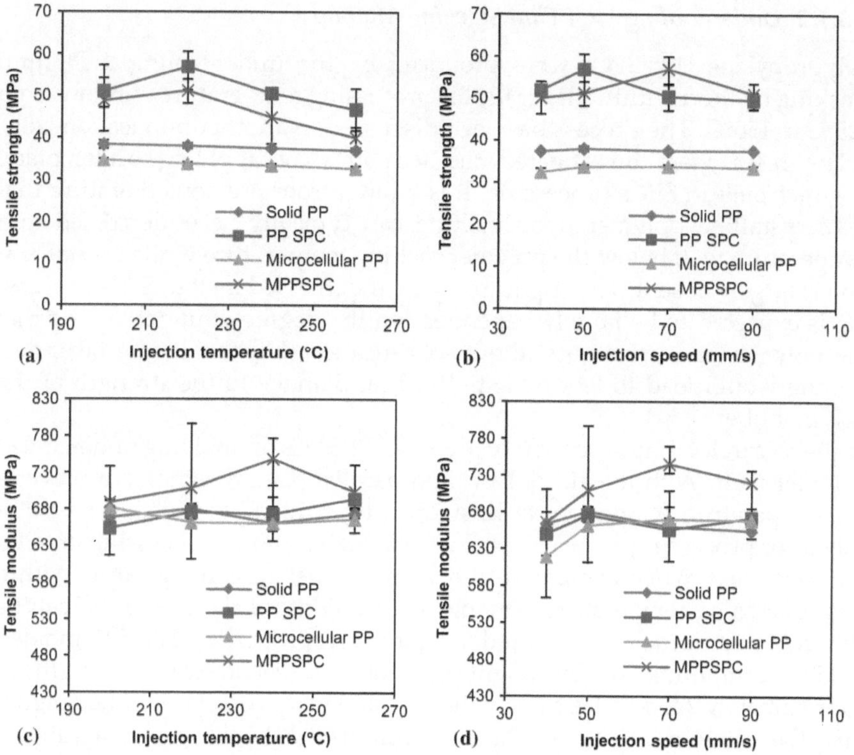

FIGURE 3.7
Tensile strength and modulus of solid PP, PP SPC, microcellular PP, and MPPSPC as a function of injection temperature and injection speed. (From Wang, J. and D. Chen., *Compos., A*, 90, 567, 2016. With permission from ScienceDirect, www.sciencedirect.com/science/article/pii/S1359835X1630272X, Figure 5.)

tensile strength. It is hence important to note that if applications involve strength, microcellular SPCs would not be the best choice. The optimum tensile strength of the MPPSPC was 60 MPa, 75% higher than that of the microcellular PP and 59% higher than that of the solid PP.

Tensile strength of 60 MPa was also higher than composites from the glass fiber (11.8%)/PP and glass fiber (21.8%)/PP. The research group further observed that in most cases the MPPSPCs had much better tensile properties than other samples. This might have been due to the interfacial bonding. The better flowability of the melt–gas solution can benefit permeability and interfacial adhesion. They commented that cell nucleation and growth created pressure in the cavity and thus further improved interfacial property of the MPPSPC. In addition, the cooling of the microcellular foam is much faster than the regular injection molding because the energy release during injection helps to cool the microcellular part quickly, which contributes to the reduction of melted fibers.

3.3.1.2 Undercooling Melt Film Stacking Method

Polypropylene [19] SPCs were produced by the undercooling melt film-stacking method within a large usable processing temperature window from 125°C to 150°C. The process has been discussed in another publication [20].

The researchers investigated the feasibility of applying undercooled polymer melt in SPCs processing. It is known from previous literature that semicrystalline polymer upon melting can typically be undercooled to a temperature well below the polymer melting temperature while crystallization is largely absent. The applicability of undercooled melt in SPCs processing is expected to be largely dependent on the degree of undercooling that the polymer can undergo without solidification. Higher degree of undercooling would lead to less potential of heat damage to the strength of the polymer fiber.

The research group successfully prepared PP SPCs by applying undercooled polymer melt. With the aid of DSC and parallel-plate rheometry, a processing temperature window of at least 25°C, from 125°C to 150°C, was established for processing PP SPCs. This is very high as far as processing of SPCs are concerned, when compared to results of other research groups. Within this processing temperature window, high fluidity of the matrix PP can be obtained without significantly reducing the fiber properties. The SPC molded at 150°C containing 50% by weight of PP fabric achieved tensile strengths of approximately 220 and 180 MPa in the weft and warp directions, much higher than the value of 30 MPa for the non-reinforced PP. Likewise, a significant improvement in storage modulus was achieved in the PP SPCs over the non-reinforced PP. The processing temperature was found to not only affect the feasibility of processing, but also affect the quality of the SPC. In particular, the tensile strength of PP SPCs decreased in both the weft and warp directions as the processing temperature decreased. The researchers opined that decrease in tensile strength may be correlated with increased viscosity and consequently reduced wetting quality of the PP fabric by the PP matrix.

Wang et al. took clues from the previous work and prepared samples with good stiffness and strength using 145±5°C. The undercooling compaction temperatures as found by the researchers were all lesser than hot compaction temperatures in traditional processing. Tensile and peel tests were carried out. The effects of compaction temperature, holding time, and compaction pressure on mechanical properties of the PP SPCs were investigated. Compaction temperature was found to be the main factor in influencing properties of the PP SPCs. At different compaction temperatures, both holding time and compaction pressure have different influences. Especially, the holding time at undercooling temperature showed different effects compared with the traditional SPCs processing. The morphological properties of the PP SPCs were also examined and good interfacial adhesion was shown.

During processing, the undercooling compaction temperature is one of the most important factors required for maximizing the adhesion and keeping the original fiber structure. The effects of compaction temperature on the tensile strength and modulus of PP SPCs can be seen in Figure 1 where the holding time has been varied. The tensile strength and modulus in the weft direction were all higher than that in the warp direction, because the weft density is higher than the warp density. Figure 3.8a shows that the tensile strength in both the warp and weft directions increased with the increase in temperature. At low compaction temperature, lower tensile strength reflects incomplete compaction and bad adhesion property. The tensile strength could be up to a maximum value of 172 MPa in the warp direction and 208 MPa in the weft direction when the compaction temperature was increased to 150°C. The highest tensile strength indicates the improvement of fiber wetting with increasing temperature. The modulus values showed a distinctly different trend compared to that of strength. The tensile modulus increased first and then decreased (Figure 3.8b–d). The maximum modulus emerged at an intermediate temperature of 135°C, when matrix fully infiltrated the void spaces between the fibers and no fibers were melted. The resulting decline in the modulus with increasing temperature is primarily due to the orientation loss of the specimen.

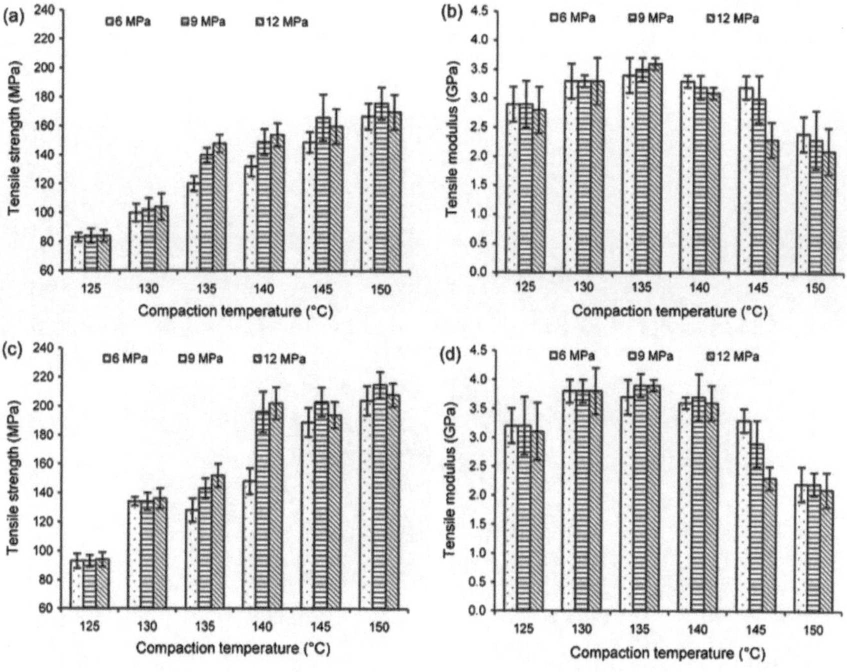

FIGURE 3.8
Tensile strength and tensile modulus of PP SPCs as a function of compaction pressure at different compaction temperatures: (a, b) tested parallel to the warp direction, and (c, d) tested parallel to the weft direction. (From Wang, J. and D. Chen., *Compos., A*, 90, 567, 2016. With permission from ScienceDirect, www.sciencedirect.com/science/article/pii/S0266353814004230#f0005, Figure 2 in reference.)

3.3.1.3 Hot Compaction of Woven Materials

A series of research studies on woven materials were carried by Ward et al. [21]. In one of the first research endeavors, they established the important parameters that control the hot compaction behavior of woven oriented polypropylene. Five commercial woven cloths (Figure 3.9), based on four different polypropylene polymers, were selected so that the perceived important variables could be studied. These include the mechanical properties of the original oriented tapes or fibers, the geometry of the oriented reinforcement (fibers or tapes), the mechanical properties of the base polymer (which were crucially dependent on the molecular weight and morphology), and the weave style. The five cloths were chosen so as to explore the boundaries of these various parameters, i.e., low and high molecular weights, circular or rectangular reinforcement (fibers or tapes), low or high tape initial orientation, coarse or fine weave. The research group commented on a vital aspect of this study. They realized that hot-compacted polypropylene could be envisaged as a composite, comprising an oriented "reinforcement" bound together by a matrix phase, formed by melting and recrystallization of the original oriented material. The crucial importance of the properties of the

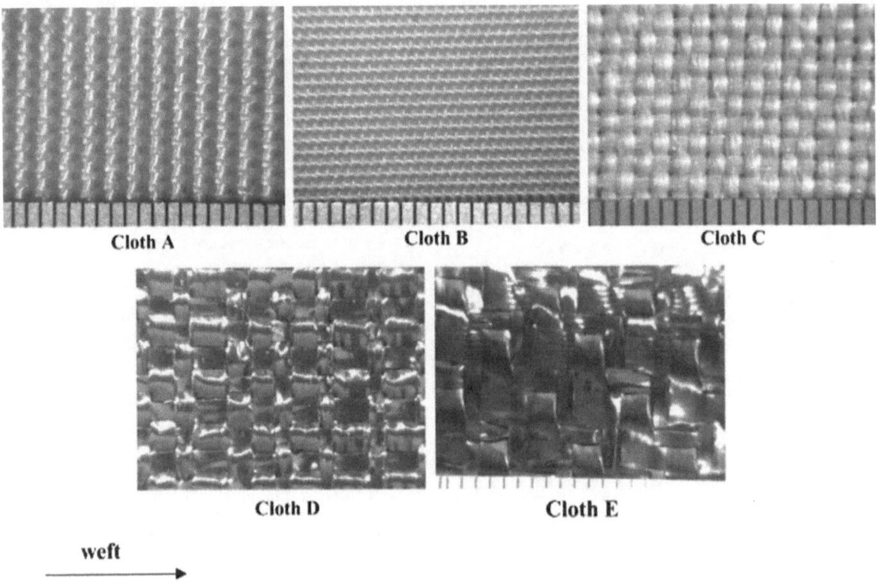

FIGURE 3.9
Photographs of the five oriented polypropylene cloths used in this study. The bottom axis is 20 mm in all cases and parallel to the weft direction. (From Hine, P. J. et al., *Polymer*, 44, 1117, 2003. With permission from ScienceDirect, www.sciencedirect.com/science/article/pii/S0032386102008091, Figure 1.)

melted and recrystallized matrix phase was established, especially the level of ductility, in controlling the properties of the hot-compacted composite.

Different varieties of cloths as shown in the figure were used as a starting material. The effects of different cloth geometries have been discussed in the article.

The research group placed the required number of woven layers in a stacked form inside a matched metal mold (125 mm^2). The group used a thermocouple between the central two layers to allow the actual assembly temperature to be measured. The whole assembly was then placed into a hot press set at the required compaction temperature, and a pressure of 2.8 MPa (400 psi) was applied. Once the assembly reached the compaction temperature, it was left for a further 10 min, during which time it was cooled to 100°C and the sample was removed.

These results indicated that as the compaction temperature increases, the percentage of melted material increases, thereby improving bonding (and stress transfer) and hence increasing modulus and strength. If too much melting occurs, a significant percentage of the oriented structure is lost and the modulus and strength decrease, although the bonding still continues to improve. The research group also made several important observations. The interlayer adhesion was found to be primarily a matrix-dependent property, although there were also obvious geometric effects due to the nature of the weave style. This was important as it is one of the works reporting weave geometry. Flat tapes again appeared to be a better style of reinforcement, as there was no preferred registering of the surfaces; hence the peel strength was the same in both the weft and warp directions. The group also observed that the ductility of the matrix was important. More ductile polymers showed higher values of peel load. Cooling rate was also found to be important. Faster cooling rates resulted in lower values of crystallinity and a more ductile matrix. They, however, observed that the only drawback to this is a small reduction in the Young's modulus of the matrix phase.

In another study using woven materials [22], fabricated materials using self-reinforced polypropylene (PP) co-extrusion tapes were used. Tapes consisting of two-component material were used, having a dense and highly crystalline-drawn PP tapes (reinforcement), coated with a thin layer of low-crystalline PP copolymer (matrix), which has a significantly lower melting temperature than the drawn tapes. The authors found a distinct gap of temperature for the crystalline and low-crystalline phase through DSC.

The consolidated process seems to make the diffraction intensity at around 18.52° and 23.8° relatively strong while it decreases the relative intensity around 21.4°. The researchers concluded that the crystalline structure of the reinforcement PP surface was selectively melted and recrystallized as the compaction temperature increased.

The researchers investigated adhesion between layers in SRC laminates indirectly by photographing the fracture surface of T-peel specimens, as shown in Figure 3.10. There were two factors determining the peel strength

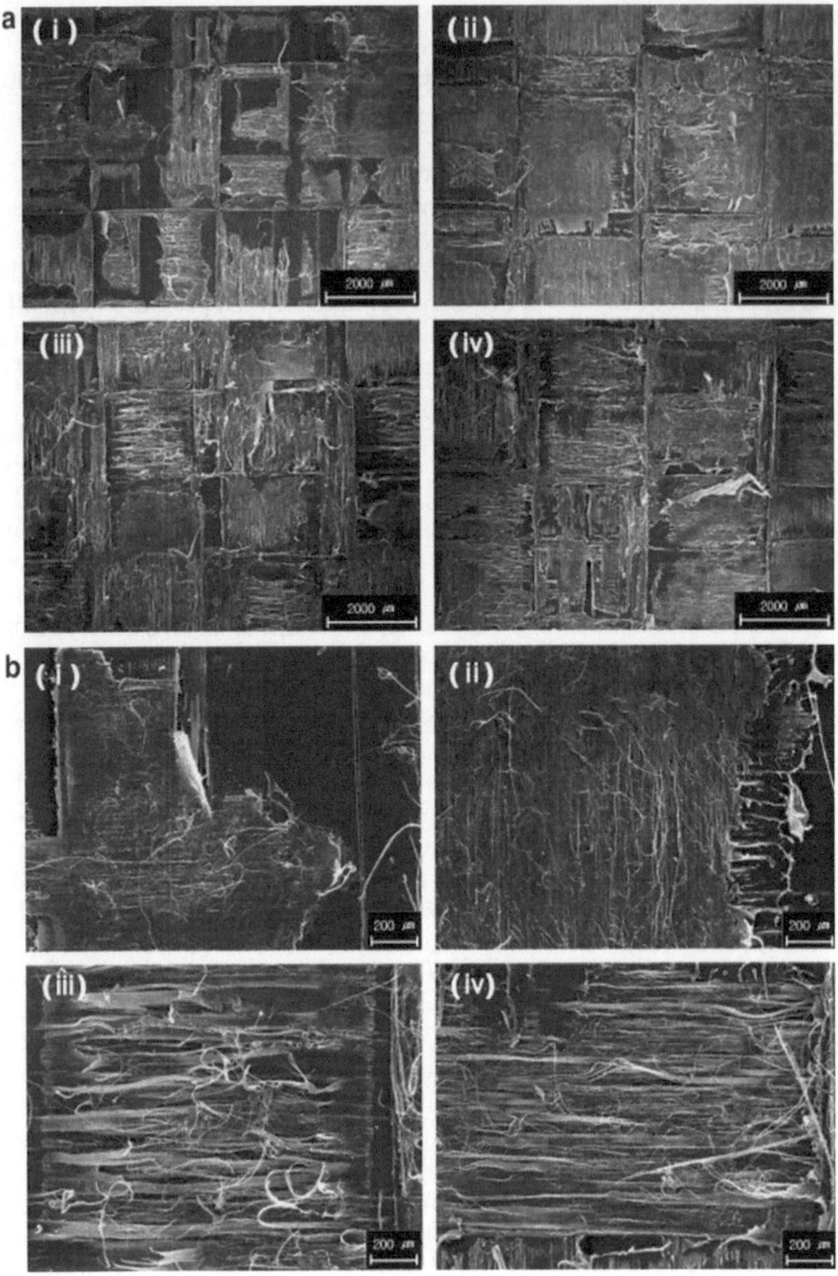

FIGURE 3.10
Fracture surfaces of T-peel test specimens: (a) low (15×) and (b) high (75×) magnification views. (i) 140°C, (ii) 147°C, (iii) 150°C, (iv) 153°C. (From Wang, J. and Chen, D., *Composi. Struct.*, 187, 403, 2018. With permission from ScienceDirect, www.sciencedirect.com/science/article/pii/S1359835X08001747, Figure 7.)

of the SRC laminates: matrix–matrix and matrix–fibers adhesion. If the consolidation temperature is not high enough to melt the surface of the coated PP copolymers, the matrix–matrix adhesion between each layer becomes weak, which is confirmed by the clear surface of the fracture surface. As shown in Figure 3.10, with rise in consolidation temperature, the facture surfaces would take a rougher appearance. However, from the theory of composite structures, the matrix–fiber adhesion is yet another important factor for efficient consolidation because it ensures effective load transfer from the matrix to the fibers.

This adhesion can be estimated by observing the pull-out debris on the fracture surface. In this regard, the authors found that as consolidation temperature approached melting temperature of the reinforcement PP, more efficient consolidation was achieved (see Figure 3.7b). Considering the two adhesion elements, the research group concluded that a consolidation temperature between 150°C and 153°C is a superior choice for producing more structurally sound composites than that between 140°C and 147°C. The best consolidation temperature was found to be 150°C.

Long-term mechanical properties of the composite were also investigated by the group. To explore anisotropic and nonlinear time-dependent behavior of SRCs, creep tests were performed using the three off-axis (0°, 22.5°, and 45°) specimens and three load levels (3, 6, and 12 MPa), from which a creep potential was determined to describe their creep behavior. For validation purposes, the authors recalculated creep strains in each direction using the creep potential and this was compared with the experiments. They concluded that an off-axis angle of 22.5° can be considered as a representative direction for describing the creep behavior of the orthotropic SRCs within a reasonable error.

3.3.1.4 Film-Stacking Method

Film stacking is one of the most popular methods of preparing composites (Figure 3.11). In a study by Barany et al. [23], self-reinforced PP composites (SRPPCs) were produced by the film-stacking method using carded mats of α-isotactic PP fibers and films of β-isotactic PP. Composite sheets were produced at various consolidation temperatures (T = 150–170°C) and constant holding time (t = 2 min). In a test series, the holding time was varied (t = 2–20 min) by keeping the consolidation temperature constant. The consolidation degree of the sheets was studied on polished sections by light microscopy and quantified by density and peel test results. Specimens cut of the sheets were subjected to in-plane static tensile and out-of-plane dynamic impact (instrumented falling weight impact, IFWI) tests. It was established that density, tensile strength, and specific perforation impact energy are the best indicators for the consolidation degree and between them linear correlations exist. Increasing the density of the composites resulted in increased tensile strength (due to lower void content and better fiber/matrix adhesion) and reduced perforation impact energy (due to hindered delamination between the constituent layers).

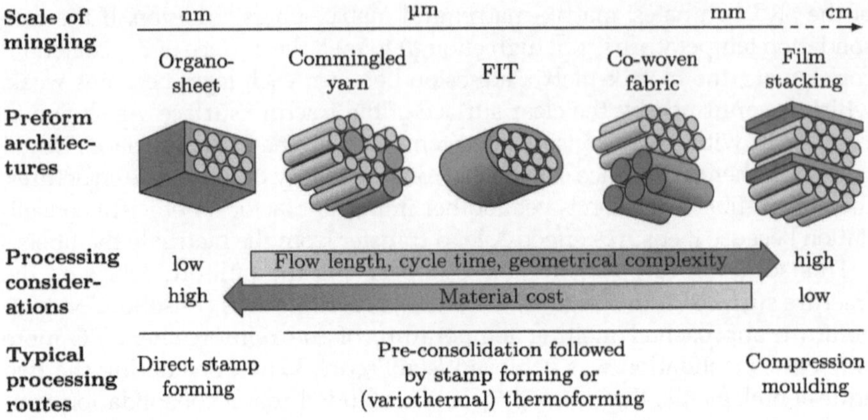

FIGURE 3.11
Schematic of different routes toward composites. (From Schneeberger, C. et al., *Compos. A*, 103, 69, 2017. With permission from ScienceDirect, www.sciencedirect.com/science/article/pii/S1359835X17303378#f0005, Figure 1.)

3.3.2 Importance of Starting Material

3.3.2.1 α- and β- Polymorphs of Isotactic PP Homopolymer and Random Copolymer

The impact behavior of SRPPCs was studied by Barany et al. [24] *α* and *β*- polymorphs of isotactic PP homopolymer and random copolymer (with ethylene) were used for matrix materials, whereas the reinforcement was a fabric woven from highly stretched split PP yarns. The composite sheets were produced by the film-stacking method and consolidated by hot pressing at 5°C and 15°C above the melting temperature of the matrix-giving PP grade. The composite sheets were subjected to static tensile, dynamic falling weight impact and impact tensile tests at room temperature. Dynamic mechanical thermal analysis (DMTA) was also performed on the related composites and their constituents. The results indicated that the *β*-modification of the PP homopolymer is more straightforward than that of the PP copolymer. Stiffness and strength were found to increase while the toughness (tensile impact strength, perforation impact energy) decreased with increasing temperature of consolidation. This was assigned to differences in the failure mode based on fractographic results.

3.3.2.2 PP Yarns and Materials α and β Crystal Forms of Isotactic PP Homopolymer

The variety of starting polymer was further investigated by the same group in a different study [25]. SRPPCs were developed and investigated by this research group. Fabric woven from highly stretched split PP yarns was used as reinforcement, whereas as matrix materials *α*- and *β*- crystal forms of isotactic PP homopolymer and random copolymer (with ethylene) were selected and used. Film-stacking method was used and compression molded at different processing temperatures, keeping the holding time and pressure constant.

The quality of the composite sheets was assessed by optical microscopy, density, and peel-strength measurements. The SRPPC specimens were subjected to static tensile, flexural and dynamic falling weight impact tests and the related results were analyzed as a function of processing temperature and polymorphic composition. The optimum processing temperature was determined on the basis of the results and found to be 20°C–25°C above the related matrix melting temperature. It was established that the β-modified PP homopolymer–based one-component SRPPCs possessed similar attractive mechanical properties as the intensively studied α-random PP copolymer–based two-component ones. Another paper investigates the impact properties of SPCs, starting with tapes. The use of highly oriented PP tapes was explored by Alcock et al. [26]. The route was to explore tapes of high tensile strength and stiffness achieved by molecular orientation during solid state drawing. Further consolidation was done to create high performance recyclable all-polypropylene (all-PP) composites. These composites had a large temperature processing window (>30°C) and a high-volume fraction of highly oriented PP (>90%). This large processing window was achieved by using coextruded, highly drawn PP tapes. This was pretty interesting because majority of the work on SPCs need to rely on a very small temperature range. The group investigated the impact resistance of these all-PP composites, and the relationship between penetrative and non-penetrative impact behavior and composite consolidation conditions. The research group reported falling weight impact together with a comparison to conventional commercial glass reinforced polypropylene composites. A model for energy absorption was also proposed by comparison with previous studies based on interfacial and tensile failure of tapes and composites.

3.3.2.3 PP Tape

In the research by Abraham et al., using tapes, different polymorphic forms of PP, in which alpha (α)-PP tapes worked as reinforcement and beta (β)-PP served as matrix, were used [27].

The schematic of the tape-winding process is shown in Figure 3.12. Initially before winding the PP tapes, a thin β-PP film layer was positioned on the surface of a steel plate. Using a typical winding machine, PP tapes were wound from a bobbin onto the same steel plate rotating at a constant speed. After laying one layer of PP tape, another layer of β-PP film was engaged and the winding direction on the steel plate was reversed.

The melting temperature of the β-modification of isotactic PP (Tm ≈ 154°C) homopolymer was found to be lower than that of the reinforcing α-PP tape

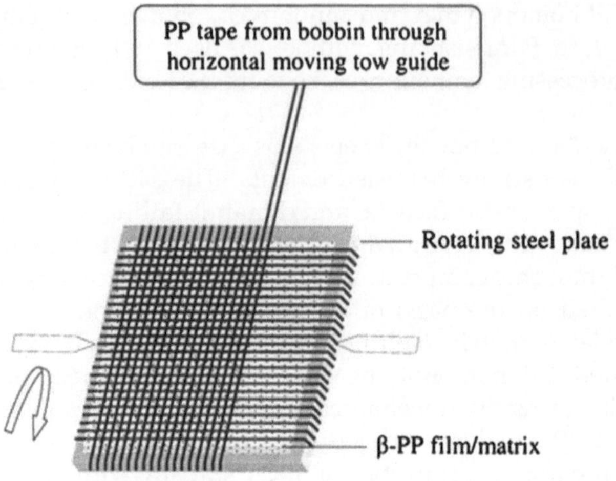

FIGURE 3.12
Schematic of winding of PP tape and film. (From Abraham, T. N. et al., *J. Materi. Sci.*, 43, 3697, 2008. With permission from Springer, https://link.springer.com/article/10.1007/s10853-008-2593-2, Figure 1.)

(Tm ≈ 165°C). The DSC trace of the all-PP composite also exhibited an interesting phenomenon of transformation of β-PP to α-form at temperatures above the melting point of the former. DMTA was used to analyze mechanical performance of the composite in a range of frequencies and temperatures. The volume fractions of matrix and reinforcement were estimated using optical microscope images. Both static flexural bending tests and DMTA revealed that α-PP tapes act as an active reinforcement for the β-PP matrix. The research group used the principle of time–temperature superposition to estimate the stiffness of the composites as a function of frequency in the form of a master curve. It was found that the Williams–Landel–Ferry model could accurately describe the change in the experimental shift factors used to create the storage modulus versus frequency master curve. Arrhenius equation was used to calculate activation energies for the α- and β- relaxations.

An alternative route was reported [28] for the manufacture of SPCs by combining the processes of hot compaction and film stacking using tapes.

The philosophy of this process is taking an assembly of oriented polymer fibers or tapes to a critical temperature, while still under pressure. The objective is to create a thin skin on the surface of each oriented element that is "selectively melted," creating a matrix phase. As found by the researchers, subsequent fast cooling results in the recrystallization of the melted material to form the matrix of the composite, with the remaining fraction of the original oriented phase acting as the reinforcement. It has to be noted that compared to other methods, matrix phase is produced around each oriented element, giving excellent wetting and infiltration.

Excellent compatibility is also found between the fiber and matrix (which are the same polymer)—the major challenge is the temperature sensitivity of the process. However, the disadvantage, when dealing with a layered composite made from multifilament bundles, is that additional material would be required in the interlayer region to fill the gaps between the rough layers of woven cloth, and it is not congenial to produce this by melting oriented fibers in the center of a fiber bundle. It is clear that a combination of hot compaction at a lower-than-optimum temperature plus an interleaved film to fill the rougher interlayer region, could help to optimize this process. The research group adopted the logic from the above philosophy in their experiments.

Assemblies of oriented fibers and tapes were produced with and without an interleaved film of the sample polymer type, over a range of processing temperatures. Temperatures in these experiments were crucial so that in some cases only the interleaved film was melted and in other cases both the film and the fiber surfaces were melted, allowing (a) a traditional film-stacking process, (b) a traditional hot-compaction technique, and (c) the combination of both to be compared. The results showed that the combination of film stacking with hot compaction gave a better overall balance of mechanical properties and a wider temperature window for processing compared to a standard hot-compaction procedure without a film. The combination process also gave much better wetting of all the oriented elements compared to a traditional film-stacking process (where only the film is melted) due to the partial melting of all the fiber surfaces.

The research group also commented that for high-modulus PE fibers, the use of an interlayer film is one of the best ways to achieve complete consolidation when using woven cloth layers. Without an interlayer film, the compaction temperature essential for manufacturing enough matrix material for a strong interlayer bond with woven cloth layers is very close to the temperature at which major crystalline melting occurs. Approach to this temperature comes at a cost of loss of mechanical properties. This makes processing very difficult (in terms of temperature control), especially for semicrystalline polymers where the melting range is narrow.

3.3.3 Advances in Testing Methods

Poszwa et al. [29] developed a method of testing polymer fibers and SPCs. The method provides fast and nondestructive detection of structural defects in fibers and delamination in SPCs; therefore, it can be applied to quality control of the fibers as well as composites. The method was based on laser scanning confocal microscopy, which uses laser to scan the surface of the specimen and detect light of wavelength longer than laser beam. The results show that the luminescence occurs only in the defected or delaminated spots of the sample. Raman and infrared spectroscopies were used to analyze the source of the luminescence. The study indicated

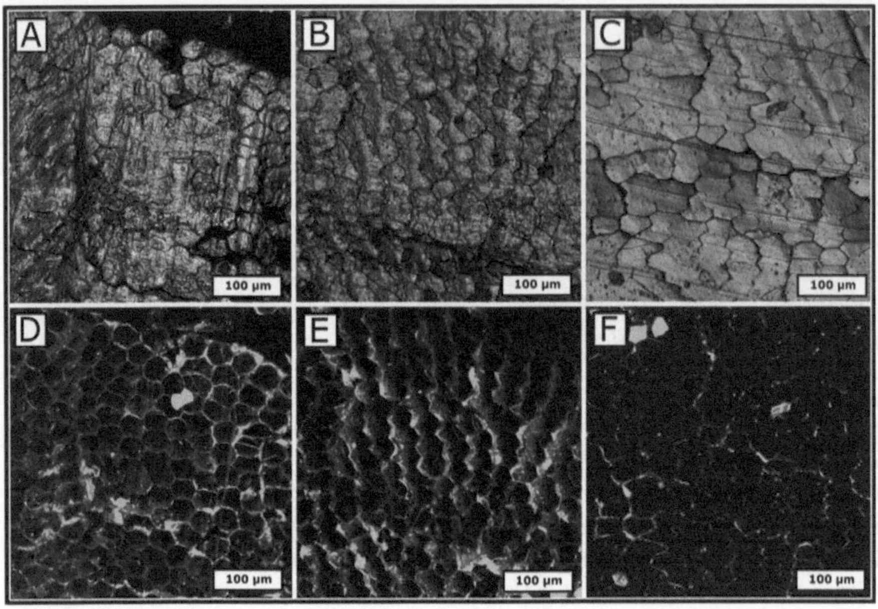

FIGURE 3.13
LSM images in material (up) and fluorescence (bottom) mode of iPP composites cross-section prepared with different process temperatures. (From Poszwa, P. et al., *Polym. Test.*, 53,174, 2016. With permission from ScienceDirect, www.sciencedirect.com/science/article/pii/S0142941816301799, Figure 6.)

that influence of impurities, additives, and effects of chemical degradation of polymer doesn't lead to luminescence of the specimen. As an explanation of the effect, mechanical stress–assisted photoluminescence was proposed.

Figure 3.13 shows mages in material (up) and iPP composites cross-section. Confocal material mode images (Figure 6a–c) show barely visible fiber structure of composites cross-section. On the other hand, PL confocal images for samples 1 (Figure 6d) show clearly visible structure of fibers cross-section. This may be a very important finding since detection of cross-section in SPCs have always been a challenge.

3.4 Conclusion

Most early studies started with PE since the high theoretical modulus of a linear PE molecule (~250 GPa) is much greater than the stiffness of the helical PP molecule (~43 GPa). This molecular modulus ultimately limits the maximum modulus achievable in a fiber of either material, and so higher properties are commonly achievable with PE fibers. An excellent review article discusses the advancements in fiber spinning [30,31] and use of woven materials in SPCs [32]. There have also been articles on long-term biological behavior of this polymer [33]. Thus, PE was the preferred choice for initial research. PE is widely used for food and medical applications, and the increasing mechanical properties by self-reinforcement do not affect this as the addition of "foreign" reinforcements might. However, both these polymers have their own advantage and disadvantage—so the choice of the starting material would very well depend on the end use. Polypropylene is essentially stiffer and more chemical and scratch-resistant, while still being very tough. Polypropylene has also higher chemical resistance in corrosive environments, as it is rugged and unusually resistant to many chemical solvents, bases, and acids [34,35]. The chemical resistance of PE is also good and a detailed resistance can be seen at a chemical resistance chart [36,37] and an excellent compatibility chart [38]. However, PE and polypropylene, both being hydrophobic in character, often require a compatibilizer when used with reinforcing agents. There have been numerous research groups that have reported modifications for reinforcement in PE [39–42] and polypropylene [43–48]. Interesting uses of polypropylene, in addition to time-dependent properties, [49] have also been reported [50,51]. However, certain uses in medical applications need reinforcement of pure polymer and avoiding compatibilizer is often a good idea.

On the basis of this chapter, Table 3.1 summarizes the mechanical properties of SPCs from polyolefins.

The highest mechanical values for PE have been reported by Marais and her group using unidirectional PE. Interestingly, this is the only group that has reported on the percentage elongation and flexure values of the composites. There were two other groups that had worked on the flexural properties but had shown no trend. Thus, there is scope for research on the rupture, flexure, and impact properties of such composites. For polypropylene, the highest values of modulus achieved have been 3.13 GPa and that of strength 168 MPa using a film-stacking technique. It is interesting to note that properties in flexure are better characterized with polypropylene. Highest strength of 110 MPa and modulus of 3.47 GPa has been achieved by the group of Andras who used woven fabrics and combined film-stacking technique in compression molding. The phenomenon of transcrystallinity and its effect on mechanical properties has also been studied well for polyolefins. There is scope for further research on impact properties and long-term mechanical properties for this class of composites. There have been studies on creep [52–54], impact [55], effect of moisture absorption [56,57], and immersion time [58] using natural fibers. However, such studies can even be carried out for single polymer systems.

TABLE 3.1

Mechanical Properties of Polyolefin-Based SPCs

Starting Material	Process	Tensile Strength (MPa)	Young's Modulus (GPa)	% Breaking Extension	Flexural Strength (MPa)	Flexural Modulus (GPa)	Ref
HDPE fibers and HDPE matrix	Prepreg, molding, and posttreatment	1300	73.9	3.5	–	60	[3]
Melt-spun high-modulus PE multifilament and PE matrix	Hot compaction	–	–	–	–	No particular trend	[4]
UHMWPE fibers	Hot compaction	–	–	–	–	No particular trend	[10]
UHMWPE fibers and UHMWPE powers	Compression molding	135	2.25	–	–	–	[11]
HDPE	Injection molding	87	3.5	–	–	–	[12]
HDPE and LDPE	Injection molding	98.3	–	–	–	–	[14]
UHMWPE fibers and HDPE	Prepreg and compression molding (solution impregnation)	1100	35	–	–	–	[16]
PP-woven fabric (reinforcement) and PP granules (matrix)	Insert-microcellular injection molding	55	0.75	–	–	–	[17]
PP-woven fabric (reinforcement) and PP granules (matrix)	Under cooling melt film stacking	172 (warp) and 208 (weft)	2.25 (both warp and weft)	–	–	–	[19, 20]
PP-woven fabric	Hot compaction	70 (warp) and 135 (weft)	2.9 (both warp and weft)	–	–	–	[21]
Carded mat of isotactic PP (b-PP)	Film stacking	99.7	2.57	17.6	–	–	[23]
Highly stretched split PP yarns–based fabric (reinforcement) and β polymorph of isotactic PP homopolymer (matrix)	Film stacking and hot pressing	100	2.9	14	–	–	[24]

(*Continued*)

TABLE 3.1 (*Continued*)

Mechanical Properties of Polyolefin-Based SPCs

Starting Material	Process	Tensile Strength (MPa)	Young's Modulus (GPa)	% Breaking Extension	Flexural Strength (MPa)	Flexural Modulus (GPa)	Ref
Highly stretched split PP yarns–based fabric (reinforcement) and α polymorph of random PP copolymer (matrix)	Film stacking and hot pressing	130	2.6	17	–	–	[24]
Highly stretched split PP yarns–based fabric (reinforcement) and β polymorph of random PP copolymer (matrix)	Film stacking and hot pressing	105	2	21	–	–	[24]
Highly stretched split PP yarns–based fabric (reinforcement) and β polymorph of isotactic PP homopolymer (matrix)	Film stacking and hot pressing	87	2.33	–	110	3.47	[25]
Highly stretched split PP yarns–based fabric (reinforcement) and α polymorph of random PP copolymer (matrix)	Film stacking and hot pressing	95	1.87	–	98	2.13	[25]
Highly stretched split PP yarns–based fabric (reinforcement) and β polymorph of random PP copolymer (matrix)	Film stacking and hot pressing	92	1.83	–	90	1.95	[25]
α-PP tapes (reinforcement) and β-PP (matrix)	Hot compaction	–	–	–	60	2.3	[27]
Woven PP cloth+interleaved PP film	Film stacking	168	3.13	–	–	–	[30]

PE, polyethylene; PP, polypropylene; HDPE, high-density polyethylene; UHMWPE, ultra-high-molecular-weight polyethylene.

References

1. https://www.britannica.com/science/polyolefin [accessed January 28, 2018].
2. Capiati, N. J., and R. S. Porter. The concept of one polymer composites modelled with high density polyethylene. *Journal of materials science* 10, no. 10 (1975): 1671–1677.
3. Marais, C., and P. Feillard. Manufacturing and mechanical characterization of unidirectional polyethylene-fiber/polyethylene-matrix composites. *Composites Science and Technology* 45, no. 3 (1992): 247–255.
4. Hine, P. J., I. M. Ward, R. H. Olley, and D. C. Bassett. The hot compaction of high modulus melt-spun polyethylene fibers. *Journal of Materials Science* 28, no. 2 (1993): 316–324.
5. Ward, I. M. The preparation, structure and properties of ultra-high modulus flexible polymers. In *Key Polymers Properties and Performance*, pp. 1–70, ed. I. M. Ward. Springer, Berlin, Heidelberg, 1985.
6. Ward, I. M., and P. J. Hine. The science and technology of hot compaction. *Polymer* 45, no. 5 (2004): 1413–1427.
7. http://nptel.ac.in/courses/116102006/6 [accessed January 28, 2018].
8. Kuo, C. J., and W. L. Lan. Gel spinning of synthetic polymer fibers. In *Advances in Filament Yarn Spinning of Textiles and Polymers*, pp. 100–112, eds D. Zhang, Woodhead Publishing, Sawston, 2014.
9. Green, E. C., Y. Zhang, H. Li, and M. L. Minus. Gel-spinning of mimetic collagen and collagen/nano-carbon fibers: Understanding multi-scale influences on molecular ordering and fibril alignment. *Journal of the Mechanical Behavior of Biomedical Materials* 65(2017): 552–564.
10. Yan, R. J., P. J. Hine, I. M. Ward, R. H. Olley, and D. C. Bassett. The hot compaction of SPECTRA gel-spun polyethylene fiber. *Journal of Materials Science* 32, no. 18 (1997): 4821–4832.
11. Deng, M., and S. W. Shalaby. Properties of self-reinforced ultra-high-molecular-weight polyethylene composites. *Biomaterials* 18, no. 9 (1997): 645–655.
12. Lai, F. S., S. P. McCarthy, D. Chiu, X. Zhu, and K. Shen. Morphology and properties of self-reinforced high density polyethylene in oscillating stress field. *Polymer* 38, no. 20 (1997): 5251–5253.
13. Peacock, A. *Handbook of Polyethylene: Structures: Properties, and Applications*. CRC Press, London, 2000.
14. Zhang, G., L. Jiang, K. Shen, and Q. Guan. Self-reinforcement of high-density polyethylene/low-density polyethylene prepared by oscillating packing injection molding under low pressure. *Journal of applied polymer science* 71, no. 5 (1999): 799–804.
15. Zhuang, X., and X. Yan. Investigation of damage mechanisms in self-reinforced polyethylene composites by acoustic emission. *Composites Science and Technology* 66, no. 3–4 (2006): 444–449.
16. Lacroix, F. V., M. Werwer, and K. Schulte. Solution impregnation of polyethylene fiber/polyethylene matrix composites. *Composites Part A: Applied Science and Manufacturing* 29, no. 4 (1998): 371–376.

17. Wang, J., and D. Chen. Microcellular polypropylene single-polymer composites prepared by insert-microcellular injection molding. *Composites Part A: Applied Science and Manufacturing* 90(2016): 567–576.
18. http://greenmolding.org/english/solution/511 [accessed February 21, 2018].
19. Wang, J., J. Chen, P. Dai, S. Wang, and D. Chen. Properties of polypropylene single-polymer composites produced by the undercooling melt film stacking method. *Composites Science and Technology* 107(2015): 82–88.
20. Dai, P., W. Zhang, Y. Pan, J. Chen, Y. Wang, and D. Yao. Processing of single polymer composites with undercooled polymer melt. *Composites Part B: Engineering* 42, no. 5 (2011): 1144–1150.
21. Hine, P. J., I. M. Ward, N. D. Jordan, R. Olley, and D. C. Bassett. The hot compaction behaviour of woven oriented polypropylene fibers and tapes. I. Mechanical properties. *Polymer* 44, no. 4 (2003): 1117–1131.
22. Kim, K. J., W. R. Yu, and P. Harrison. Optimum consolidation of self-reinforced polypropylene composite and its time-dependent deformation behavior. *Composites Part A: Applied Science and Manufacturing* 39, no. 10 (2008): 1597–1605.
23. Bárány, T., J. Karger-Kocsis, and T. Czigány. Development and characterization of self-reinforced poly (propylene) composites: carded mat reinforcement. *Polymers for Advanced Technologies* 17, no. 9–10 (2006): 818–824.
24. Bárány, T., A. Izer, and J. Karger-Kocsis. Impact resistance of all-polypropylene composites composed of alpha and beta modifications. *Polymer Testing* 28, no. 2 (2009): 176–182.
25. Izer, A., T. Bárány, and J. Varga. Development of woven fabric reinforced all-polypropylene composites with beta nucleated homo-and copolymer matrices. *Composites Science and Technology* 69, no. 13 (2009): 2185–2192.
26. Alcock, B., N. O. Cabrera, N. M. Barkoula, and T. Peijs. Low velocity impact performance of recyclable all-polypropylene composites. *Composites Science and Technology* 66, no. 11–12 (2006): 1724–1737.
27. Abraham, T. N., S. Siengchin, and J. Karger-Kocsis. Dynamic mechanical thermal analysis of all-PP composites based on β and α polymorphic forms. *Journal of Materials Science* 43, no. 10 (2008): 3697–3703.
28. Hine, P. J., R. H. Olley, and I. M. Ward. The use of interleaved films for optimising the production and properties of hot compacted, self reinforced polymer composites. *Composites Science and Technology* 68, no. 6 (2008): 1413–1421.
29. Poszwa, P., K. Kędzierski, B. Barszcz, and A. B. Nowicka. Fluorescence confocal microscopy as effective testing method of polypropylene fibers and single polymer composites. *Polymer Testing* 53(2016): 174–179.
30. Zhang, W., Z. Hu, Y. Zhang, C. Lu, and Y. Deng. Gel-spun fibers from magnesium hydroxide nanoparticles and UHMWPE nanocomposite: the physical and flammability properties. *Composites Part B: Engineering* 51(2013): 276–281.
31. Schneeberger, C., J. C. H. Wong, and P. Ermanni. Hybrid bicomponent fibers for thermoplastic composite preforms. *Composites Part A: Applied Science and Manufacturing* 103(2017): 69–73.
32. Gao, C., L. Yu, H. Liu, and L. Chen. Development of self-reinforced polymer composites. *Progress in Polymer Science* 37, no. 6 (2012): 767–780.

33. Sen, S. K., and S. Raut. Microbial degradation of low density polyethylene (LDPE): a review. *Journal of Environmental Chemical Engineering* 3, no. 1 (2015): 462–473.
34. https://www.industrialspec.com/images/editor/polypropylene-chemical-compatibility-chart-2016-from-ism.pdf [accessed February 21, 2018].
35. https://www.calpaclab.com/polypropylene-chemical-compatibility-chart/ [accessed February 21, 2018].
36. http://www.cdf1.com/technical%20bulletins/Polyethylene_Chemical_Resistance_Chart.pdf [accessed February 21, 2018].
37. https://www.ineos.com/globalassets/ineos-group/businesses/ineos-olefins-and-polymers-usa/products/technical-information--patents/ineos-hdpe-chemical-resistance-guide.pdf [accessed February 21, 2018].
38. https://www.spilltech.com/wcsstore/SpillTechUSCatalogAssetStore/Attachment/documents/ccg/POLYETHYLENE.pdf [accessed February 21, 2018].
39. Lai, S. M., F. C. Yeh, Y. Wang, H. C. Chan, and H. F. Shen. Comparative study of maleated polyolefins as compatibilizers for polyethylene/wood flour composites. *Journal of Applied Polymer Science* 87, no. 3 (2003): 487–496.
40. Abdelmouleh, M., S. Boufi, M. N. Belgacem, and A. Dufresne. Short natural-fiber reinforced polyethylene and natural rubber composites: effect of silane coupling agents and fibers loading. *Composites science and technology* 67, no. 7–8 (2007): 1627–1639.
41. Wang, M., and W. Bonfield. Chemically coupled hydroxyapatite–polyethylene composites: structure and properties. *Biomaterials* 22, no. 11 (2001): 1311–1320.
42. Bengtsson, M., P. Gatenholm, and K. Oksman. The effect of crosslinking on the properties of polyethylene/wood flour composites. *Composites Science and Technology* 65, no. 10 (2005): 1468–1479.
43. Asumani, O. M. L., R. G. Reid, and R. Paskaramoorthy. The effects of alkali–silane treatment on the tensile and flexural properties of short fiber non-woven kenaf reinforced polypropylene composites. *Composites Part A: Applied Science and Manufacturing* 43, no. 9 (2012): 1431–1440.
44. Fuad, M. A., Z. Ismail, Z. A. M. Ishak, and A. K. M. Omar. Application of rice husk ash as fillers in polypropylene: effect of titanate, zirconate and silane coupling agents. *European Polymer Journal* 31, no. 9 (1995): 885–893.
45. Demjen, Z., and B. Pukanszky. Effect of surface coverage of silane treated CaCO3 on the tensile properties of polypropylene composites. *Polymer Composites* 18, no. 6 (1997): 741–747.
46. Coutinho, F., T. H. Costa, and D. L. Carvalho. Polypropylene–wood fiber composites: effect of treatment and mixing conditions on mechanical properties. *Journal of Applied Polymer Science* 65, no. 6 (1997): 1227–1235.
47. Gassan, J., and A. K. Bledzki. The influence of fiber-surface treatment on the mechanical properties of jute-polypropylene composites. *Composites Part A: Applied Science and Manufacturing* 28, no. 12 (1997): 1001–1005.
48. Cantero, G., A. Arbelaiz, R. Llano-Ponte, and I. Mondragon. Effects of fiber treatment on wettability and mechanical behaviour of flax/polypropylene composites. *Composites science and technology* 63, no. 9 (2003): 1247–1254.
49. Wang, J., and D. Chen. Flexural properties and morphology of microcellular-insert injection molded all-polypropylene composite foams. *Composite Structures* 187 (2018): 403–410.

50. Kim, K. J., W. R. Yu, and P. Harrison. Optimum consolidation of self-reinforced polypropylene composite and its time-dependent deformation behavior. *Composites Part A: Applied Science and Manufacturing* 39, no. 10 (2008): 1597–1605.
51. Abraham, T. N., S. Siengchin, and J. Karger-Kocsis. Dynamic mechanical thermal analysis of all-PP composites based on β and α polymorphic forms. *Journal of Materials Science* 43, no. 10 (2008): 3697–3703.
52. Lee, S. Y., H. S. Yang, H. J. Kim, C. S. Jeong, B. S. Lim, and J. N. Lee. Creep behavior and manufacturing parameters of wood flour filled polypropylene composites. *Composite Structures* 65, no. 3–4 (2004): 459–469.
53. Sain, M. M., J. Balatinecz, and S. Law. Creep fatigue in engineered wood fiber and plastic compositions. *Journal of Applied Polymer Science* 77, no. 2 (2000): 260–268.
54. Xu, Y., Q. Wu, Y. Lei, and F. Yao. Creep behavior of bagasse fiber reinforced polymer composites. *Bioresource Technology* 101, no. 9 (2010): 3280–3286.
55. Bledzki, A. K., and O. Faruk. Creep and impact properties of wood fiber–polypropylene composites: influence of temperature and moisture content. *Composites Science and Technology* 64, no. 5 (2004): 693–700.
56. Taib, R. M., Z. M. Ishak, H. D. Rozman, and W. G. Glasser. Effect of moisture absorption on the tensile properties of steam-exploded acacia mangium fiber–Polypropylene osites. *Journal of Thermoplastic Composite Materials* 19, no. 5 (2006): 475–489.
57. Balatinecz, J. J., and B. D. Park. The effects of temperature and moisture exposure on the properties of wood-fiber thermoplastic composites. *Journal of Thermoplastic Composite Materials* 10, no. 5 (1997): 476–487.
58. Mat Taib, R., S. Ramarad, Z. M. Ishak, and H. D. Rozman. Effect of immersion time in water on the tensile properties of acetylated steam-exploded Acacia mangium fibers-filled polyethylene composites. *Journal of Thermoplastic Composite Materials* 22, no. 1 (2009): 83–98.

4

Single-Polymer Composites from Polyamides

4.1 Introduction

Polyamide (PA), as defined by *Encyclopaedia Britannica* [1], is "any polymer in which repeating units in the molecular chain are linked together by amide groups." Amide groups have the general chemical formula CO-NH (Figure 4.1).

Polyamides are a prominent class amongst the semicrystalline polymers [2]. Polyamides are produced either by the reaction of a diacid with a diamine (polyamide 66; PA66) or by ring-opening polymerization of lactams (polyamide 6; PA6) (Figure 4.2a and b).

These classes of polymers are either all aliphatic or all aromatic. The aliphatic PAs are produced on a much larger scale and are the most important class of engineering thermoplastics. They are amorphous or only moderately crystalline when injection molded, but the degree of crystallinity can be much increased for fiber and film applications by orientation via mechanical stretching. The increase in crystallinity has been confirmed by wide-angle diffraction studies. Till date, the two most important PAs are poly(hexamethyleneadipamide) (Nylon 6,6) and polycaprolactam (Nylon 6). The heat deflection temperature of PA-6,6 is typically between 180°C and 240°C, which exceeds that of polycarbonate and polyester.

There is also a class of aromatic PAs, often called aramids (Figure 4.3). They are characterized by higher strength; better solvent, flame, and heat resistance; and greater dimensional stability than the all-aliphatic amides.

However, they are much more expensive and more difficult to produce. Polyamides have very good mechanical properties, are particularly tough, and have excellent sliding and wear characteristics. A detailed discussion on the properties of the PAs are available on the polymer database website [5]. Nylon as a polymer has been extensively used in composite applications and even in advanced research applications in composites [6–9].

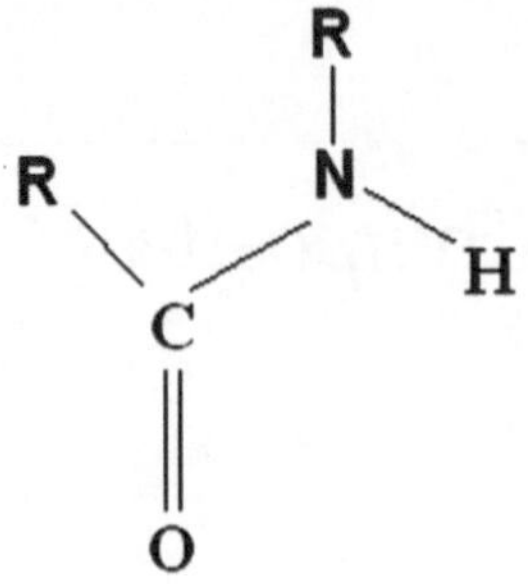

FIGURE 4.1
General structure of an amide.

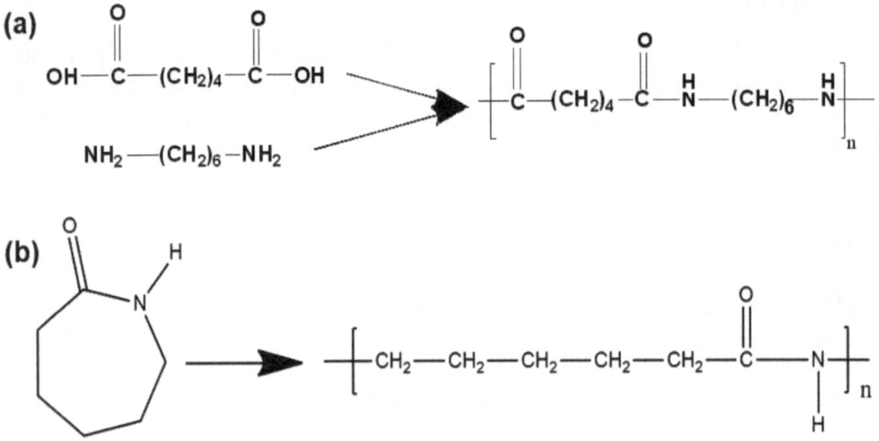

FIGURE 4.2
(a) Nylon 66 and (b) Nylon 6 formation. (Adapted from http://nylon66membrane.com/Preparation-of-Nylon-66.html. With permission.)

In the field of single-polymer composites (SPCs), Ward and his group were the first to report a study on PA [10], where an assembly of oriented elements, often in the form of a woven cloth, is held under pressure. This assembly is then taken to a critical temperature so that only a small fraction of the surface of each oriented element is melted. This material on cooling recrystallizes to form the matrix of the single polymer composite. The research group had already worked on polyolefins and polyesters and extended the technology to study the feasibility on PAs. This method is consequently a way of producing novel high-volume-fraction polymer/polymer composites in which the two phases are chemically the same material. Oriented nylon multifilaments that are available on a commercial scale were used as the starting material.

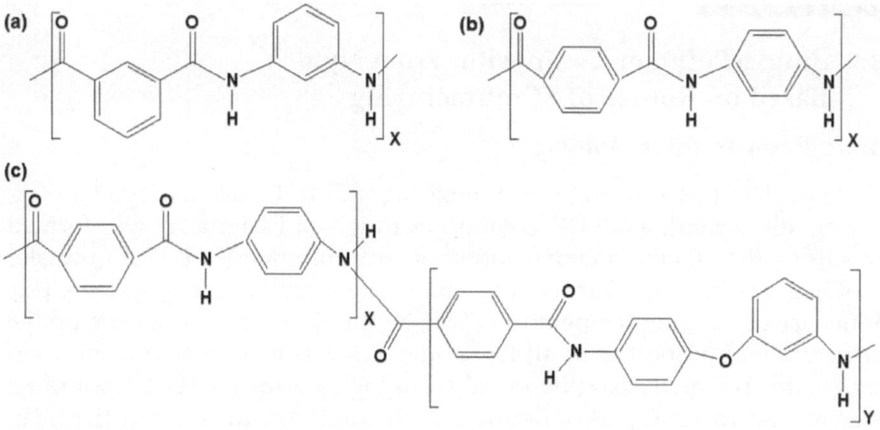

FIGURE 4.3
Aromatic PAs (a) poly(m-phenylene isophthalamide) [Nomex, Conex, etc.] (b) poly (p-phenylene terephthalamide) [Kevlar, Twaron, etc.] (c) co-poly [p-phenylene/3,4-oxydiphenylene terephthalamide] [Technora]. (Adapted from https://sites.google.com/site/grupodepolimeros/aromatic-polyamides---aramids. With permission.)

An important feature of this work, not earlier scrutinized during the use of hot compaction for other oriented polymers, was the sensitivity of the properties to absorbed water. This was investigated with a change in properties calculated immediately after hot-compaction processing and 2 weeks later when 2% water had been absorbed by the compacted nylon sheets. The researchers found that water uptake had a higher effect on properties that influence interactions of local chains like modulus and yield strength and had less effect on properties that was a function of large-scale properties of molecular network, such as, strength. The initial modulus decreased from 4.2 to 2.8 GPa, with only 2.2% absorbed moisture. The researchers wanted to isolate the effect of the constituents in the SPCs. The mechanical behavior study yielded the stress–strain curves of nylon matrix samples tested dry and with 1% or 2% absorbed water.

It was found that when Nylon 6,6 is dry, it is linear up to a sharp yield point at a stress of up to 90 MPa. However, with increase in moisture content, the initial Young's modulus and yield stress fall significantly. Typical values of the modulus, as reported by the researchers, were 3.1 GPa for the dry sample and 1.91 GPa for the sample equilibrated at 50% RH (2.7% w/w water uptake). With 2% absorbed water, the behavior is more nonlinear, and this makes it difficult to pick out a clear yield point. It is clear that these changes in the matrix behavior will have a momentous effect on the properties of the compacted sheet. However, apart from performance at elevated temperature, the majority of the measured properties of the hot-compacted nylon sheets were comparable to those of hot-compacted polypropylene and poly(ethylene terephthalate).

4.2 Single-Polymer Composites from Nylons Based on Routes of Manufacturing

4.2.1 Resin Transfer Molding

Gong et al. [11] used resin transfer molding (RTM) (Figure 4.4) for manufacturing all-polyamide (all-PA) composites in which PA6 matrix was formed in situ by the anionic polymerization of ε-caprolactam (CL). Resin transfer molding has been used in various composite manufacturing systems [12]. Influence of molding temperature (TM), a critical process parameter, on the structure and properties of all-PA composites was investigated using thermogravimetric analysis, differential scanning calorimetry (DSC), scanning electron microscope (SEM), tensile and flexural test. Increase in the melting temperature (TM) resulted in the decrease of CL conversion and the enhancement of fiber/matrix interface bonding. By comparing the mechanical properties of all-PA composites prepared at different TM (140°C–200°C), an optimal TM (180°C) was found in this temperature range. As a whole, the complete consolidation of all-PA composites and the very high reinforcing effect of PA66 fibers on PA6 matrix were secured by high-CL conversion, low-void fraction, and strong interface performance, though in a wide TM range.

One of the major challenges was the control of voids. In engineering all-PA composites by the research group using RTM, the reactive mixture was permeated into mold by nitrogen and saturated PA66 plain cloth was used to eject the air out. However, the movement of reactive mixture in the mold was intricate. The weave structure of PA66 fiber reinforcement phase affected the flow rate and various flow rates were formed as a result of the varying

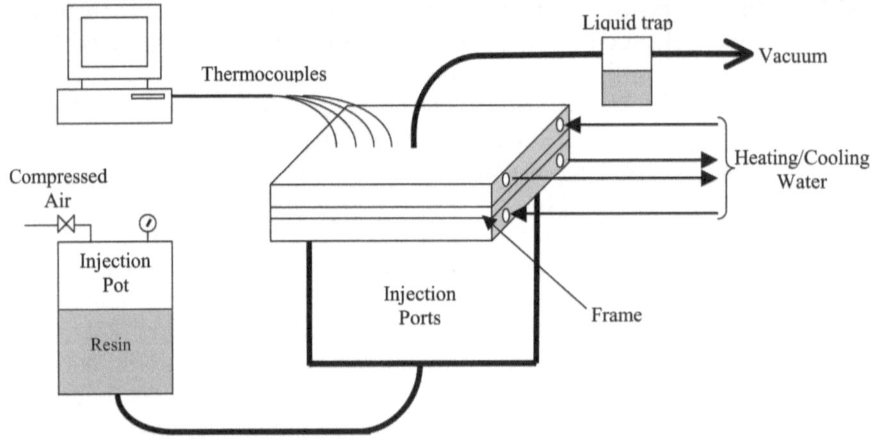

FIGURE 4.4
Schematic of RTM, 1: Cope 2: Drag 3: Clamp 4: Mixing chamber 5: Fiber preform 6: Heated mold 7: Resin 8: Curative. (From Rouison, D. et al., Compos. Sci. Technol., 64, 629, 2004. With permission from ScienceDirect, www.sciencedirect.com/science/article/pii/S0266353803002963.)

starting material. It was observed that with difference in reactive mixture, flow rates met each other, air could be entrapped, and void was retained in all-PA composites after the PA6 matrix was formed in situ by anionic polymerization. It is well-known that existence of voids reduces the strength and surface quality of the composites. Therefore, obtaining low-void fraction was crucial for acquiring high performance of composites. It was observed that to achieve high-performance PA6 matrix by anionic polymerization of CL to successfully bond PA66 fibers together, high-CL monomer conversion is required. This was one important observation by the group.

Thermal behavior of the SPCs was also investigated. It was found that all-PA composites prepared at 160°C showed two melting endotherms (214°C and 254°C). When compared with the melting behaviors of PA66 fibers and pure PA6 prepared at 160°C, it was observed that higher melting peak during heating all-PA composites corresponded to the melting of PA66 fibers, and lower melting peak corresponded to the melting of PA6 matrix formed by anionic polymerization of CL. Tensile fracture morphology of all-PA composites was investigated by SEM. When the melting point was lower than 200°C, PA66 fibers stuck out of the fracture surface. The researchers commented that this morphology showed the fiber pull-out and debonding failure mechanism. However, at 200°C, PA66 fibers fractured almost in the cross-section, which showed that bond between fibers and matrix was better and failure mode transformed into fiber fracture. The interfacial adhesion was thus found to be better up to a temperature of 200°C.

4.2.2 Film-Stacking Technique

Gong et al. [13] used polyamide 66 (PA66) fiber and polyamide 6 (PA6) matrix to manufacture all-PA composites using various processing conditions. In these all-PA composites, the reinforcement and matrix share same molecular structure unit (–CONH–$(CH_2)_5$–). Because of the chemical similarity of the two components, good bonding at the fiber/matrix interface could be expected. Effects of processing temperature and cooling rate on the structure and physical properties of composites were investigated by SEM, DMA, DSC analyses, and static tensile test. Fiber/matrix interface strength benefited from elevated processing temperatures. The static tensile results showed that maximum tensile strength was observed in the processing temperature range of 225°C–245°C. At different cooling rates, crystallization temperature of PA6 in the composites was increased compared to the pure PA6 because of the nucleation effect of PA66 fiber surface to the PA6 matrix. A study of the matrix microstructure in a single fiber–polymer composite gave proof of the transcrystalline growth at the fiber–matrix interface, the reasons behind which were the similar chemical compositions and lattice structures between PA6 and PA66.

Film-stacking technique was used for this research (Figure 4.5) and for various other SPC systems [14]. PA66 plain weave was used as the reinforcement phase and PA6 as the matrix. Plain weave of PA66 was selected to

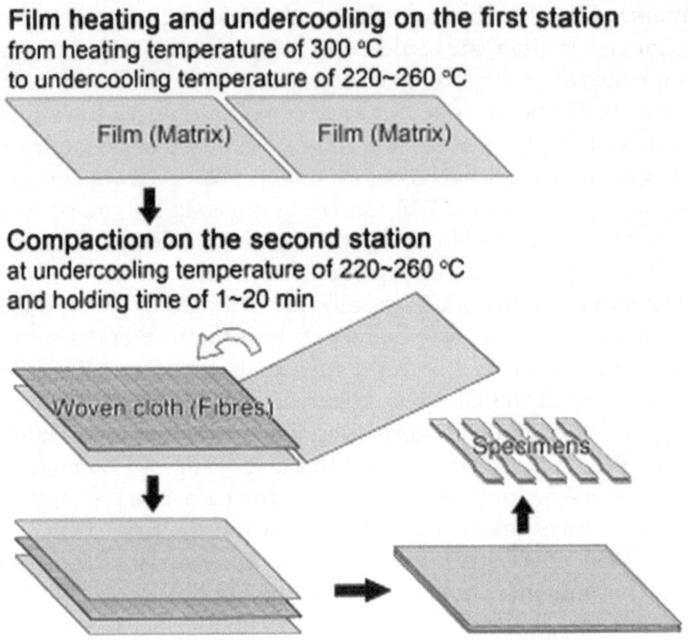

FIGURE 4.5
Film-stacking technique in composite manufacturing. (From Wang, J. et al., *Composi. Sci. Technol.*, 91, 50, 2014. With permission from ScienceDirect, www.sciencedirect.com/science/article/pii/S0266353813004624.)

balance the properties in both warp and weft directions of the final composites. The difference in the melting points of PA66 and PA6 was essential for the successful preparation of the composites and guaranteed a wide temperature processing window. The researchers also commented on the mechanism of bonding. The effect of processing temperature on interface bonding is related to physical and chemical reasons. First, the decrease of PA6 matrix viscosity as the temperature increased improved the wetting out of PA6 melt on the fiber surface and promoted the physical diffusion effect between fiber and matrix. Second, PA66 fiber and PA6 matrix both have sufficient –CONH– which could form hydrogen bonds between them. Increase in temperature promoted the formation of amide hydrogen bonds on the interface and strengthened the interface bonding. They studied the dynamic mechanical properties and the effect of cooling rate on the crystallization behavior of composite materials. Transcrystallinity was clearly observed and the reader is referred to the original article to find details of transcrystallization.

Bhattacharya et al. [15] demonstrated the use of two basic techniques for the preparation of SPCs, hot compaction, and film stacking. Single polymer composite was manufactured from polyamide 6 (PA 6). The starting materials were PA 6 high-tenacity yarn, which was used as reinforcement, and PA 6 film, prepared via melt quenching—used as matrix material. The polymorphic modifications of PA 6 differed in their melting temperatures. The prepared SPC was characterized by the formation of layered structure and demonstrated superior mechanical properties due to good wetting. Tensile modulus was seen to improve by 200% and the ultimate tensile strength by 300%–400% as compared to the isotropic matrix film.

4.2.3 Film-Casting Technique

Polymer film casting is a process used for making flexible plastic components that are typically in the shape of a single or multi-lumen tube and are generally utilized in the medical industry. This manufacturing technology is unique because the process does not necessitate conventional extrusion or injection-molding technologies, yet it readily incorporates components and features traditionally produced by these processes.

A very different and detailed approach for SPC preparation (Figure 4.6) was used by a group of researchers [16]. All-PA composites were manufactured in the form of a flat laminate on the fabric substrate by a method of isothermal immersion-precipitation. Film casting has been used by various researchers like Giuseppe [17].

Seven solutions of PA production waste in formic acid were made by dissolving different amounts of PA in 100 g (82 ml) formic acid at room temperature. It was found that in low concentrations (less than ≈30% w/w), PA66 readily dissolved in formic acid at room temperature, but for higher concentrations, the solution required to be agitated for a longer time. To assure the completion of dissolution and have the same agitation condition for all the solutions, the sealed solution flasks were put in a shaker to obtain a

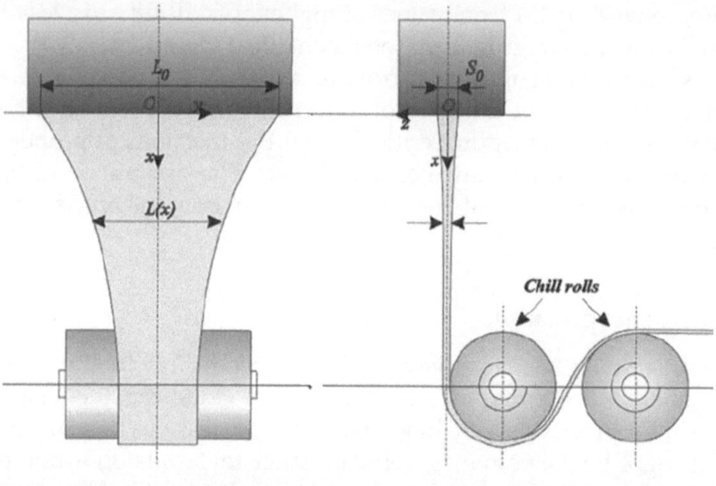

FIGURE 4.6
A schematic of film-casting method. (From Titomanlio, G. and Lamberti, G., *Rheol. Acta.*, 43, 146, 2004. With permission from Springer, https://link.springer.com/article/10.1007/s00397-003-0329-4.)

homogeneous solution called dope. The dopes were cooled to room temperature, and after centrifugation for 10 min to remove the bubbles, the dopes were casted on a PA fabric. Once the casting process was done, after waiting for different intervals, the glass plate was immersed in a distilled water coagulation bath at room temperature to induce polymer precipitation. From then on, the waiting time, which is the time between applying the solution to the fabric and immersion in the coagulant, is called gelling time. After 1 h coagulation (in the water bath), the composites obtained were first washed with distilled water and then held under light press between two sheets of filter papers and dried. The composites were named according to their corresponding solution concentrations and gelling time. Interestingly, this method doesn't rely on partial melting of polymers.

The researchers commented on the morphological properties of the composites. Scanning electron micrographs of the cross-section showed that practically all specimens demonstrated good adhesion between the fabric and formed film. The researchers commented that the boundaries between the fiber of the fabric and the formed film were not clear, having faded due to the adhesion of the two components. At higher concentrations of the solvent, the dissolving power of the dope was found to be greater; hence, it could penetrate more into the fabric and dissolve a larger part of the fabric. It is obvious that a larger part of the cross-sectional area of the fabric was dissolved and the polymer chains were found to be inter-diffused into each other. Although higher surface dissolution helped to create a better adhesion between the fabric and the formed film, there is a possibility of degradation and a change in fabric structure from a fibrous form to a film form—which is not the ideal thing to happen. Fibers are spun and drawn polymers and have a high crystallinity and thus good strength, while films are amorphous. Therefore, converting PA from values of higher crystallinity to a less crystalline form is not encouraging from the mechanical viewpoint.

This has been a challenge for majority of the research groups, especially in the solvent route. The optimization of concentration where the best mechanical properties are obtained are critical—and the mechanical properties are seen to vary over a small range of dissolution. The orientation of the reinforcing phase is often critical and in quest of a perfect interphase; often the orientation gets hampered.

4.2.4 Microencapsulation

Microencapsulation has been described as a process in which tiny particles or droplets are surrounded by a coating to give small capsules (Figure 4.7) of many useful properties [18]. In general, this process has been used to contain food ingredients, enzymes, cells, or other materials on a micrometric scale. Microencapsulation can also be used to enclose solids, liquids, or gases inside a micrometric wall made of hard or soft soluble film, so as to reduce dosing frequency and prevent the degradation of pharmaceuticals [19].

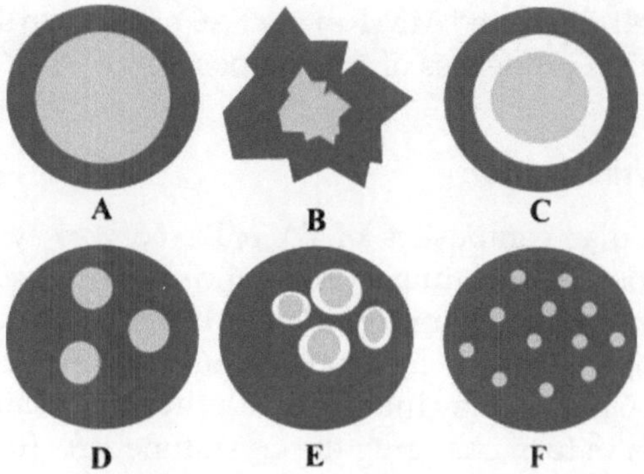

FIGURE 4.7
Microencapsulation process. (A) Simple, (B) Irregular, (C) Multiwall, (D) Multi-core, (E) Aggregate and (F) Matrix. (From Goncalves, A. et al., *Trends Food Sci. Technol.*, 51, 76, 2016. With permission from ScienceDirect, www.sciencedirect.com/science/article/pii/S092422441530131X.)

Vasileva et al. [20] disclosed a new two-stage strategy for the preparation of all-PA laminate composites based on PA6 matrices reinforced by high volume fractions of PA66 textile structures and three different types of nanoclays. The task of manufacturing was completed in two stages. PA6 microcapsules (MC) were loaded with montmorillonite (MMT) nanoclays in the first stage. In the second stage, the MMT-loaded MC obtained with controlled molecular weight, composition, and granulometry are compression molded in the presence of PA66 textile structures to produce the final dually reinforced laminate composites. The mechanical properties showed significant improvements. Mechanical tests in tension, flexure, and impact for selected composites in this study showed up to 73% increase of the Young's modulus, up to 142% increase of the stress at break, and more than a fivefold increase of the notched impact resistance. The mechanical behavior of the dually reinforced composites was deliberated in combination with the morphology of samples studied by optical and electron microscopy, and the crystalline structure of matrix as revealed by DSC and synchrotron X-ray diffraction. The researchers commented that the microencapsulation strategy toward dually reinforced polymer composites has big potential in combining matrix and reinforcements. For a detailed discussion, the reader is referred to the full version of the paper.

One of the main objectives of this unique study was to prepare all-PA laminate composites with a high-volume fraction of PA66 textile plies. Matrix using PA6 was additionally reinforced by MMT nanoclays. This was achieved by impregnation of the textile plies by MMT-loaded PA6 MC and ensuing compression molding. The researchers commented that such dual reinforcement was not considered before in any attempts to produce PA laminates. They justified their work to be inspired from previous studies that have shown that the PA66 textile reinforcement can significantly increase the mechanical properties of PA6. There also exist many reports suggesting that the presence of well-dispersed MMT in the PA6 matrix can strongly enhance [21,22] all mechanical properties of the composite.

4.2.5 In situ Polymerization

A series of molecular composites of PA 6/PA 66 was synthesized via in situ polymerization [23]. The impact resistance of PA 6 was improved dramatically by incorporating a minor amount of PA 66 (2–10 wt.-%), without decreasing the tensile strength. Inserting PA 66 macromolecules at a molecular level into a PA 6 matrix may interfere with the arrangement of the hydrogen bonds of PA 6, in turn changing the crystalline structure and impeding the crystallization of PA 6.

The researchers explained the basic theory for attempting to manufacture the SPCs. If one type of PA can be completely dissolved in ε-caprolactam monomer and a homogeneous transparent solution is attained, this PA is thought to be thermodynamically miscible with the monomer ε-caprolactam. However, If the PA/ε-caprolactam solution exhibits translucent character, the system is only partially miscible. During ring-opening polymerization, since ε-caprolactam contains an amide ring, no other groups would be released, thus making the chemical composition of PA 6 polymer almost the same as its monomer. In addition, the strong hydrogen bonding between PA and the ε-caprolactam monomer and between the PAs plays an important role, which overcomes the phase separation problem caused by the reduction of ΔSmix after polymerization. Therefore, it is expected that if a selected PA is miscible with ε-caprolactam monomer, it will also exhibit good miscibility with PA 6, and possibly form a molecular level composite.

Izod notched impact strength of PA 6/PA 66 showed that there was significant improvement when a small amount of PA 66, around 4 wt.-%, was incorporated. The impact strength at room temperature increased two times compared to that of pure PA 6. The notched impact strength at –40°C was also greatly improved. Wide-angle X-ray diffraction (WAXD) was performed to study the crystal forms. The WAXD patterns of PA 6/PA 66 revealed that the peaks of the $\alpha 1$ crystal form of PA 6 decreased, those of the $\alpha 2$ crystal form increased with increasing PA 66, and the γ crystal form diminished and disappeared when PA 66 content was increased. The researchers commented that PA 66 macromolecules significantly influence the crystalline structure of the PA 6 matrix. Fourier-transform infrared data showed that there was no phase separation. Hence the researchers were able to create molecular composites of PA 6/PA 66 via in situ polymerization. The impact resistance of PA 6 was improved dramatically without decreasing the tensile strength by incorporating a small amount of PA 66 (2–10 wt.-%). The specific hydrogen bonding interaction between PA 66 and PA 6 plays a critical role. PA 6/PA 66 molecular composites could be synthesized successfully due to the presence of hydrogen bonding interactions. PA 66 did not crystallize and undergo phase separation in molecular composites as the PA 66 macromolecules were closely surrounded by PA 6.

Single-polymer composites based on PA6 were prepared by in-mold activated anionic ring-opening polymerization [24] (AAROP) of caprolactam in the presence of PA6 textile fibers. The researchers studied the influence of the reinforcing fibers content, their surface treatment, as well as of the temperature of AAROP upon the crystalline structure, morphology, and mechanical properties of the composites. This group also observed the presence of oriented transcrystalline layer on the surface of the reinforcing fibers. Its orientation and polymorph structure were determined by synchrotron wide-angle X-ray scattering. They studied the mechanical behavior in tension of the SPCs when composites showed a well-expressed growth of the stress at break (70%–80%) and deformation at break (up to 150%–190%) in composites with 15%–20 wt% of reinforcements. The best mechanical properties were found in SPCs whose reinforcing fibers were solvent-pretreated prior to AAROP so as to remove the original finish. In these samples, a stronger adhesion at the fiber/matrix interface was proved by scanning electron microscopy of cryofractured samples.

Single-polymer composites were manufactured through in situ polymerization in another work by Gong et al. [25]. After melting in a flask, 100 g CL monomer was exposed to vacuum conditions at 140°C for a duration of 15 min to remove water. NaOH (catalyst) was then added into the CL melt and the mixture was exposed to a vacuum. Subsequently, the mixture was cooled under dry nitrogen. The schematic of the RTM set up is shown in Figure 4.8.

Using nitrogen pressure of 0.02 MPa, the reactive mixture was permeated into a matched stainless steel mold (200 × 100 × 0.6 mm). Ten layers of PA6 plain clothes were plied in the mold. The mold was placed in a hot press at various temperatures—140°C, 160°C, 180°C, and 200°C.

When molding temperature was lower than 160°C, increasing the temperature had a positive influence on the strength of the composites. With increasing the TM further (higher than 160°C), the strength of SPCPA drastically decreased to 81 MPa at 200°C (Figure 4.9). With textile volume fractions

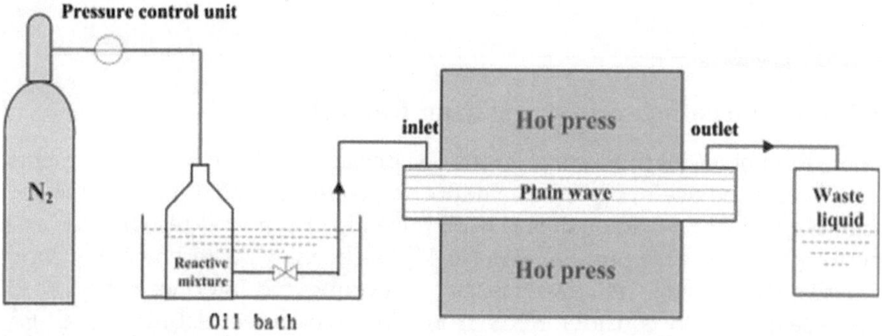

FIGURE 4.8
Structure of the RTM set up used by the researchers. (From Gong, Y. et al., *Compos. A*, 41, 1006, 2010. With permission from ScienceDirect, www.sciencedirect.com/science/article/pii/S1359835X10001065, Figure 8.)

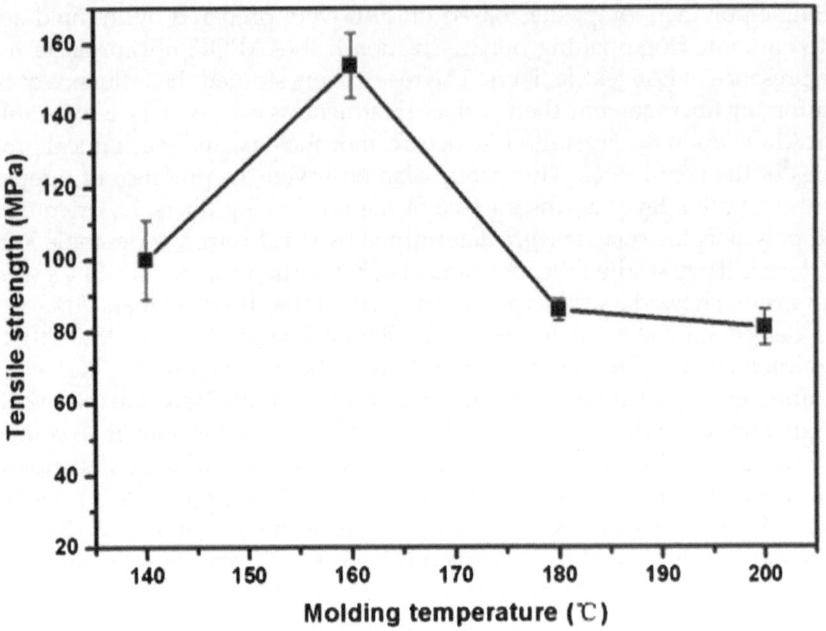

FIGURE 4.9
Tensile strength of composites prepared at different molding temperatures. (From Gong, Y. et al., *Compos. A*, 41, 1006, 2010. With permission from ScienceDirect, www.sciencedirect.com/science/article/pii/S1359835X10001065.)

close to 0.7, the tensile strengths of the FS samples reached 180 MPa and of those by RTM–155 MPa, the value of the neat PA6 matrix being 68 to 70 MPa. Such significant improvement was attributed to good adhesion and possible chemical bonding at the matrix–fiber interface.

4.3 Comparisons and Concluding Remarks

There has been a lot of research in the last decade on Nylon 6 reinforcements [26–29] and the use of Nylon 6 as nanocomposites [30–32]. Nylon 66 [33–36] and its use in nanocomposites [37–43] have also been well researched. The performance of these popular PAs [44,45] has been evaluated by a few comparative studies [46]. However, issues of recycling and interface have forced the researchers to consider SPCs from PAs. Good recyclability of all-PA composites as compared to those reinforced by glass or carbon fibers and the absence of compatibilizers/surface modifiers in the system would make them a preferred choice in certain applications. An overall comparative statement of mechanical properties has been given in Table 4.1.

TABLE 4.1

Mechanical Properties of Polyamide-Based SPCs

Starting Material	Process	Tensile strength (MPa)	Young's modulus (GPa)	% Breaking extension	Flexural strength (MPa)	Remarks	Reference
Woven Nylon 6,6 multifilaments	Hot compaction	170	4.1	–	–	–	[10]
Nylon 6,6 plain fabric (reinforcement) and Nylon 6,6 plain woven fabric (Nylon 6 as matrix)	Resin transfer molding (anionic polymerization of e-caprolactam)	156	–	–	130		[11]
Nylon 6,6 plain fabric (reinforcement) and Nylon 6 film (matrix)	Resin transfer molding	192	–	–	–	–	[12]
Nylon 6 yarn (reinforcement) and Nylon 6 film (matrix)	Hot compaction and film stacking	359.07	5.31	–	–	With catalyst	[13]
Nylon 6,6 plain woven fabric and ε-caprolactam (Nylon 6 as matrix)	Microencapsulation and activated anionic ring-opening polymerization of ε-caprolactam in solution	85.7 (warp), 128.1 (weft)	1.93 (warp), 2.37 (weft)	20.0 (warp), 22.4 (weft)	–	9 piles, without nanoclays	[15]
Nylon 6,6 plain woven fabric and ε-caprolactam (Nylon 6 as matrix)	Microencapsulation and activated anionic ring-opening polymerization of ε-caprolactam in solution	115 (warp), 152.2 (weft)	2.85 (warp), 2.86 (weft)	19.5 (warp), 24.3 (weft)	–	9 piles, with Cloisite 15A	[15]
Nylon 6,6 plain woven fabric and ε-caprolactam (Nylon 6 as matrix)	Microencapsulation and activated anionic ring-opening polymerization of ε-caprolactam in solution	89.5 (warp), 139.6 (weft)	2.35 (warp), 2.70 (weft)	15.4 (warp), 24.6 (weft)		9 piles, with Cloisite 20A	[15]

(Continued)

TABLE 4.1 (*Continued*)

Mechanical Properties of Polyamide-Based SPCs

Starting Material	Process	Tensile strength (MPa)	Young's modulus (GPa)	% Breaking extension	Flexural strength (MPa)	Remarks	Reference
Nylon 6,6 plain woven fabric and ε-caprolactam (Nylon 6 as matrix)	Microencapsulation and activated anionic ring-opening polymerization of ε-caprolactam in solution	87.1 (warp), 160 (weft)	2.37 (warp), 2.88 (weft)	19.2 (warp), 26.1 (weft)	–	9 piles with Nanomer I.24 TL	[15]
Nylon 6,6 and melted ε-caprolactam monomer	Molding of mixtures of raw materials	68.5	–	567	–	Nylon 6,6 content 8 wt%	[21]
Nylon 6 and Nylon 6,6	Blending and injection molding	77.6	–	33	–	Nylon 6,6 content 12 wt%	[21]
Nylon 6 filament and e-caprolactam	In-mold activated anionic ring-opening polymerization	122.5	1.74	38.1	–	Fiber content s20 wt% and with treatment in acetone	[22]
Nylon 6 fiber and ε-caprolactam (Nylon 6 as matrix)	In situ anionic polymerization of ε-caprolactam	154	–	–	155 MPa (Modulus-3GPa)	–	[23]

Thus, the highest tensile strength and modulus of 360 MPa of 5.3 GPa have been using the film-stacking method. The highest flexural strength of 155 MPa has been achieved through the route of in situ anionic polymerization. However, compared to tensile properties, only few attempts have been made to study flexural properties of the composite, and more research is required in this direction. In a study on the Izod notched impact strength, there was significant improvement when a small amount of PA 66 was incorporated in PA6. The impact strength at room temperature increased two times compared to that of pure PA 6. Researchers have attempted interesting routes like in-mold activated anionic ring-opening polymerization, microencapsulation inspired from other fields [47], which have been discussed in detail in this chapter. However, so far systematic comparative studies on the overall mechanical properties of PA SPCs in tension, flexure, and impact are missing.

References

1. https://www.britannica.com/science/polyamide [accessed January 22, 2018].
2. http://www.ensinger-online.com/en/materials/engineering-plastics/polyamides/ [accessed January 22, 2018].
3. http://nylon66membrane.com/Preparation-of-Nylon-66.html [accessed January 22, 2018].
4. https://sites.google.com/site/grupodepolimeros/aromatic-polyamides---aramids [accessed January 22, 2018].
5. http://polymerdatabase.com/polymer%20classes/Polyamide%20type.html [accessed January 22, 2018].
6. Xu, Z., and C. Gao. In situ polymerization approach to graphene-reinforced nylon-6 composites. *Macromolecules* 43, no. 16 (2010): 6716–6723.
7. Liu, T., I. Y. Phang, L. Shen, S. Y. Chow, and W. D. Zhang. Morphology and mechanical properties of multiwalled carbon nanotubes reinforced nylon-6 composites. *Macromolecules* 37, no. 19 (2004): 7214–7222.
8. Zhang, W. D., L. Shen, I. Y. Phang, and T. Liu. Carbon nanotubes reinforced nylon-6 composite prepared by simple melt-compounding. *Macromolecules* 37, no. 2 (2004): 256–259.
9. Zheng, L. Y., L. X. Zhao, and J. J. Zhang. Tribological study of three-dimensionally braided carbon fiber–nylon 6 composites against 316L stainless steel. *Current Applied Physics* 7(2007): e120–e124.
10. Hine, P. J., and I. M. Ward. Hot compaction of woven nylon 6, 6 multifilaments. *Journal of Applied Polymer Science* 101, no. 2 (2006): 991–997.
11. Gong, Y., and G. Yang. All-polyamide composites prepared by resin transfer molding. *Journal of Materials Science* 45, no. 19 (2010): 5237–5243.
12. Rouison, D., M. Sain, and M. Couturier. Resin transfer molding of natural fiber reinforced composites: cure simulation. *Composites Science and Technology* 64, no. 5 (2004): 629–644.

13. Gong, Y., and G. Yang. Manufacturing and physical properties of all-polyamide composites. *Journal of Materials Science* 44, no. 17 (2009): 4639–4644.
14. Wang, J., J. Chen, and P. Dai. Polyethylene naphthalate single-polymer-composites produced by the undercooling melt film stacking method. *Composites Science and Technology* 91(2014): 50–54.
15. Bhattacharyya, D., P. Maitrot, and S. Fakirov. Polyamide 6 single polymer composites. *Express Polymer Letters* 3, no. 8 (2009): 525–532.
16. Jabbari, M., M. Skrifvars, D. Åkesson, and M. J. Taherzadeh. Introducing all-polyamide composite coated fabrics: A method to produce fully recyclable single-polymer composite coated fabrics. *Journal of Applied Polymer Science* 133, no. 7 (2016).
17. Titomanlio, G., and G. Lamberti. Modeling flow induced crystallization in film casting of polypropylene. *Rheologica Acta* 43, no. 2 (2004): 146–158.
18. Gonçalves, A., B. N. Estevinho, and F. Rocha. Microencapsulation of vitamin A: a review. *Trends in Food Science and Technology* 51(2016): 76–87.
19. Singh, M. N., K. S. Y. Hemant, M. Ram, and H. G. Shivakumar. Microencapsulation: A promising technique for controlled drug delivery. *Research in Pharmaceutical Sciences* 5, no. 2 (2010): 65.
20. Vasileva Dencheva, N., D. Manso Vale, and Z. Zlatev Denchev. Dually reinforced all-polyamide laminate composites via microencapsulation strategy. *Polymer Engineering and Science* 57, no. 8 (2017): 806–820.
21. Kojima, Y., A. Usuki, M. Kawasumi, A. Okada, T. Kurauchi, and O. Kamigaito. Synthesis of nylon 6–clay hybrid by montmorillonite intercalated with ϵ-caprolactam. *Journal of Polymer Science Part A: Polymer Chemistry* 31, no. 4 (1993): 983–986.
22. Fornes, T. D., P. J. Yoon, H. Keskkula, and D. R. Paul. Nylon 6 nanocomposites: the effect of matrix molecular weight. *Polymer* 42, no. 25 (2001): 09929–09940.
23. Li, Y., and G. Yang. Studies on molecular composites of polyamide 6/polyamide 66. *Macromolecular Rapid Communications* 25, no. 19 (2004): 1714–1718.
24. Dencheva, N., Z. Denchev, A. S. Pouzada, A. S. Sampaio, and A. M. Rocha. Structure–properties relationship in single polymer composites based on polyamide 6 prepared by in-mold anionic polymerization. *Journal of Materials Science* 48, no. 20 (2013): 7260–7273.
25. Gong, Y., A. Liu, and G. Yang. Polyamide single polymer composites prepared via in situ anionic polymerization of ε-caprolactam. *Composites Part A: Applied Science and Manufacturing* 41, no. 8 (2010): 1006–1011.
26. Liu, T., I. Y. Phang, L. Shen, S. Y. Chow, and W. D. Zhang. Morphology and mechanical properties of multiwalled carbon nanotubes reinforced nylon-6 composites. *Macromolecules* 37, no. 19 (2004): 7214–7222.
27. Xu, Z., and C. Gao. In situ polymerization approach to graphene-reinforced nylon-6 composites. *Macromolecules* 43, no. 16 (2010): 6716–6723.
28. Zhang, W. D., L. Shen, I. Y. Phang, and T. Liu. Carbon nanotubes reinforced nylon-6 composite prepared by simple melt-compounding. *Macromolecules* 37, no. 2 (2004): 256–259.
29. Cho, J. W., and D. R. Paul. Nylon 6 nanocomposites by melt compounding. *Polymer* 42, no. 3 (2001): 1083–1094.
30. Fornes, T. D., and D. R. Paul. Modeling properties of nylon 6/clay nanocomposites using composite theories. *Polymer* 44, no. 17 (2003): 4993–5013.

31. Fornes, T. D., and D. R. Paul. Crystallization behavior of nylon 6 nanocomposites. *Polymer* 44, no. 14 (2003): 3945–3961.
32. Fong, H., W. Liu, C. S. Wang, and R. A. Vaia. Generation of electrospun fibers of nylon 6 and nylon 6-montmorillonite nanocomposite. *Polymer* 43, no. 3 (2002): 775–780.
33. Lü, J., and X. Lu. Elastic interlayer toughening of potassium titanate whiskers–nylon66 composites and their fractal research. *Journal of Applied Polymer Science* 82, no. 2 (2001): 368–374.
34. Kyu, T., T. I. Chen, H. S. Park, and J. L. White. Miscibility in poly-p-phenylene terephthalamide/nylon 6 and nylon 66 molecular composites. *Journal of Applied Polymer Science* 37, no. 1 (1989): 201–213.
35. Xu, X., B. Li, H. Lu, Z. Zhang, and H. Wang. The effect of the interface structure of different surface-modified nano-SiO2 on the mechanical properties of nylon 66 composites. *Journal of Applied Polymer Science* 107, no. 3 (2008): 2007–2014.
36. Clark Jr, R. L., M. D. Craven, and R. G. Kander. Nylon 66/poly (vinyl pyrrolidone) reinforced composites: 2: Bulk mechanical properties and moisture effects. *Composites Part A: Applied Science and Manufacturing* 30, no. 1 (1999): 37–48.
37. Li, L., C. Y. Li, C. Ni, L. Rong, and B. Hsiao. Structure and crystallization behavior of Nylon 66/multi-walled carbon nanotube nanocomposites at low carbon nanotube contents. *Polymer* 48, no. 12 (2007): 3452–3460.
38. Shen, L., I. Y. Phang, L. Chen, T. Liu, and K. Zeng. Nanoindentation and morphological studies on nylon 66 nanocomposites. I. Effect of clay loading. *Polymer* 45, no. 10 (2004): 3341–3349.
39. Yu, Z. Z., C. Yan, M. Yang, and Y. W. Mai. Mechanical and dynamic mechanical properties of nylon 66/montmorillonite nanocomposites fabricated by melt compounding. *Polymer International* 53, no. 8 (2004): 1093–1098.
40. Dasari, A., Z. Z. Yu, M. Yang, Q. X. Zhang, X. L. Xie, and Y. W. Mai. Micro-and nano-scale deformation behavior of nylon 66-based binary and ternary nanocomposites. *Composites Science and Technology* 66, no. 16 (2006): 3097–3114.
41. Shen, L., T. Liu, and P. Lv. Polishing effect on nanoindentation behavior of nylon 66 and its nanocomposites. *Polymer Testing* 24, no. 6 (2005): 746–749.
42. Yu, Z. Z., M. Yang, Q. Zhang, C. Zhao, and Y. W. Mai. Dispersion and distribution of organically modified montmorillonite in nylon-66 matrix. *Journal of Polymer Science Part B: Polymer Physics* 41, no. 11 (2003): 1234–1243.
43. Xu, X., B. Li, H. Lu, Z. Zhang, and H. Wang. The effect of the interface structure of different surface-modified nano-SiO2 on the mechanical properties of nylon 66 composites. *Journal of Applied Polymer Science* 107, no. 3 (2008): 2007–2014.
44. Chavarria, F., and D. R. Paul. Comparison of nanocomposites based on nylon 6 and nylon 66. *Polymer* 45, no. 25 (2004): 8501–8515.
45. Vlasveld, D. P. N., S. G. Vaidya, H. E. N. Bersee, and S. J. Picken. A comparison of the temperature dependence of the modulus, yield stress and ductility of nanocomposites based on high and low MW PA6 and PA66. *Polymer* 46, no. 10 (2005): 3452–3461.
46. Fornes, T. D., and D. R. Paul. Structure and properties of nanocomposites based on nylon-11 and-12 compared with those based on nylon-6. *Macromolecules* 37, no. 20 (2004): 7698–7709.
47. Gonçalves, A., B. N. Estevinho, and F. Rocha. Microencapsulation of vitamin A: a review. *Trends in Food Science and Technology* 51(2016): 76–87.

5

Single-Polymer Composites from Polyesters

5.1 Introduction

Polyester, as defined by *Encyclopaedia Britannica* [1], is a class of synthetic polymers built up from multiple chemical repeating units linked together by ester (CO–O) groups. Shellac, a form of natural polyester, as secreted by the insect lac, was used by the ancient Egyptians for preserving mummies. Berzelius was the first scientist who synthesized polyesters of polybasic acids and polyvalent alcohols by reacting tartaric acid and glycerol in 1847.

In the late 1920s, it was the famous chemist W. H. Carothers from DuPont who made a vast number of polyesters varying in molecular weight between 2,500 and 5,000 by condensation reactions of dicarboxylic acids and diols. Carothers and Arvin synthesized polyesters in 1930s, having a molecular weight of ~4,000 by reacting octadecanedioic acid with polydioxanone (dicarboxylic acids with 5% excess of diols). Polyester is generally formed by step-growth polycondensation from dicarboxylic acid or its diester and diol. The by-products produced are water or methanol depending upon the reactant used, namely, dicarboxylic acid or its dimethyl ester. In a typical polycondensation reaction for the synthesis of polyester, a number of catalysts are used. They are mainly compounds of germanium, titanium, antimony, aluminum, and so on. Ring-containing polyesters are the greater and commercially more important group amongst all the classes of polyester. The most important member of this class is polyethylene terephthalate (PET), a stiff, strong polymer that has been spun into fibers known by such trademarks as Dacron and Terylene and which has the highest share of the market. There has been a lot of development including that of bio-based polyesters (Figure 5.1).

There have been several studies on self-reinforced polyester composites by different groups, the research findings of which will be dealt with in the subsequent sections. The starting material was found to vary for the preparation of these materials.

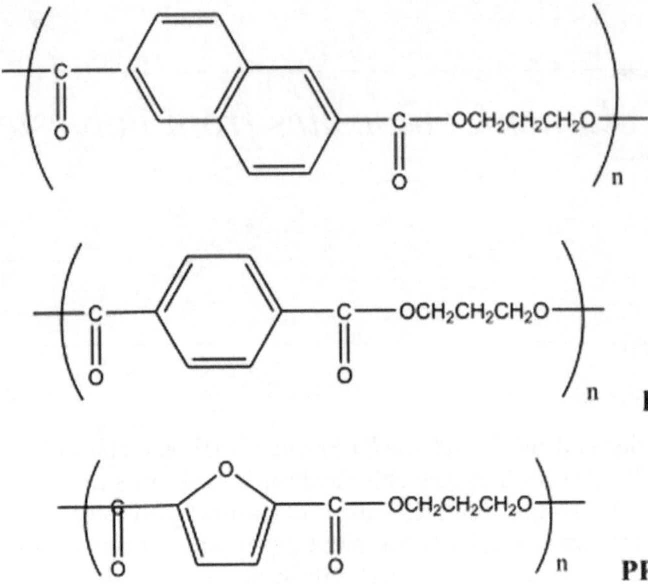

FIGURE 5.1
Schematic of bio-based polyester. (From Papageorgiou, G. Z. et al., *Polymer*, 62, 28, 2015. With permission from ScienceDirect, www.sciencedirect.com/science/article/pii/S003238611 5001391#sch1, scheme 1.)

5.2 Single-Polymer Composites from Different Starting Materials

Self-reinforced composites from polyester can be classified in several ways, based on the starting material.

5.2.1 Fibers

Fibers were the starting material for a process by Rasburn et al. [3] explaining the successful compaction of PET fibers. Different kinds of orientation of fibers have been used in the manufacture of composites (Figure 5.2).

In this research, fiber was wound unidirectionally using a specialized winding apparatus and subsequently placed between mold plates, with a surface area of 55 x 55 mm^2. The complete assembly of fibers and mold was then transferred to a preheated compression-molding machine. Compaction temperature was decided based on the melting endotherm of the original fiber, determined by differential scanning calorimetry (DSC). The measurement of mechanical properties shows that a very high proportion of the original fiber properties are preserved and that resultant compacted samples

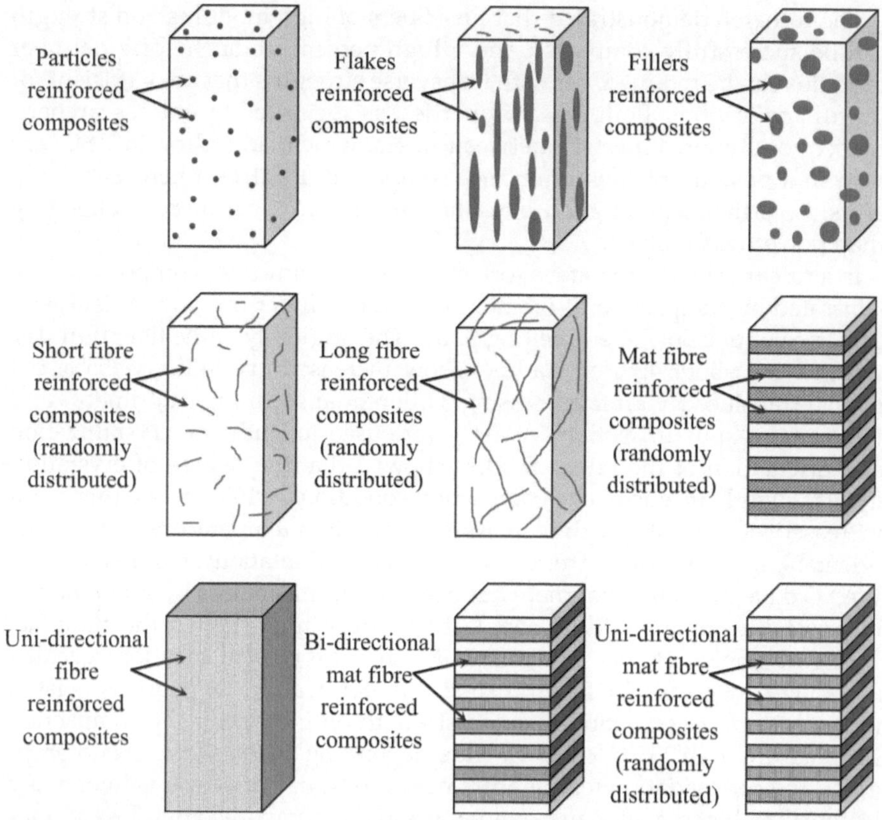

FIGURE 5.2
Types of fiber, particle, and filler-reinforced composites. (From Nirmal, U. et al.,*Tribol. Int.*, 83, 77, 2015. With permission from ScienceDirect, www.sciencedirect.com/science/article/pii/S0301679X14003843, Figure 10.)

have a good degree of constancy. The researchers performed the electron microscopy studies of etched samples that revealed the important role played by compaction temperature on the developed structure of resultant samples.

The longitudinal moduli of the compacted fiber samples again parallel the behavior found with the melt-spun polyethylene fibers. At low compaction temperatures, the lower values, contrasted to the original fiber modulus, reflect incomplete compaction and suggest the presence of lateral voids. Modulus emerges to be the maximum at intermediate temperatures, when filling of voids by molten polymer resulted in a coherent material, with a minimal amount of fiber having been melted in the process. Modulus is highly sensitive to temperature ranges where the orientation of polymers gets disturbed. A decline in the modulus with increasing compaction temperature was also observed, which was due to a loss of orientation of the melted and recrystallized part of the sample.

The research demonstrated that PET fibers of high modulus and strength can be successfully compacted and a high percentage of the original fiber modulus can be retained, yet with transverse strengths that are sufficient for useful applications. Both these properties, as suggested by the researchers, reflect good interconnection between fibers, which, according to DSC, are held in a position of constraint once compacted. Electron microscopy suggests, nevertheless, that good interfibrillar adhesion was more challenging than polyethylene fibers.

In a research by Rojanapitayakorn et al. [5], self-reinforced composites were fabricated by compaction of oriented PET fibers under pressure at temperatures near, but below, their melting point. The originally white fiber bundles, which were about 40% crystalline, show increased crystallinity (55%) but optical translucency after processing. Differential scanning calorimetry and wide-angle X-ray diffraction (WAXD) were used to study the crystallization and orientation of the fibers. Studies showed that the degree of crystallinity was slightly insensitive to compaction conditions, while the melting point increased significantly with increasing compaction temperature. Crystalline orientation, determined from the Hermans orientation parameter from WAXD data, indicated that there has been no significant loss in orientation of the crystalline fraction due to compaction. This is important in the manufacturing of single-polymer composites. Mechanical characterization revealed a stepwise decrease in flexural modulus (9.4–8.1 GPa) and an associated increase in transverse modulus and strength on increasing the compaction temperature from 255°C to 259°C. Thus, depending on the choice of end properties, the consolidation temperature needs to be optimized. It is interesting to note that this transition in behavior was also accompanied by a loss of optical transparency and a change in the distribution of amorphous fraction from fine intrafibrillar domains to coarse interfibrillar domains as observed with electron microscopy. The researchers strongly felt that the mechanical properties of PET compactions are influenced more by orientation of the amorphous phase rather than that of the crystalline phase. The impact properties of compacted materials showed remarkable impact resistance after compaction, with impact toughness declining as compaction temperature was increased.

A very interesting trend between the compaction temperature and melting isotherms and crystallinity was reported by the group.

While the melting point of the compacted fibers improved with increasing compaction temperature below the melting point of the original fibers (Figure 5.3), it is interesting to note that the melting endotherm of the compacted samples became broader than that for the original fibers. In addition, for annealing temperatures below 259°C, the endotherm exhibited a peak or shoulder on the lower temperature side of the major peak. However, there was a drastic decrease of crystallinity and peak melting temperature when the composites were taken to compaction temperature beyond 262°C. In fact, the DSC thermogram (not shown here) showed that the crystalline structure has been lost.

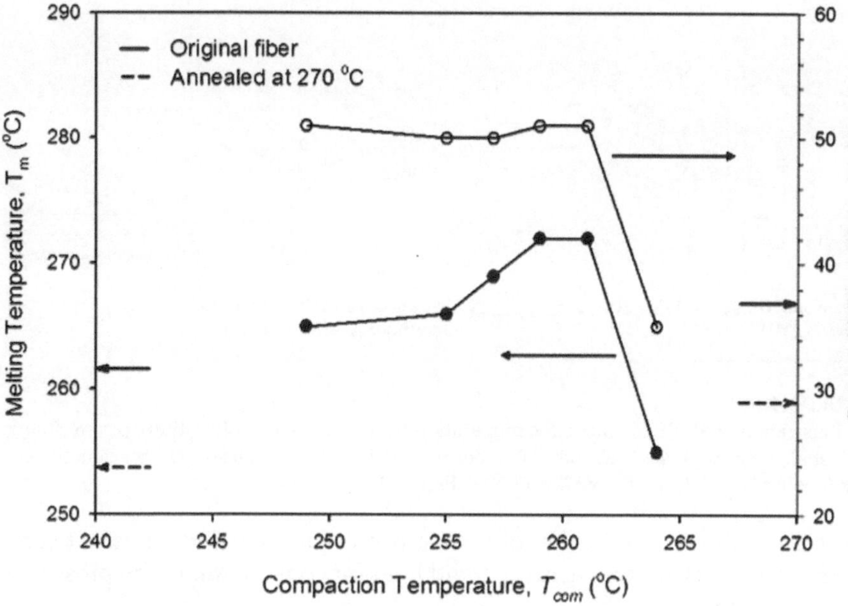

FIGURE 5.3
Melting temperature (●) and degree of crystallinity (○) for PET fiber compactions as a function of the compaction temperatures. Dashed arrows denote the melting temperature and degree of crystallinity for unoriented PET (annealed at 270°C), while solid arrows denote the melting point and degree of crystallinity for an original unconstrained PET fiber. (From Rojanapitayakorn, P. et al., *Polymer*, 46, 761, 2005. With permission from ScienceDirect, www.sciencedirect.com/science/article/pii/S003238610401136X, Figure 4.)

5.2.2 Tapes

In a work by Zhang et al. [6], all-PET composites used the method of film stacking of oriented PET tapes. Initial experiments were done using PET tapes and co-PET films (Figure 5.4) to determine the temperature window, including DSC and T-peel tests. The tensile properties of PET tape, co-PET film, and all-PET composites were testified and contrasted with a commercial coextruded PURE® polypropylene tape. The research group studied the effect of compaction pressures and temperatures on tensile properties of all-PET composites to determine optimum processing parameters for harmonizing good interfacial adhesion between tapes and residual tensile properties of PET tapes.

In the experimental part, unidirectional composite laminates were produced by winding tape from a bobbin onto a steel frame (dimension: 120 mm × 140 mm) using a winding machine. To achieve a high fiber volume fraction with sufficient matrix, nine layers of co-PET films wound between six layers of PET tapes as a multiple sandwich structure were used. A high fiber volume fraction of 70% was achieved.

The frame was then placed in a mold, and the assembly of PET tapes and co-PET films was compacted in a hot press into a unidirectional composite

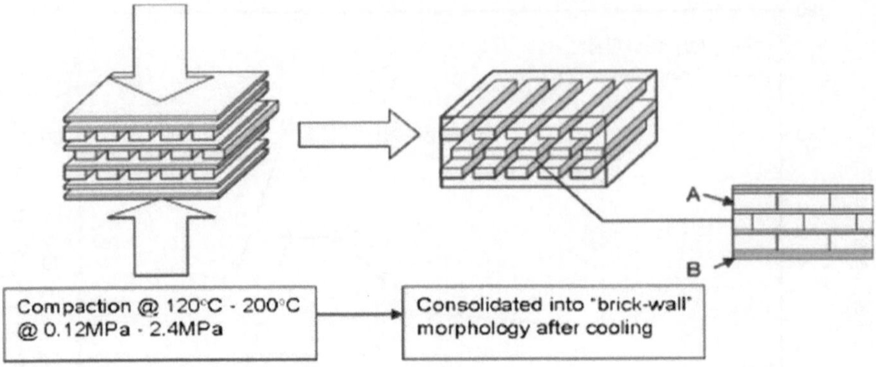

FIGURE 5.4
An illustration of all-PET composites preparation (A: PET tape; B: co-PET film). (From Zhang, J. M. et al., *Compos. A*, 40, 1747, 2009. With permission from ScienceDirect, www.sciencedirect.com/science/article/pii/S1359835X09002486, Figure 3.)

sheet through the application of heat and pressure. After the desired compaction temperature was achieved and held for consolidation, the press was rapidly cooled.

All-PET composites were manufactured using a film-stacking technique. This technique includes a heating process to enable diffusion between reinforcement (PET tape) and matrix (co-PET film). This may compromise mechanical properties of PET tapes at elevated temperatures. To investigate the thermal stability of PET tapes, free shrinkage measurements and static tensile tests at elevated temperatures were carried out.

Temperature was found to be the single most important parameter in the experiments. A higher compaction temperature has been found to improve stress transfer between the tapes, but molecular relaxation of tapes gives a reduction in longitudinal properties at compaction temperatures above 180°C. The researchers found that among the range of compaction temperatures selected, 160°C–180°C (Figure 5.5) was the optimum compaction temperature for composite consolidation. This temperature gave a good balance between longitudinal and transverse properties.

Overall, it has to be noted that film-stacking method based on co-PET films allowed for a large processing temperature window in composite preparation. A large temperature window allows more confidence in manufacturing. From the research, it was found that satisfactory interfacial bonding was achieved when the PET/co-PET assembly was fused in a hot press at 160°C–180°C and a compaction pressure of around 0.6 MPa.

5.2.3 Double-Covered Uncommingled Yarn

Hybrid yarn has been used extensively in composites. They can be manufactured by different ways comprising commingling and uncommingling techniques, which are frequently used approaches in hybrid yarn production.

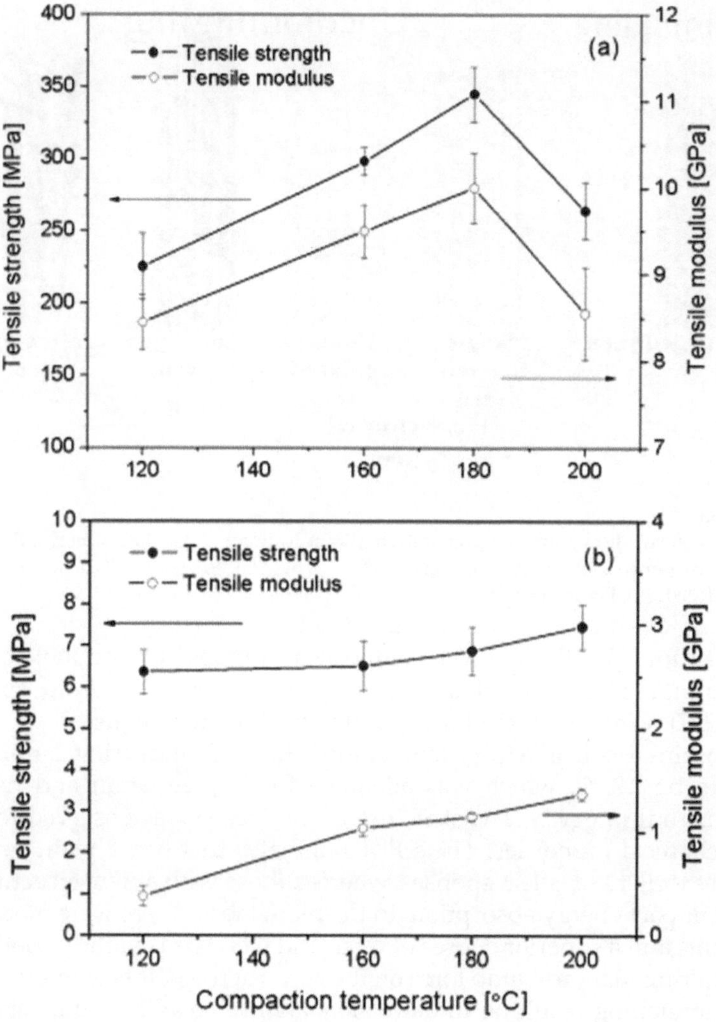

FIGURE 5.5
Longitudinal (a) and (b) transverse mechanical properties of unidirectional all-PET composites (compaction pressure: 0.6 MPa) vs. compaction temperature. (From Zhang, J. M. et al., *Compos. A*, 40, 1747, 2009. With permission from ScienceDirect, www.sciencedirect.com/science/article/pii/S1359835X09002486.)

Commingling is the technique used to produce even distribution of fiber and matrix along the surface of the yarn. It can be done by different types of technologies, such as air jet spinning, water blending, and electrostatic charge spinning. Uncommingled is an uneven distribution of the matrix around reinforcing yarn. Uncommingled yarn can be subsequently classified according to their morphological structures: (a) single-covered yarn, (b) double-covered yarn, (c) core-spun yarn, and (d) stretch-broken yarn (Figure 5.6).

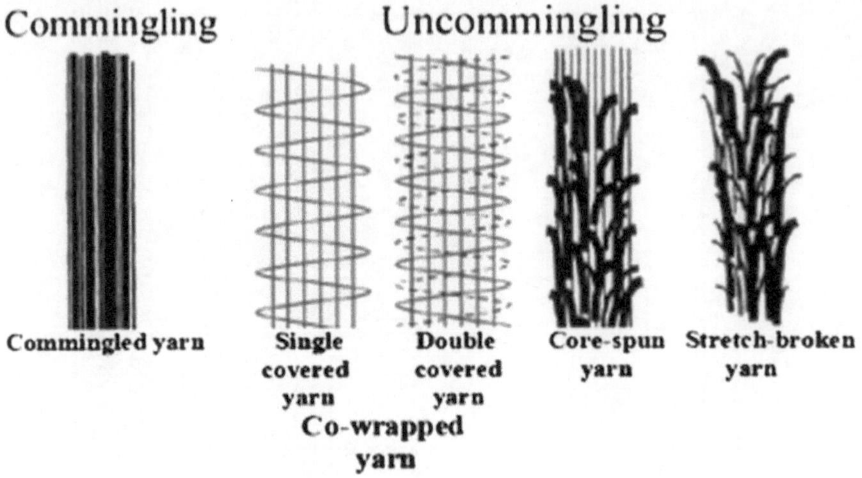

FIGURE 5.6
Types of preforms for composite preparation. (From Kannan, T. G. et al., *Compos. B*, 43, 2836, 2012. With permission from ScienceDirect, www.sciencedirect.com/science/article/pii/S1359836812002867, Figure 1.)

Self-reinforced PET (srPET) composites composed of double-covered uncommingled yarn [8] that was prepared through co-wrap spinning (Figure 5.7) were hot pressed using a film-stacking technique.

The optimal consolidation temperature for manufacturing lamina was found to be 240°C, which was adequate for impregnation and reinforcing structural integrity. The structure of the weave was observed to affect the mechanical properties. The srPET laminates that had a plain structure showed excellent tensile responses, whereas those with a twill structure had highest impact energy absorption. In the plain fabric, there were more interlace points and a superior degree of wrap and weft yarns in the periodic unit cell, resulting in more clamp force on the yarns at the points of interlacement during stretching resulting in more elongation. The srPET composites prepared using basket-weaving [9] (Figure 5.7) structural fabric broke apart and resulted in the lowest impact energy, which can be attributed to the larger gap between two interweaving points that creates a weak path.

Dynamic mechanical measurements over a wide range of temperatures are useful for determining viscoelastic behavior and valuable insights into the relationship among the structure, morphology, and properties of polymers and related composites. Dynamic mechanical analysis was employed to verify the reinforcing effect of the fibers. A strong influence from the weaving structure was observed for the relaxation behavior of the composites. T_α shifted to a higher temperature because the reinforcing of the fibers and the structural integrity was more effective. The researchers commented on the mechanical properties of the composites. The stress–strain curves for srPET laminae signify substantial yielding and post-yield strain hardening.

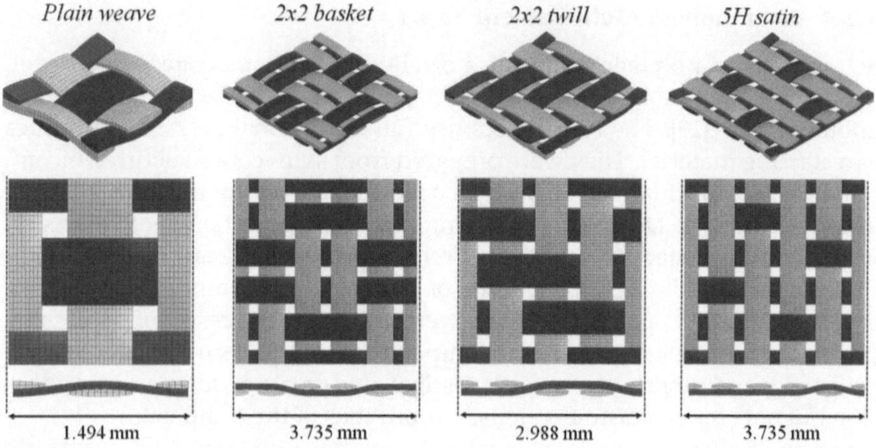

FIGURE 5.7
Different weave forms. (From Erol, O. et al., *Compos. A*, 101, 554, 2017. With permission from ScienceDirect, www.sciencedirect.com/science/article/pii/S1359835X17302750#f0010, Figure 2.)

In the above study, the effects reveal structural homogeneity of composites. Good tensile properties of the laminae prepared at high consolidation temperatures indicated that the srPET lamina structure was more compact at higher consolidation temperatures and had lower viscosity and improved impregnation. The slope between the yield point and the failure point represents the reinforcing efficiency of the srPET laminae. The post-yield modulus has been found to be susceptible to thermal degradation of the polyester matrix and the resulting poor interfacial adhesion. The failure mechanism was also investigated by the authors. They reported some differences between the reinforcing yarns and the polymer matrix. In the srPET lamina prepared at 250°C, the reinforcing yarns were melted and well incorporated with the copolymerized PET (mPET) multifilaments matrices. The highest value of tensile modulus reported was 5.2 GPa, which was higher by 87% and 49%, respectively, than the values obtained at 235°C and 240°C. According to the researchers, high modulus can be attributed to substantial impregnation and structural integrity caused by strong bonding. Brittle failure was observed in the samples consolidated at 250°C, suggesting that safe temperatures could not go beyond this limit.

The researchers also commented on the phenomenon of strain hardening. The tensile stress–strain curves of the srPET composites indicate substantial yielding and post-yield strain hardening, both indicative of the reinforcing effect and structural homogeneity of the srPET composites. The flexural properties were also discussed by the authors; the best properties in terms of flexure were exhibited by the fabric with basket or matt weave and the twill weave had shown the weakest results. The impact properties were also investigated. The absorbed impact energy of the srPET laminates prepared was found to be 30 times greater than that of pure polyester resin (28 J/m). This is a significant improvement in terms of the composite behavior.

5.2.4 Bicomponent Multifilament Yarns

Self-reinforced polymer composites or all-polymer composites were developed with acceptable mechanical lightweight and interfacial properties in another study [10]. The researchers used unidirectional all-PET composites as a starting material. They were prepared from skin–core structured bicomponent PET multifilament yarns by a collective process of filament winding and hot pressing. During hot pressing, the thermoplastic copolyester skin was partially melted to weld high-strength polyester cores, creating an all-PET composite. Physical properties of the resulting composites including thickness, density, and void content were reported. The result of processing parameters, that is, consolidation temperature and pressure on mechanical properties and morphology, was investigated to create good interfacial adhesion but not compromising on tensile properties of the composite.

Figure 5.8 shows a simple schematic of how the bicomponent PET yarns are converted into an all-PET composite after hot pressing. DSC measurements were done on bicomponent PET yarns and all-PET composites at a constant heating and cooling rate of 10°C min^{-1}. The data obtained describes the melting temperatures of two polyester phases, which are shown as peak endotherms.

It is quite interesting to see that using a bicomponent structure (Figure 5.9) gives a good range of operation for the single-polymer composites. Longitudinal and transverse strengths were investigated.

The longitudinal tensile modulus, as observed by the researchers, declines (Figure 5.10) from 10.8 to 8.5 GPa with rising temperatures from 200°C to 240°C, which is caused by loss of orientation due to molecular relaxation at elevated temperatures. Modulus is more disturbed by loss of orientation. The transverse modulus recovers as it is a function of the bonding that betters with temperature. Tensile strength in the longitudinal direction augments by 44% from a consolidation temperature of 200°C to 210°C. There is a drastic decrease by 52% from 210°C to 240°C. According to the researchers, this is due to a combined effect of two factors (1) the effect of temperature on the residual

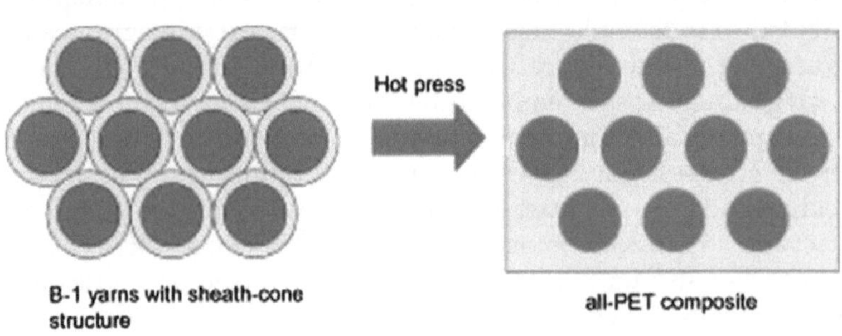

FIGURE 5.8
Schematic of processing all-PET composites based on bicomponent PET yarns. (From Zhang, J. M. and Peijs, T., *Compos. A*, 41, 964, 2010. With permission from ScienceDirect, www.sciencedirect.com/science/article/pii/S1359835X10001004, Figure 3.)

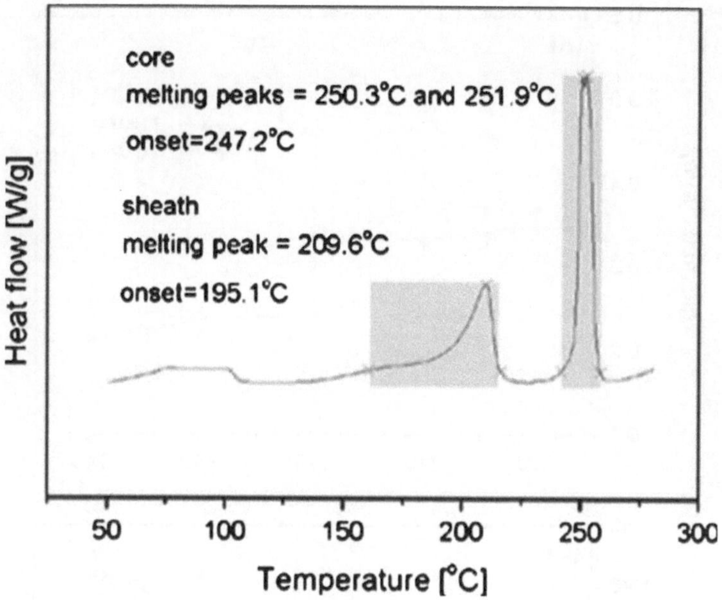

FIGURE 5.9
DSC curves of bicomponent PET yarn, showing the melting peaks of the copolyester sheath and polyester core. (From Zhang, J. M. and Peijs, T., *Compos. A*, 41, 964, 2010. With permission from ScienceDirect, www.sciencedirect.com/science/article/pii/S1359835X10001004, Figure 5.)

polyester fiber mechanical properties and (2) the essential formation of higher matrix phase to facilitate stress transfer between remaining fiber cores. In terms of transverse properties, tensile modulus increases dramatically from 200°C to 220°C and remains stable between 220°C and 240°C. Similarly, the researchers observed that tensile strength augments from 200°C to 230°C and remains constant up to 240°C, which is coherent with their failure modes.

5.2.5 Woven Sheets

In another study by Hine [11], single-polymer composites were manufactured from hot-compacted sheets of woven PET multifilaments. Time as a variable often plays an important role; though for the class of single-polymer composites, temperature has been found to be more crucial. In this research, investigation of the several processing parameters showed that a key aspect was the time spent at the compaction temperature. This is popularly referred to as the "dwell time." Temperature sensitivity of the polymer played an important role in consolidation. Molecular weight measurements showed that hydrolytic degradation occurred rapidly at the temperatures required for successful compaction. This resulted in embrittlement of the resulting materials with increasing dwell time. The dwell time of 2 min was found to be optimum because this gave the required percentage of melted material to bind the

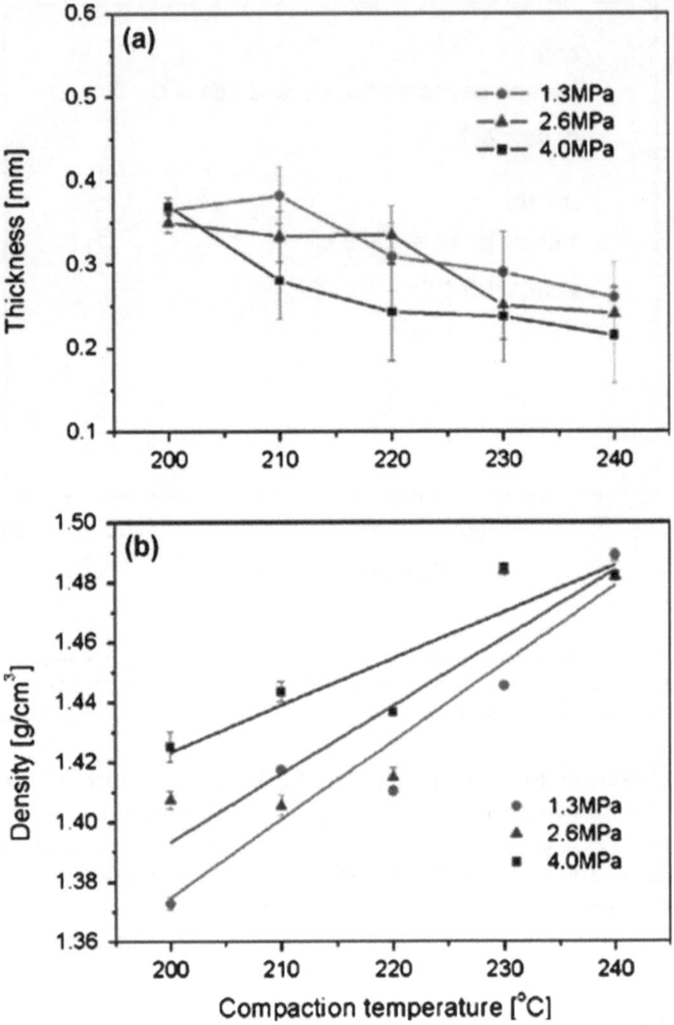

FIGURE 5.10
Longitudinal and transverse tensile strength (a) and modulus (b) vs. consolidation temperature at a consolidation pressure Pc=2.6MPa. (From Zhang, J. M. and Peijs, T., *Compos. A*, 41, 964, 2010. With permission from ScienceDirect, www.sciencedirect.com/science/article/pii/S1359835X10001004, Figure 6.)

structure together, resulting only in a small reduction in molecular weight. Different techniques of characterization including mechanical tests, differential scanning calorimetry and scanning electron microscopy were used to examine the mechanical properties and morphology of the optimum compacted sheets. The properties of hot-compacted sheets as self-reinforced composites, as found by the researchers, was a combination of the properties of the two components, a) the original oriented multifilaments and b) melted and recrystallized matrix. Other key findings included the importance of obtaining high ductility in the melted and recrystallized phases, promoted by using a high molecular weight or by suppressing crystallinity during processing.

Samples were prepared over a range of temperatures from 253°C to 259°C, all using a 2-min dwell. Measurements were made of the peel strength and tensile modulus. As the compaction temperature increased, the peel strength also increased. The researchers opined that the level of interlayer bonding developed as a consequence of melting, which resulted in increasing percentage of the original oriented phase. The in-plane modulus, meanwhile, increased to a plateau value of about 5 GPa and then remained at this level between 255°C and 259°C. Above 259°C the samples were found to be completely melted and the modulus reverted to that of the original isotropic polymer at around 2.4 GPa. The importance of temperature window is highlighted through these experiments. The authors made a very important observation of the increase in bonding over this temperature range of 255°C and 259°C. They found this to be the highest in compacted composites investigated. One can therefore conclude that PET makes a better matrix material for a hot-compacted PET composite than polyolefins as a consequence of its higher cohesive strength. In another research by Rasburn [12], measurement of mechanical properties showed that a very high proportion of original fiber properties were retained and that compacted samples had a good degree of coherence. Electron microscopy studies of suitably etched samples showed the effect of compaction temperature on the structure of the compacted samples.

5.3 Comparative Study and Conclusions

Single-polymer composites from polyester have been manufactured from different starting materials like fiber and different forms of yarn and fabric (Table 5.1). Tensile modulus of all-PET composites from film stacking

TABLE 5.1

Comparative Mechanical Properties of Different Composites When Compared to Polyester Single-Polymer Composites

Nature of Composite	Tensile Modulus (GPa)	Tensile Strength (MPa)	Strain to Failure (%)	Ref.
Unidirectional all-PET (film stacking of tapes)	10	350	10.2	Zhang [6]
Unidirectional all-PET (bicomponent yarns;)	10.4	245	9.8	Zhang [10]
All-PET 0°/90° sheet (hot compaction)	5.8	130	11.4	Hine [11]
Unidirectional all-PET (hot compaction)	13.3	–	–	Rasburn [12]

and bicomponent yarns have reached quite high values of around 13.3 GPa, while the highest tensile strength of 350 MPa was reported from film stacking of tapes. Mechanical properties have been found to be very sensitive to temperature. Overall, researchers have been successful in development of single-polymer composites from polyester. Though polyester composites in general are being characterized for flexural [13–15] impact properties [16,17] and studies on aging [18] have been carried out, they have majorly been with natural fibers as reinforcements. These kinds of studies can be carried out on single-polymer composites as well those leading to findings in flexure, impact, and long-term behavior of manufactured composites.

References

1. https://www.britannica.com/science/polyester (accessed February 23, 2018).
2. Papageorgiou, G. Z., D. G. Papageorgiou, V. Tsanaktsis, and D. N. Bikiaris. Synthesis of the bio-based polyester poly (propylene 2, 5-furan dicarboxylate). Comparison of thermal behavior and solid state structure with its terephthalate and naphthalate homologues. *Polymer* 62(2015): 28–38.
3. Rasburn, J., P. J. Hine, I. M. Ward, R. H. Olley, D. C. Bassett, and M. A. Kabeel. The hot compaction of polyethylene terephthalate. *Journal of Materials Science* 30, no. 3 (1995): 615–622.
4. Nirmal, U., J. Hashim, and M. M. Ahmad. A review on tribological performance of natural fibre polymeric composites. *Tribology International* 83(2015): 77–104.
5. Rojanapitayakorn, P., P. T. Mather, A. J. Goldberg, and R. A. Weiss. Optically transparent self-reinforced poly (ethylene terephthalate) composites: molecular orientation and mechanical properties. *Polymer* 46, no. 3 (2005): 761–773.
6. Zhang, J. M., C. T. Reynolds, and T. Peijs. All-poly (ethylene terephthalate) composites by film stacking of oriented tapes. *Composites Part A: Applied Science and Manufacturing* 40, no. 11 (2009): 1747–1755.
7. Kannan, T. G., C. M. Wu, and K. B. Cheng. Effect of different knitted structure on the mechanical properties and damage behavior of Flax/PLA (Poly Lactic acid) double covered uncommingled yarn composites. *Composites Part B: Engineering* 43, no. 7 (2012): 2836–2842.
8. Wu, C. M., P. C. Lin, and C. T. Tsai. Fabrication and mechanical properties of self-reinforced polyester composites by double covered uncommingled yarn. *Polymer Composites* 37, no. 12 (2016): 3331–3340.
9. Erol, O., B. M. Powers, and M. Keefe. Effects of weave architecture and mesoscale material properties on the macroscale mechanical response of advanced woven fabrics. *Composites Part A: Applied Science and Manufacturing* 101(2017): 554–566.
10. Zhang, J. M., and T. Peijs. Self-reinforced poly (ethylene terephthalate) composites by hot consolidation of Bi-component PET yarns. *Composites Part A: Applied Science and Manufacturing* 41, no. 8 (2010): 964–972.

11. Hine, P. J., and I. M. Ward. Hot compaction of woven poly (ethylene terephthalate) multifilaments. *Journal of Applied Polymer Science* 91, no. 4 (2004): 2223–2233.
12. Rasburn, J., P. J. Hine, I. M. Ward, R. H. Olley, D. C. Bassett, and M. A. Kabeel. The hot compaction of polyethylene terephthalate. *Journal of Materials Science* 30, no. 3 (1995): 615–622.
13. Sathishkumar, T. P., P. Navaneethakrishnan, and O. Shankar. Tensile and flexural properties of snake grass natural fiber reinforced isophthallic polyester composites. *Composites Science and Technology* 72, no. 10 (2012): 1183–1190.
14. Ahmed, K. S., and S. Vijayarangan. Tensile, flexural and interlaminar shear properties of woven jute and jute-glass fabric reinforced polyester composites. *Journal of materials processing technology* 207, no. 1–3 (2008): 330–335.
15. Dhakal, H. N., Z. Y. Zhang, and M. O. W. Richardson. Effect of water absorption on the mechanical properties of hemp fibre reinforced unsaturated polyester composites. *Composites science and technology* 67, no. 7–8 (2007): 1674–1683.
16. Rout, J., M. Misra, S. S. Tripathy, S. K. Nayak, and A. K. Mohanty. The influence of fibre treatment on the performance of coir-polyester composites. *Composites Science and Technology* 61, no. 9 (2001): 1303–1310.
17. Devi, L. Uma, S. S. Bhagawan, and S. Thomas. Mechanical properties of pineapple leaf fiber-reinforced polyester composites. *Journal of Applied Polymer Science* 64, no. 9 (1997): 1739–1748.
18. Pothan, L. A., S. Thomas, and N. R. Neelakantan. Short banana fiber reinforced polyester composites: mechanical, failure and aging characteristics. *Journal of Reinforced Plastics and Composites* 16, no. 8 (1997): 744–765.

6

PLA-Based Single-Polymer Composites

Poly(lactic acid) (PLA) is a biocompatible, biodegradable, and thermoplastic polyester synthesized from renewable resources. Among the family of biomass-derived biodegradable polymers, PLA has relatively high strength and modulus. The conversion route from lactic acid to poly(lactic acid) is demonstrated in Figure 6.1 [1].

Conversion of the lactic acid to dimer form and ring-opening polymerization allows for generation of higher molecular weight polymers with a greater control of polymer molecular weight. However, owing to the general stereochemistry of central carbon, this polymer has been found to exist in both DL and L lactic forms (Figure 6.2) [2]. The subtle difference in stereochemistry has a drastic impact on mechanical properties and degradation behavior [3].

An excellent review of the polymer from the application point of view has been attempted [4] in addition to a review on environmental issues related to the polymer [5]. Polylactic acid suffers from high brittleness and low heat deflection temperature (HDT). This has been the main reason for PLA not having gained full market recognition as an engineering resin. Various fibers have been thus used to reinforce polylactic acid. Improved HDTs are often observed in natural fiber-reinforced PLA composites, often due to increased values of modulus and crystallinity. However, it is important to note that although the usage of natural fibers to reinforce PLA at the onset appeared to be an environmentally friendly approach owing to their renewability, there are some issues with respect to end-of-life scenarios for these class of composites. In the case of mechanical recycling, their relatively poor thermal stability may lead to severe additional thermal degradation of the composites during subsequent reprocessing steps [6]. "Self-reinforced polymer" (SRP) composites or "all-polymer" composites have thus also been tried with polylactic acid. This class of composite polymer matrix is reinforced with oriented fibers or tapes of the same polymer.

6.1 PLA Self-Reinforced Composites Based on Composite Manufacturing

Mai et al. [7] optimized a tape extrusion process to ensure superior mechanical properties. A screw extruder was used to obtain PLA-extruded film. The

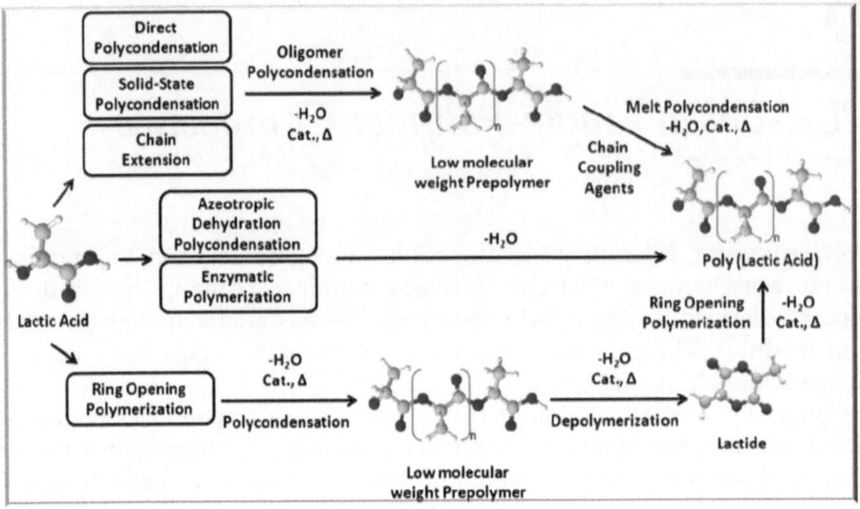

FIGURE 6.1
Conversion scheme of lactic acid to polylactic acid. (From Farah, S. et al., *Adv. Drug Deliv. Rev.*, 107, 367, 2016. With permission from ScienceDirect, www.sciencedirect.com/science/article/pii/S0169409X16302058#f0010, Figure 2.)

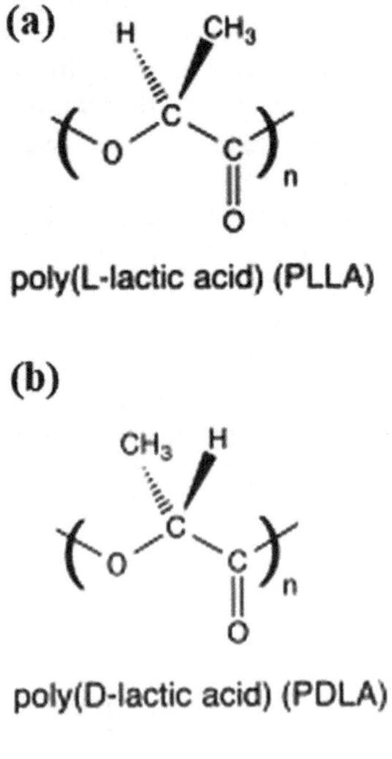

FIGURE 6.2
Poly(lactide) in L or DL racemic form. (From Tsuji, H. and Hayakawa, T., *Polymer*, 55, 721, 2014. With permission from ScienceDirect, www.sciencedirect.com/science/article/pii/S0032386113011695#fig1, Figure 1.)

film was quenched by winding on a chill roll, followed by post-drawing on heated rollers to create an oriented tape. The tapes were drawn in a two-step solid-state drawing process below the melting temperature. The first drawing step provided some initial orientation, but ultimate drawing was performed in the second step.

Unidirectional (UD) and bidirectional (BD) laminates with a thickness of ~1.6 mm (67 vol.% of tape) were manufactured by stacking in a $[0]_{20}$ or $[0, 90]_5$ lay-up configuration in a mold (Figure 6.3). The number in between brackets indicates tape orientation within each layer, while the number layers of repeating groups are shown by subscript.

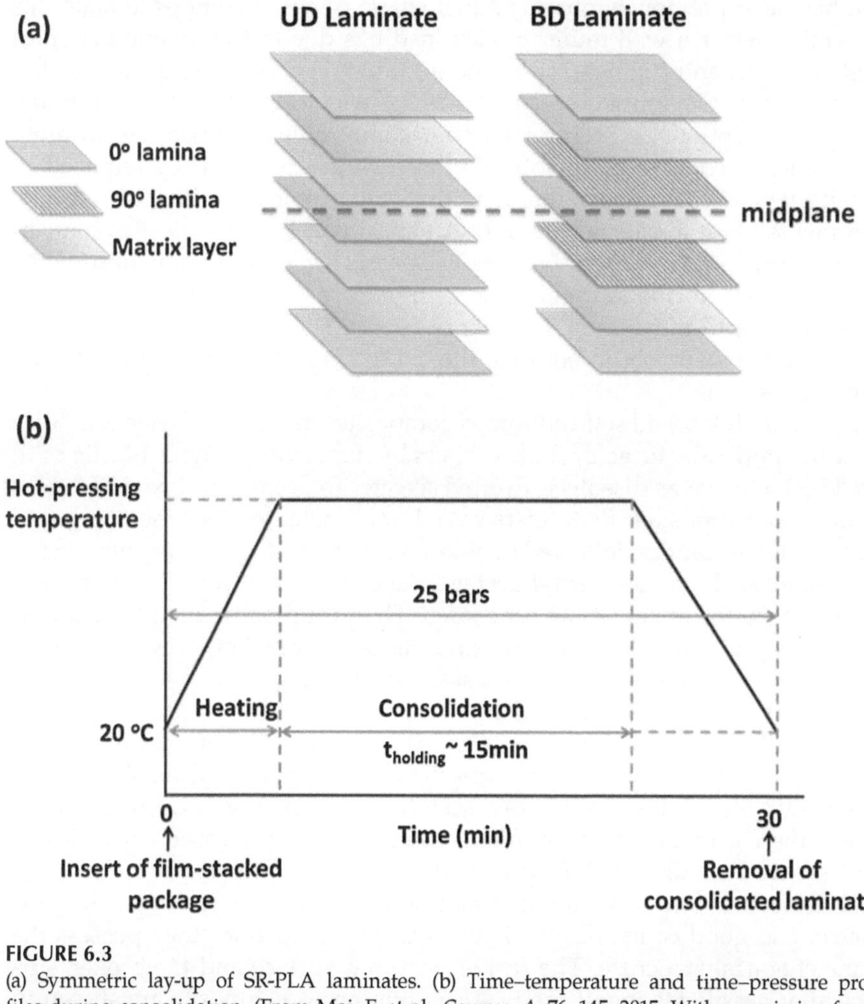

FIGURE 6.3
(a) Symmetric lay-up of SR-PLA laminates. (b) Time–temperature and time–pressure profiles during consolidation. (From Mai, F. et al., *Compos. A*, 76, 145, 2015. With permission from ScienceDirect, www.sciencedirect.com/science/article/pii/S1359835X15001918#b0020, Figure 2.)

The researchers observed that peel strength improved with compaction temperature. At 150°C, good fusion bonding was not achieved, and compression-molded sample could be easily peeled apart. A remarkable increase in peeling strength was observed when temperature was raised to 160°C or above. It is interesting to note that a compaction temperature of 170°C led to a peel force of 0.95 N/mm, 10 times higher than the peel force achieved at 150°C. This is because T_m of the matrix was found to be around 155°C as measured by differential scanning calorimetry (DSC), and at 170°C the matrix fully melted, which resulted in superior bonding. The tensile and impact properties are reported by the researchers. Impact strength was interesting, as relatively poor impact strength of PLA generally prevents a broader field of application of these materials.

It has been previously observed that single polymer composite materials generally perform well under impact loadings due to their combination of high strength and high stiffness, and additional energy absorption mechanisms such as delamination. Isotropic PLA was found to exhibit very low energy absorption. Self-reinforced PLA composites absorbed significantly more energy compared to isotropic PLA (Figure 6.4). Energy required to break a tape is 12.7 times higher than that of isotropic PLA film. The authors commented that tensile failure of PLA tapes in SR-PLA composites contributed greatly to total energy absorption. General processes during impact like delamination, fibrillation, and tape pull-out are additional processes that absorbed significant amounts of energy.

The researchers observed that the HDT improved for self-reinforced composites.

Uniaxial PLA-based self-reinforced composites have been developed [8] by bonding poly (l-lactic acid) (PLLA) fibers by depositing poly(D, l-lactic acid) (PDLLA) which was dissolved in ethyl acetate. To prepare uniaxial PLA self-reinforced composites, PLA fibers were impregnated by the research group using PDLLA solution followed by winding them under tension onto a rotating mandrel. To remove ethyl acetate, the composites were dried at room temperature for certain time schedules. The group subsequently removed this assembly from the mandrel and the structure hot pressed at 50°C under 2 tons for 5 min, to achieve a smooth surface.

Microstructure, thermal and mechanical properties of the developed self-reinforced composite have been characterized using scanning electron microscopy (SEM), dynamic mechanical analysis, and tensile testing. The results showed that after ethyl acetate treatment, the PLLA fibers could retain their geometry and orientation well, had an increased crystallinity, and showed a slightly damaged surface. Electron microscopy and thermal analysis showed that the interface between the PLA fiber and the PLA matrix had good compatibility, which was expected since they possess the same chemical structure. The freeze-fractured surface and the broken surface after tensile testing of the PLA self-reinforced composite are shown in Figure 6.5. It can be seen that the PLLA fibers are properly coated by PDLLA

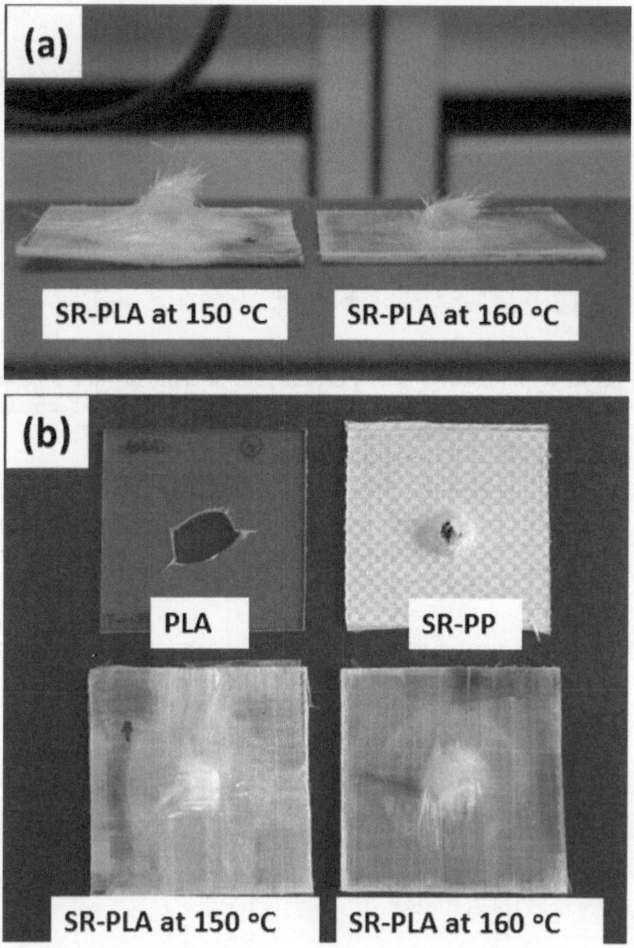

FIGURE 6.4
(a) Side image of out-of-plane deformation of SR-PLA BD composite laminates. (b) Front image of typical impact penetration damage of different materials. (From Mai, F. et al., *Compos. A*, 76, 145, 2015. With permission from ScienceDirect, www.sciencedirect.com/science/article/pii/S1359835X15001918#b0090, Figure 2.)

matrix, indicative of a strong interface formed between PLLA fibers and PDLLA matrix.

The modulus, tensile strength, and elongation, as well as the thermal stability of uniaxial PLA self-reinforced composites, were all significantly greater than those of the PDLLA film. The dynamic mechanical properties were also evaluated (Figure 6.6). The storage modulus values and retention of storage modulus was found to be higher for the composites. The composite was found to retain storage modulus values up to 140°C.

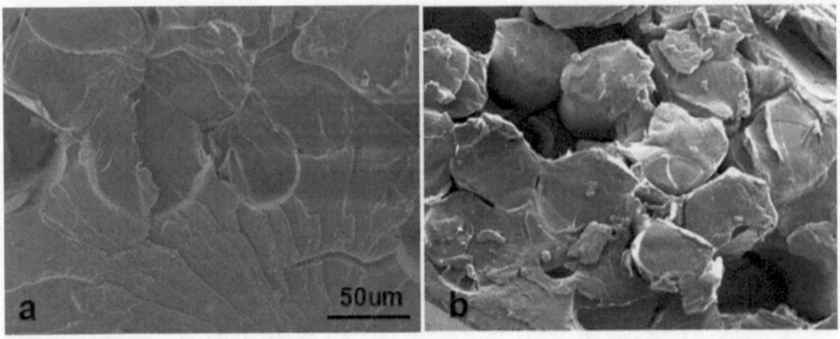

FIGURE 6.5
Fractured surface showing coating of PDLLA on PLA fibers and minimum fiber pull-out. (From Gao, C. et al., *Composi. Sci. Technol.* 117, 392, 2015. With permission from ScienceDirect, www.sciencedirect.com/science/article/pii/S0266353815300361, Figure 7.)

The results showed that the researchers developed a method to prepare PLA composites with enhanced mechanical properties, without the problem of narrow processing temperature window normally required when using traditional thermal processing techniques.

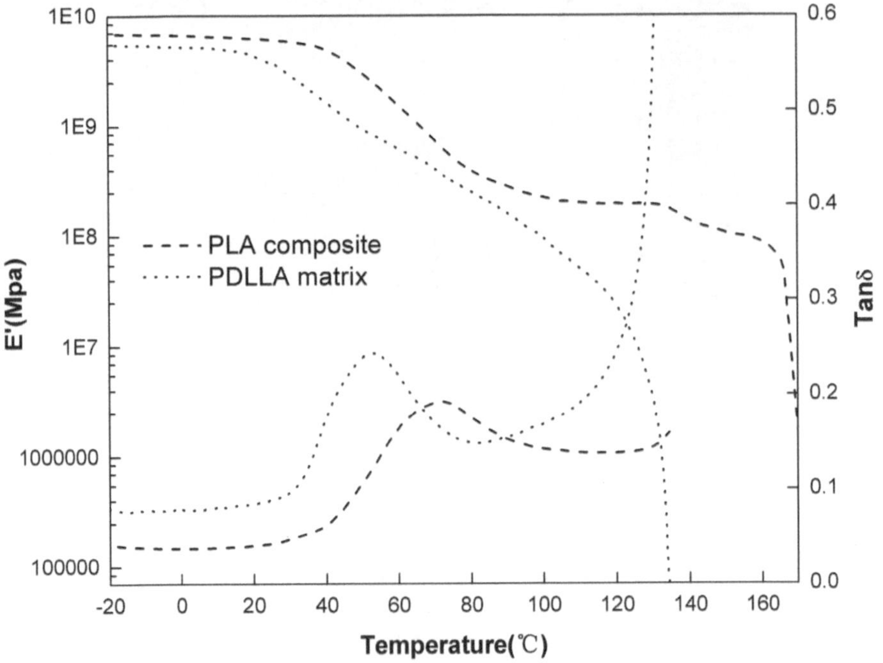

FIGURE 6.6
DMA showing storage modulus (E′) and tanδ of PDLLA film and PLA self-reinforced composite. (From Gao, C. et al., *Composi. Sci. Technol.* 117, 392, 2015. With permission from ScienceDirect, www.sciencedirect.com/science/article/pii/S0266353815300361, Figure 8.)

6.1.1 Importance of Temperature

Li et al. [9] demonstrated the preparation of PLA single-polymer composites (SPCs) on the foundation of PLA's gradually crystallizing structure. In general practice, owing to its slow crystallizing nature, PLA can be processed with usual polymer-processing techniques resulting in end products with varied crystallinities, highly crystalline fibers to amorphous films. In the study by this research group, amorphous PLA sheets and crystalline PLA fibers/fabrics were first laminated and subsequently compression molded to form SPC at a processing temperature substantially lower than PLA's melting temperature. The researchers found that processing temperature played a profound role in affecting the fiber–matrix bonding properties. For a temperature of 120°C or below, good fusion bonding was not obtained. The compressed sample could be easily peeled apart at interface. The researchers observed a drastic increase in peeling force at platen temperature of 130°C. With increasing processing temperature, a drastic improvement in the interfacial bonding ensued at temperature of around 135°C, which exhibited lower boundary of the process window. For SPC specimens molded at 130°C, very long fiber pull-out was observed at the surface of break, and the pull-out length was found to be much higher than the thickness of the SPC sheet. Interestingly, as platen temperature increased to 140°C, the fiber pull-out at surface decreased in length. This indicated the development of a strong bonding between the fiber and the matrix. The compression-molded SPC displayed enhanced mechanical properties; particularly, the tearing strength of the fabric-reinforced SPC was almost an order higher than that of the nonreinforced PLA.

In another study by Liu et al. [10], polylactide (PLA) SPCs were prepared using PLA nonwovens made of core–sheath PLA fibers as raw materials by hot pressing. The core and sheath materials used were poly(l-lactide) (PLLA) and PLA with d-lactide of about 10 mol%, respectively. The melt-processing window of PLA–SPCs reached more than 40°C. The effects of hot-pressing temperature on the crystallinity, the size distribution of crystallites, the lamellar thickness, and mechanical properties of PLA–SPCs were significant, while those of hot-pressing pressure were small. The strong interfacial adhesion between matrix and reinforcement led to high mechanical properties of PLA–SPCs.

JIa et al. [11] studied fully biodegradable PLA fiber-reinforced PLA and poly (butylene succinate) (PBS) matrix composites to reduce the impact on the environment by long-lasting plastics-based composites. A film-stacking method was used (Figure 6.7). Untwisted PLA yarn was used as the reinforcement, while PLA film was used as the matrix. For the sake of comparison, a PBS film was also used.

The time, temperature, and pressure cycle have been shown in Figure 6.8. For each type of composite, three different structures were studied by the researchers. They were UD composites with fiber area density of 40 gsm

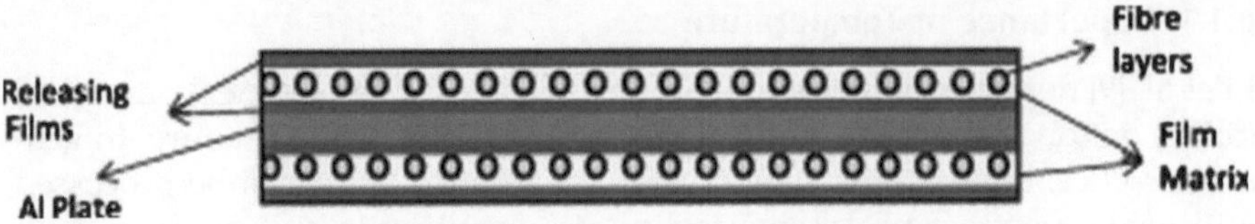

FIGURE 6.7
Film stacking. (From Jia, W. et al., *Compos. B*, 62,104, 2014. With permission from ScienceDirect, www.sciencedirect.com/science/article/pii/S1359836814001012, Figure 1.)

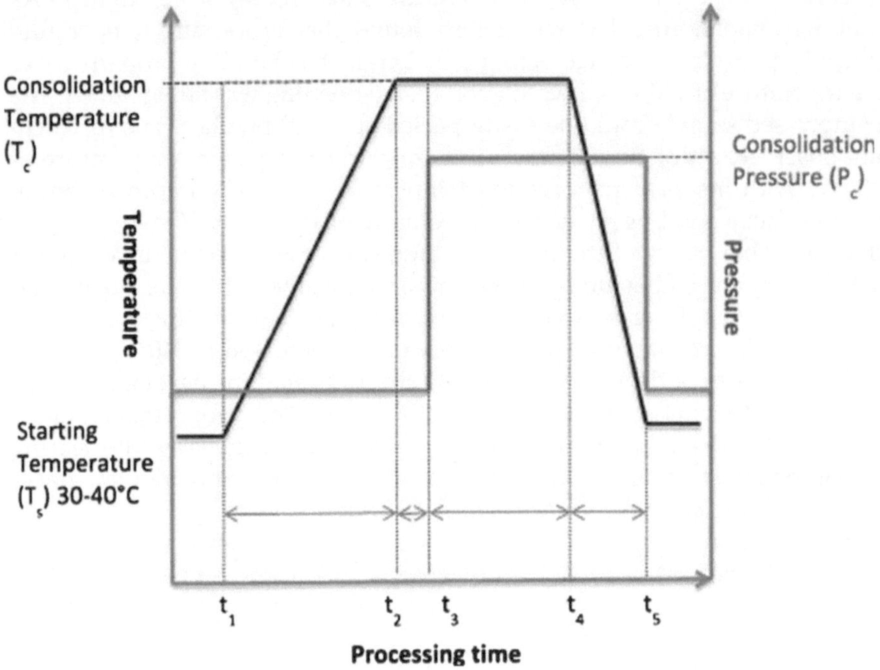

FIGURE 6.8
Time–temperature cycle in experimentation. (From Jia, W. et al., *Compos. B*, 62,104, 2014. With permission from ScienceDirect, www.sciencedirect.com/science/article/pii/S1359836814001012#f0005, Figure 2.)

and BD composites with fiber area densities of 20 gsm (1.5-mm yarn interval) and 30 gsm (1.0-mm yarn interval).

Comparison of the PLA-SRC and PLA–PBS composites revealed that tensile strength and Young's modulus of PLA-SRC are higher than that of PLA–PBS by 12%–40% and 39%–54%, respectively. The authors cited two reasons: first, the PLA film at the onset had better tensile strength and Young's modulus than that of the PBS film; second, PLA-SRC had better interfacial adhesion due to greater chemical similarity.

Wu et al. [12] obtained three-dimensional (3D)–braided single poly (lactic acid) composites (PLA–SPCs) by combining 3D and five-direction braiding

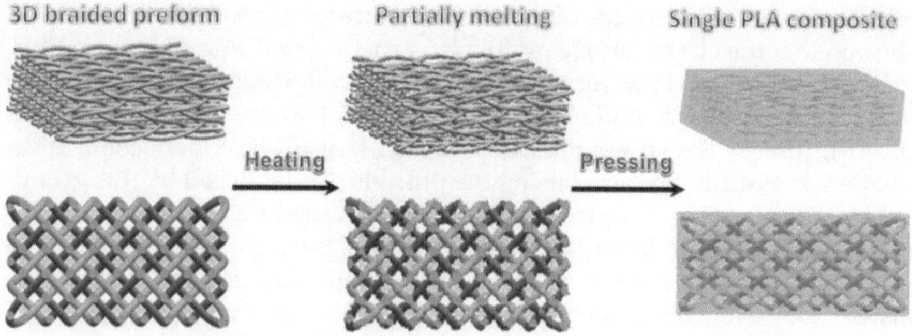

FIGURE 6.9
Schematic of using braided preform for composites. (From Wu, N. et al., *Compos. B*, 52, 106, 2013. With permission from ScienceDirect, www.sciencedirect.com/science/article/pii/S1359836813001066, Figure 3.)

technique (Figure 6.9) and hot-compression technical process. Preforms were preheated in the specially designed die system to partially melt all the fibers. In the next stage, the researchers used the preforms and consolidated them under pressure (from 7.8 to 10 MPa) at temperatures ranging from 130°C up to 150°C. Under the controlled processing conditions, one part of fiber body formed the matrix, while the other part retained its fibrous form.

At the same consolidation temperature, the maximum bending stress values resulted to be considerably dependent on fiber volume fraction of PLA–SPCs, while values of bending modulus were largely subjected to fiber content in the length direction. The researchers found that an increase in consolidation pressure gave rise to superior fusion of neighboring fibers. This resulted in increased values of stress and modulus. Like other SPCs, consolidation temperature played an important role. With an increase in temperature, fusion bonding improved. Increase in temperature also led to higher values of bending stress and modulus.

6.1.2 Importance of Time

Self-reinforced poly(lactic acid) (SR-PLA) nanocomposites was developed for the first time by using electrospun PLA fiber mats as the starting material [13] followed by hot compaction.

PLA pellets (PLLA) were used by the researchers with intrinsic viscosity of 1.28 dL g^{-1} and crystallinity of 36.7% (determined by DSC analysis). They were converted into electrospun materials using a PLA solution of 8% w/v in 7:3 v/v dichloromethane (DCM) and dimethylformamide (DMF) as a solvent system. For details of layering of PLA and stacking, the reader is requested to refer to the original work. The hot-pressing time was varied from 10 to 60s to optimize processing conditions. The SR-PLA nanocomposite films were found to be transparent. They were analyzed using DSC, SEM, and tensile testing. SEM images of the fracture surfaces of SR-PLA film samples showed that the characteristic multilayer structure of PLA electrospun fiber mats in nanocomposites was retained after hot compaction. The researchers also found the PLA nanofiber cores to be intact, particularly at shorter hot-pressing times. This observation confirmed that SR-PLA nanocomposites can be successfully prepared using the procedure established by this group. They confirmed that hot-pressing time significantly affected the composite microstructure. With longer hot-pressing times, the appearance of the electrospun fibers started to vanish as more of the fibers melted and turned into matrix. However, the research group found that the multilayer structure of stacked PLA fiber materials was still obvious in all SR-PLA samples, but less identifiable at higher hot-pressing times. DSC results showed that the SR-PLA films produced had considerably higher crystallinity with more stable crystallites and also higher glass transition temperature when compared to bulk isotropic PLA films.

The researchers found that tensile properties of the SR-PLA nanocomposite films were significantly enhanced as compared to isotropic PLA films, which were found to be brittle and weak. All SR-PLA films were considerably more ductile and tougher than the isotropic films. Thus, a much better material has been produced through the process of self-reinforcement in terms of mechanical properties. Of particular interest was the notable increase in toughness of the SR-PLA nanocomposite films as a result of their unique microstructure and related failure mechanisms.

6.2 Application of PLA Self-Reinforced Composites

Majola et al. evaluated the strength and strength retention of self-reinforced (SR) absorbable polylactic acid composite rods after intramedullary and subcutaneous implantation in rabbits [14]. The researchers manufactured rods made of poly-l-lactic acid (SR-PLLA) and of poly-dl-lactic acid + poly-l-lactic acid composite (SR-PDLLA/PLLA). There was a variation in molecular mass. (Mv) of PLLA was 260.000 and that of PDLLA used was 100.000. The bending and shear strengths were measured after a follow-up of 1–48 weeks. Researchers reported values of 250–271 MPa as the initial bending strength of SR-PLLA rods and the shear strength was found to be 94–98 MPa. After intramedullary and subcutaneous implantation of 12 weeks, the bending strength of the SR-PLLA implants was found to be 100 MPa. At 36 weeks the bending strength had decreased to the level of the strength of cancellous bone (10–20 MPa). However, there was no alteration in shear strength during 12 weeks hydrolysis. The initial bending strength of SR-PDLLA/PLLA implants was observed to be 209 MPa, and during subsequent testing, the implants lost their bending and shear strength sooner than SR-PLLA implants. This investigation gave important clues to continue studies with fixation of experimental cortical bone osteotomies with SR-PLLA intramedullary rods.

Another research [15] demonstrated the preparation of impact-resistant all-PLA composites by the method of film stacking of highly crystalline PLA fibers with fully amorphous PLA films. The film-stacking technique was used by the researchers for manufacturing matrix films, resulting in fully amorphous films even after processing (as was confirmed from DSC measurements). The research group felt the need of finalizing the temperature of processing. On the basis of the theory of single polymer composite processing, final temperature is a function of (a) the initial melting temperature of crystalline phase of the reinforcing PLA yarn (152°C), and (b) temperature-dependent viscosity (fluidity) of the matrix materials. Accordingly, the group finalized a processing temperature of 140°C for composite preparation. It was also observed that applied fire-retardant (FR) materials noticeably increased the fluidity of PLA, as established from melt flow indices of flame-retarded films at the temperature of composite processing.

The authors used a modified method of filament winding. Filament winding [16,17] is one of the composite fabrication techniques (Figure 6.10) involving wrapping pre-tensioned, resin-saturated continuous filaments around

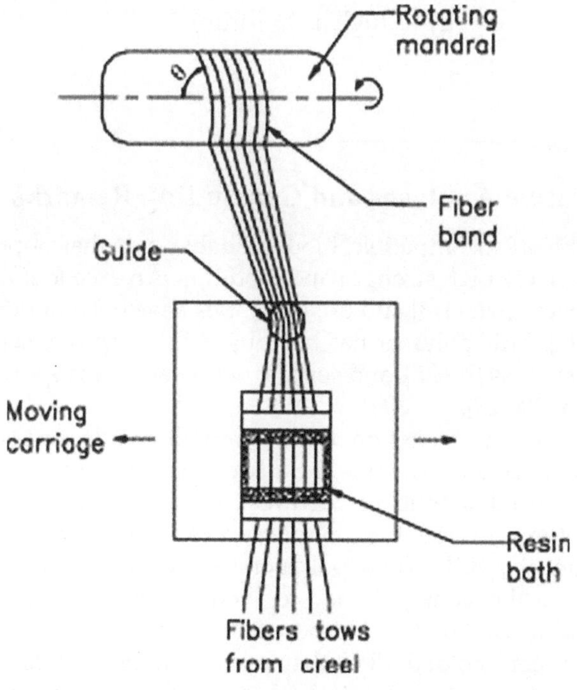

FIGURE 6.10
Schematic of filament winding. (From Abdalla, F. H. et al., *Mater. Des.* 8, 234, 2007. With permission from ScienceDirect, www.sciencedirect.com/science/article/pii/S0261306905001743#fig1, Figure 1.)

a mandrel (male mold). The mandrel is usually circular in cross-section; however, other cross sections are achievable with the correct programming.

The researchers used 11 plies of 65 ± 5 μm thick matrix films and 10 PLA filament layers. They were laminated onto a 6 mm thick aluminum core (300 × 300 mm) using a modified filament-winding process to attain cross-ply configuration, as shown in Figure 6.11. The filament-wound laminated packages were consolidated by compression molding in a laboratory hot press (custom construction).

In this research, flammability of PLA-SRCs was efficiently reduced by incorporation of ammonium polyphosphate–based flame-retardant additive and montmorillonite clays in a weight ratio of 10:1 in matrix. Interestingly, a very low value of 16 wt% FR was found to be adequate for achieving self-extinguishing behavior, and for achieving significant reduction in peak heat-release rate and total heat emission. Generally, introduction of FR materials come at the cost of reduction of mechanical properties. However, in this research, FR additives improved tensile properties compared to the FR-free all-PLA composite. Flexural properties, however, suffered with addition of FRs. The stiffness of PLA-SRCs increased steadily with the FR contents of their matrix layers. The researchers explained that the prominent energy absorption capacity was likely due to the improved fiber–matrix bonding (impact perforation energy as high as 16 J/mm).

6.3 Comparative Analyses and Concluding Remarks

Polylactic acid is a biocompatible, biodegradable, and thermoplastic polyester having relatively high strength and modulus. An excellent review of this polymer has been given [18] and opportunities have been underlined in certain articles [19]. This polymer has been used in composite systems with a fair degree of success [20–25], and some articles have also reported advanced characterization [26,27].

Inspired by previous works on SR-PLA and all-cellulose nanocomposites, researchers have tried to combine the advantages of self-reinforcement and some have even tried nanofiber structures to create SR-PLA nanocomposites with a target of improvement in toughness and thermal properties. PLA has been used to develop SRP (where polymer matrix is reinforced with oriented fibers or tapes of the same polymer) to overcome its limitations so that it's used as an engineering material/product.

The compression-molded SPC displayed enhanced mechanical properties; particularly, the tearing strength of the fabric-reinforced SPC. Peel strength was found to improve drastically with compaction temperature at 170°C as the matrix is fully melted, which resulted in superior bonding.

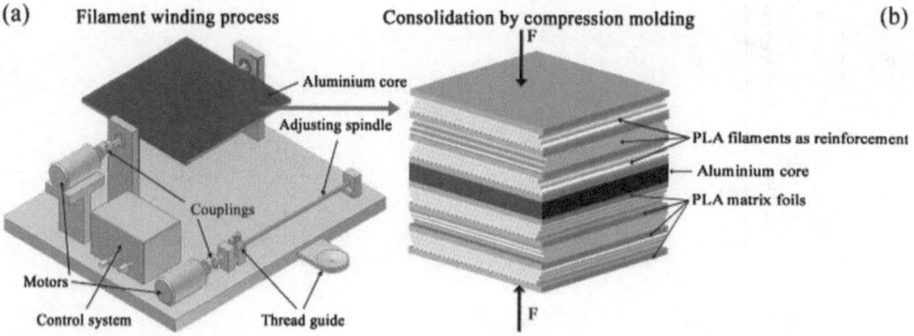

FIGURE 6.11
Preparation of cross-ply SR-PLA composites (a) with filament winding and (b) with subsequent compression molding. (From Bocz, K. et al., *Compos. A*, 70, 27, 2015. With permission from ScienceDirect, www.sciencedirect.com/science/article/pii/S1359835X14003881, Figure 1.)

Self-reinforced PLA composites absorbed significantly more energy compared to isotropic PLA.

Tensile strength and Young's modulus of PLA-SRC are higher than PLA–PBS by 12%–40% and 39%–54%, respectively. PLA-SRC had better interfacial adhesion due to greater chemical similarity. Maximum bending stress values are dependent on fiber volume fraction of PLA–SPCs. Bending modulus was largely subjected to fiber content in length direction.

The effects of hot-pressing temperature on the crystallinity, the size distribution of crystallites, lamellar thickness, and mechanical property of PLA–SPCs were significant. Hot-pressing time significantly affected the composite microstructure in case of nanocomposites. Multilayer structure of stacked PLA fiber materials was still obvious in all SR-PLA samples, but less identifiable at higher hot-pressing times. DSC results showed that the SR-PLA films produced had considerably higher crystallinity with more stable crystallites and also higher glass transition temperature when compared to bulk isotropic PLA film. All SR-PLA films were considerably more ductile and tougher than the isotropic films.

As evident from Table 6.1, the maximum strength and modulus were obtained using PLA nonwoven as the core material. Majority of the research addressed tensile properties—there is enough scope to work on the flexure and impact properties of SPCs developed from PLA. Keeping in mind the predictions of this polymer [28], promise of research [29], and even a zone-wise prediction of growth [30,31], this polymer has a lot of scope for research. There is an excellent article that reviews the possibilities of this polymer [32] and another article on philosophies of compatibilization and some futuristic applications [33–38]. Further research on SPCs based on this polymer can get important clues from this study.

TABLE 6.1

Comparative Mechanical Properties of SPCs from PLA

Starting Material	Process	Tensile Strength (MPa)	Young's Modulus (GPa)	% Breaking Extension	Flexural Strength (MPa)	Reference
PLA amorphous		59	1.28	7	106	i
PLLA fibers PDLLA matrix	Coating matrix on the fibers.	44		5.4	88	i
Amorphous PLA sheets Crystalline PLA fibers/fabrics	Lamination Compression Molding done at temperature lower than melting point of PLA	58.6	0.1	4		ix
PLA nonwoven with core material used poly(l-lactide) (PLLA) and sheath material used PLA with d-lactide	Hot pressing Hot pressing temperature is significant, while effect of pressing pressure is small	48.03 ± 2.29	3.29 ± 0.14			xi
PLA yarns (reinforcement and PLA film as matrix)	Film stacking Untwisted PLA yarn was used as the reinforcement while PLA film as the matrix	27.95 ± 3.68	1.8 ± 0.15			xi
Braided preforms fibers formed matrix as well retained fibrous form	3D braiding technique using hot-compression process	95				xii
Extruded PLA film Laminate UD and BD laminates (67 vol.% of tape)	Stacking in symmetrical lay-up configuration in a mold	61				xiii

(Continued)

TABLE 6.1 (*Continued*)

Comparative Mechanical Properties of SPCs from PLA

Starting Material	Process	Tensile Strength (MPa)	Young's Modulus (GPa)	% Breaking Extension	Flexural Strength (MPa)	Reference
Electrospun PLA fiber mats (PLA pellets [PLLA] dissolved in a PLA solution of 8% w/v in DCM and DMF as a solvent)	Hot compaction. With longer hot-pressing times, the appearance of the electrospun fibers started to vanish as more of the fibers melted and turned into matrix	51				xiii
Plies of thick matrix films and PLA filament layers	Filament Winding Lamination on Aluminium core Compression molding					xvii
Plies of thick matrix films PLA filament layers ammonium polyphosphate–based flame-retardant additive/ montmorillonite clays in matrix	Filament Winding, Lamination on Aluminium core using Compression molding					xv

References

1. Farah, S., D. G. Anderson, and R. Langer. Physical and mechanical properties of PLA, and their functions in widespread applications—A comprehensive review. *Advanced Drug Delivery Reviews* 107(2016): 367–392.
2. Tsuji, H., and T. Hayakawa. Hetero-stereocomplex formation between substituted poly (lactic acid) s with linear and branched side chains, poly (l-2-hydroxybutanoic acid) and poly (d-2-hydroxy-3-methylbutanoic acid). *Polymer* 55, no. 3 (2014): 721–726.
3. http://normalpolymer.com/ (accessed on January 21, 2018).
4. Farah, S., D. G. Anderson, and R. Langer. Physical and mechanical properties of PLA, and their functions in widespread applications—A comprehensive review. *Advanced Drug Delivery Reviews* 107(2016): 367–392.
5. Dubey, S. P., V. K. Thakur, S. Krishnaswamy, H. A. Abhyankar, V. Marchante, and J. L. Brighton. Progress in environmental-friendly polymer nanocomposite material from PLA: Synthesis, processing and applications. *Vacuum* 146(2017): 655–663.
6. Cabrera, N., B. Alcock, J. Loos, and T. Peijs. Processing of all-polypropylene composites for ultimate recyclability. *Proceedings of the Institution of Mechanical Engineers, Part L: Journal of Materials: Design and Applications* 218, no. 2 (2004): 145–155.
7. Mai, F., W. Tu, E. Bilotti, and T. Peijs. Preparation and properties of self-reinforced poly (lactic acid) composites based on oriented tapes. *Composites Part A: Applied Science and Manufacturing* 76(2015): 145–153.
8. Gao, C., L. Meng, L. Yu, G. P. Simon, H. Liu, L. Chen, and S. Petinakis. Preparation and characterization of uniaxial poly (lactic acid)-based self-reinforced composites. *Composites Science and Technology* 117(2015): 392–397.
9. Li, R., and D. Yao. Preparation of single poly (lactic acid) composites. *Journal of Applied Polymer Science* 107, no. 5 (2008): 2909–2916.
10. Liu, Q., M. Zhao, Y. Zhou, Q. Yang, Y. Shen, R. H. Gong, F. Zhou, Y. Li, and B. Deng. Polylactide single-polymer composites with a wide melt-processing window based on core-sheath PLA fibers. *Materials & Design* 139(2018): 36–44.
11. Jia, W., R. H. Gong, and P. J. Hogg. Poly (lactic acid) fibre reinforced biodegradable composites. *Composites Part B: Engineering* 62(2014): 104–112.
12. Wu, N., Y. Liang, K. Zhang, W. Xu, and L. Chen. Preparation and bending properties of three dimensional braided single poly (lactic acid) composite. *Composites Part B: Engineering* 52(2013): 106–113.
13. Somord, K., O. Suwantong, N. Tawichai, T. Peijs, and N. Soykeabkaew. Self-reinforced poly (lactic acid) nanocomposites of high toughness. *Polymer* 103(2016): 347–352.
14. Majola, A., S. Vainionpää, P. Rokkanen, H. M. Mikkola, and P. Törmälä. Absorbable self-reinforced polylactide (SR-PLA) composite rods for fracture fixation: Strength and strength retention in the bone and subcutaneous tissue of rabbits. *Journal of Materials Science: Materials in Medicine* 3, no. 1 (1992): 43–47.
15. Bocz, K., M. Domonkos, T. Igricz, Á. Kmetty, T. Bárány, and G. Marosi. Flame retarded self-reinforced poly (lactic acid) composites of outstanding impact resistance. *Composites Part A: Applied Science and Manufacturing* 70(2015): 27–34.

16. http://www.nuplex.com/composites/processes/filament-winding (accessed on January 21, 2018).
17. Abdalla, F. H., S. A. Mutasher, Y. A. Khalid, S. M. Sapuan, A. M. S. Hamouda, B. B. Sahari, and M. M. Hamdan. Design and fabrication of low cost filament winding machine. *Materials & Design* 28, no. 1 (2007): 234–239.
18. Nampoothiri, K. M., N. R. Nair, and R. P. John. An overview of the recent developments in polylactide (PLA) research. *Bioresource Technology* 101, no. 22 (2010): 8493–8501.
19. Murariu, M., L. Bonnaud, P. Yoann, G. Fontaine, S. Bourbigot, and P. Dubois. New trends in polylactide (PLA)-based materials: "Green" PLA–Calcium sulfate (nano) composites tailored with flame retardant properties. *Polymer Degradation and Stability* 95, no. 3 (2010): 374–381.
20. Oksman, K., M. Skrifvars, and J. F. Selin. Natural fibres as reinforcement in polylactic acid (PLA) composites. *Composites Science and Technology* 63, no. 9 (2003): 1317–1324.
21. Serizawa, S., K. Inoue, and M. Iji. Kenaf-fiber-reinforced poly (lactic acid) used for electronic products. *Journal of Applied Polymer Science* 100, no. 1 (2006): 618–624.
22. Bax, B., and J. Müssig. Impact and tensile properties of PLA/Cordenka and PLA/flax composites. *Composites Science and Technology* 68, no. 7–8 (2008): 1601–1607.
23. Suryanegara, L., A. N. Nakagaito, and H. Yano. The effect of crystallization of PLA on the thermal and mechanical properties of microfibrillated cellulose-reinforced PLA composites. *Composites Science and Technology* 69, no. 7–8 (2009): 1187–1192.
24. Islam, M. S., K. L. Pickering, and N. J. Foreman. Influence of accelerated ageing on the physico-mechanical properties of alkali-treated industrial hemp fibre reinforced poly (lactic acid)(PLA) composites. *Polymer Degradation and Stability* 95, no. 1 (2010): 59–65.
25. Huda, M. S., L. T. Drzal, A. K. Mohanty, and M. Misra. Effect of fiber surface-treatments on the properties of laminated biocomposites from poly (lactic acid)(PLA) and kenaf fibers. *Composites Science and Technology* 68, no. 2 (2008): 424–432.
26. Awal, A., M. Rana, and M. Sain. Thermorheological and mechanical properties of cellulose reinforced PLA bio-composites. *Mechanics of Materials* 80(2015): 87–95.
27. Csikós, Á., G. Faludi, A. Domján, K. Renner, J. Móczó, and B. Pukánszky. Modification of interfacial adhesion with a functionalized polymer in PLA/wood composites. *European Polymer Journal* 68(2015): 592–600.
28. http://www.mynewsdesk.com/us/pressreleases/global-polylactic-acid-pla-market-to-see-strong-growth-and-business-scope-from-2017-to-2022-2053457 (accessed February 21, 2018).
29. http://cms.natureworksllc.com/~/media/The_Ingeo_Journey/EcoProfile_LCA/EcoProfile/NTR_Eco_Profile_Industrial_Biotechnology_032007_pdf.pdf (accessed February 21, 2018).
30. https://www.marketsandmarkets.com/Market-Reports/polylacticacid-387.html (accessed February 21, 2018).
31. http://www.insidertradings.org/2018/01/16/united-states-polylactic-acid-pla-market-research-on-chemical-segmentation-to-2022/ (accessed February 21, 2018).

32. Nagarajan, V., A. K. Mohanty, and M. Misra. Perspective on polylactic acid (PLA) based sustainable materials for durable applications: Focus on toughness and heat resistance. *ACS Sustainable Chemistry & Engineering* 4, no. 6 (2016): 2899–2916.
33. Wang, Y., K. Chen, C. Xu, and Y. Chen. Supertoughened biobased poly (lactic acid)–epoxidized natural rubber thermoplastic vulcanizates: fabrication, co-continuous phase structure, interfacial in situ compatibilization, and toughening mechanism. *The Journal of Physical Chemistry B* 119, no. 36 (2015): 12138–12146.
34. Liu, Z., Y. Luo, H. Bai, Q. Zhang, and Q. Fu. Remarkably enhanced impact toughness and heat resistance of poly (L-Lactide)/thermoplastic polyurethane blends by constructing stereocomplex crystallites in the matrix. *ACS Sustainable Chemistry & Engineering* 4, no. 1 (2015): 111–120.
35. Zeng, J. B., K. A. Li, and A. K. Du. Compatibilization strategies in poly (lactic acid)-based blends. *Rsc Advances* 5, no. 41 (2015): 32546–32565.
36. Lu, X., X. Wei, J. Huang, L. Yang, G. Zhang, G. He, M. Wang, and J. Qu. Supertoughened poly (lactic acid)/polyurethane blend material by in situ reactive interfacial compatibilization via dynamic vulcanization. *Industrial & Engineering Chemistry Research* 53, no. 44 (2014): 17386–17393.
37. Cao, Z. Q., X. R. Sun, R. Y. Bao, W. Yang, B. H. Xie, and M. B. Yang. Carbon Nanotube Grafted Poly (L-lactide)-block-poly (D-lactide) and Its Stereocomplexation with Poly (lactide) s: The Nucleation Effect of Carbon Nanotubes. *ACS Sustainable Chemistry & Engineering* 4, no. 5 (2016): 2660–2669.
38. Chen, Y., D. Yuan, and C. Xu. Dynamically vulcanized biobased polylactide/natural rubber blend material with continuous cross-linked rubber phase. *ACS Applied Materials & Interfaces* 6, no. 6 (2014): 3811–3816.

7

All-Cellulose Composites: Concepts, Raw Materials, Synthesis, Phase Characterization, and Structure Analysis

7.1 Introduction

Petroleum-based polymer composite materials, typically glass fibers or carbon fibers–reinforced composite embedded into unsaturated polyester or epoxy resin, show excellent mechanical and thermal properties. They are widely used in various applications, including structural commodities, aerospace vehicles, sports, packaging, house furnishings, smart devices, and so on [1]. However, these composite materials are not biodegradable and cause environmental problems when they are disposed by combustion [2]. Therefore, of late, attention has been increasingly focused on environmentally friendly, biodegradable, and plant-derived composites based on renewable resources, which are designated as "green" composites [3–4].

Cellulose-based natural fibers like jute, flax, hemp, sisal, and so on have been well recognized as potential reinforcements for bio-composites. Being lighter in weight, the specific strength and modulus of bio-composites are higher than that of traditional glass, carbon, or metal-based composites [5]. In spite of these benefits, they also have certain drawbacks in terms of structural inhomogeneity and poor interfacial adhesion between hydrophilic cellulosic reinforcement and hydrophobic matrix phases, resulting in inefficient load transfer from matrix to reinforcement [6]. The strength and Young's modulus of cellulose-reinforced composites hardly exceed 300 MPa and 30 GPa, respectively [7]. To overcome the issue of interfacial adhesion, many different approaches have been attempted. The most common approach involves chemical or physical modification of fiber or matrix to enhance compatibility and improve interfacial bonding [5].

Next alternative approaches to environmentally friendly polymer composites have focused on so-called "self-reinforced polymer composites" or "single-polymer composites" based on single-material ecodesign concepts [3]. Following the success of synthetic polymer–based all-polymer composites such as all-polypropylene composite or all-polyethylene composite, cellulose-based self-reinforced bio-composites have been developed on similar lines recently. In this case, the matrix is formed from dissolved and precipitated cellulose, while the reinforcement component is formed from the undissolved cellulose [1]. Nishino et al. coined the term "all-cellulose composites" (ACC) for this new class of materials [8]. Due to the inherent compatibility and excellent interfacial bonding between matrix and fibers, a powerful reinforcing effect can be anticipated in these types of composites. Nishino et al. reported tensile strength and storage modulus (at 300°C) values of up to 480 MPa and 45 GPa, respectively, for ramie fiber–reinforced ACC [8]. However, highest tensile strength achieved till date has been 910 MPa for unidirectionally aligned Bocell fiber ACCs [3] and Young's modulus up to 28 GPa for unidirectionally aligned ramie fiber–based ACCs [9]. These remarkable properties support the assumption that interfacial properties are greatly improved in a single polymer–based bio-composite. Another big advantage of this self-reinforced cellulose composite is complete recyclability, either for reuse or for end-of-life disposal [6].

All-cellulose composite is a fully cellulose-based composite and it is one of the most interesting classes of self-reinforced composites. However, before illuminating directly on ACCs, an initial fundamental discussion on cellulose will be useful.

7.2 Cellulose: Chemistry and Overview

7.2.1 Solid-State Structures of Native Cellulose

Cellulose ($C_6H_{11}O_5$) is the most abundantly available biopolymer in the world [1] and it consists of D-anhydroglucose units (AGU) as repeat units joined by 1, 4-β-D-glycosidic linkages or acetal linkages at C1 and C4 positions [2]. The spatial arrangement or stereochemistry of these acetal linkages has been very important. In cellulose, every second AGU ring is rotated by 180° in the plane to provide preferred bond angles of the acetal oxygen bridges. In this way, two adjacent structural AGUs define the disaccharide cellobiose unit [10–12]. The molecular structure of cellulose has been shown in Figure 7.1.

Cellulose chain consists of a D-glucose unit with an original C4-OH group at the nonreducing end and the other end structure (reducing end) is terminated with an original C1-OH group, which is in equilibrium with the aldehyde [12–14], as shown in Figure 7.1. Each repeating unit of cellulose contains three hydroxyl groups at positions C2, C3 (secondary, equatorial), and C6 (primary). Relative to the O5-C5, C4-C5 bond, the CH_2OH side group is arranged in a trans-gauche (T-G) position. Due to the hydrogen bond–formation ability of hydroxyl groups in cellulose, it forms a hierarchical structure. This H-bonding ability also conducts a major role in directing the crystalline

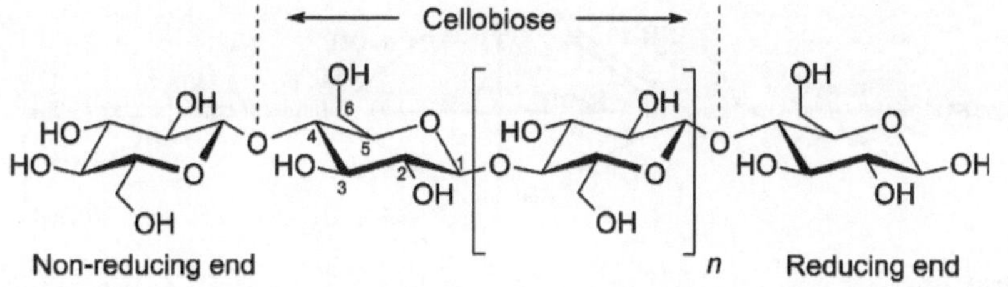

FIGURE 7.1
Molecular structure of cellulose. (From Pinkert, A. et al., *Chem. Rev.*, 109, 6712, 2009. With permission.)

packing of cellulose and governs properties such as hydrophilicity, chirality, degradability, and chemical reactivity [13]. As a result of super-molecular solid structure of cellulose, it forms a microcrystalline structure with fractions of high-ordered crystalline regions and low-ordered amorphous regions.

Intense research has been conducted on the hierarchical structure of cellulose since more than hundred years and often some controversy arises with insights of cellulose structure. As a first approximation, X-ray diffraction was used to determine the crystal structure of a native cellulose (cellulose I) and described by monoclinic sphenoid structures where the molecular chains are orientated along fiber direction [10,13]. Later, using high-resolution cross-polarization magic angle spinning (CP/MAS^{13}C-NMR) spectroscopy it was found that native crystalline cellulose I consists of two allomorphs, named I_α and I_β. This result was confirmed later by electron diffraction analysis. Generally, the Valonia (algae) and bacterial celluloses are rich in I_α, whereas ramie and cotton celluloses are dominated by I_β. So, the ratio of I_α/I_β depends on origin of the cellulose [13,15]. Although I_α and I_β have triclinic and monoclinic unit cells, respectively, the crystal structures are almost similar and both have parallelly aligned chains. The main difference between these two crystal structures lies in the fact that some chains in one structure are shifted by half the repeat distance relative to those in the other one. The I_α phase is considered to be less stable than the I_β phase, because I_α can be irreversibly converted to I_β by annealing in the solid state [16].

7.2.2 Polymorphism of Cellulose

Apart from native cellulose (i.e., cellulose I), it may also occur in other forms (cellulose II, cellulose III, cellulose IV). Among these polymorphic structures, the most stable structure is cellulose II due to the presence of an antiparallel chain orientation [17]. Cellulose I, with a parallel chain arrangement, is thermodynamically less stable than cellulose II. This less stable cellulose I can be converted to cellulose II by mercerization or alkali treatment. Therefore, cellulose fibers synthesized from regenerated cellulose (known as man-made fibers) mainly consist of cellulose II. Cellulose III_I and cellulose III_{II} can be formed from cellulose I and cellulose II, respectively, by liquid ammonia

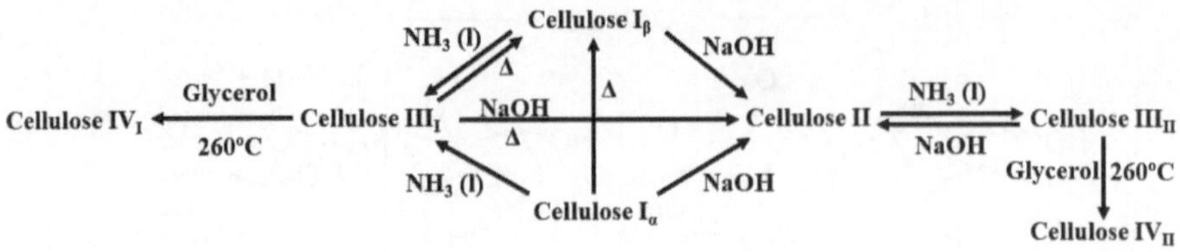

FIGURE 7.2
Transformation of cellulose into its several polymorphs. (From Kaplan, D.L., *Biopolymers from Renewable Resources*, Springer, Berlin, 1998. With permission.)

treatment. By a heat treatment of cellulose III with glycerol at about 260°C, it can be converted to cellulose IV_I or cellulose IV_{II} [16,18], as shown in Figure 7.2.

7.2.3 Physical and Chemical Properties of Cellulose

The super-molecular structure of cellulose, its orientation, and number of repeating units (AGU unit) per molecule determine many of its physical and chemical properties. The degree of polymerization (DP) may be as high as 10,000, although it varies depending on the type of natural fiber and the treatment given to raw materials [10–11]. The DP of bacterial cellulose is quite high and for regenerated cellulosic fibers varies only from 250 to 500. For wood pulp, the DP values typically stand between the 300–1700 range, whereas depending on treatment, the DP values of cotton and other fibers varied between 800–10,000 [13]. The properties of cellulose, especially physical properties, have been observed to be a linear function of DP.

Among different allomorphs, cellulose I is the strongest, having theoretical tensile strength of about 17.8 GPa [6,10]. Nishino et al. measured the elastic modulus of the crystalline regions of cellulose polymorphs in the direction parallel to the chain axis by X-ray diffraction. They reported that the elastic modulus values of crystalline cellulose I is about 138 GPa, which is the highest. However, a drastic change may occur in the crystal modulus depending on the crystal modification. Nishino et al. reported that the values of elastic modulus for cellulose II, III_I, III_{II}, and IV were 88, 87, 58, and 75 GPa, respectively [18]. Whereas Gassan et al. have done calculations on the elastic properties of natural fibers based on some experimental data and theoretical models. According to their calculation, modulus of pure cellulose varied from 74–168 kN/mm^2 (or GPa) with respect to fibril axis [19]. These values are comparable with the elastic modulus values of high-performance synthetic fibers such as Kevlar (156 GPa), Vectran (126 GPa), Technora (88 GPa), and also with aluminium (70 GPa) and glass fibers (76 GPa) [8]. The average Young's modulus for amorphous cellulose as calculated by Chen et al. using force field model was around 10.3 GPa [20]. Lyons reported a value of 1.582–1.585 gm/cc as the density of native cellulose in the continuous crystal lattice [21]. Variation in mechanical properties among different allomorphs of cellulose is mainly due to the difference in their molecular structure. In spite of good mechanical properties of cellulose, the strength and modulus of cellulosic fiber–reinforced composites remain far below the potential provided by cellulose. It is mainly due to the heterogeneous structure and composition of plant fibers [10]. Cellulose is resistant to strong alkali (17.5 wt%), but it is easily hydrolyzed by acid treatment and quantitatively decomposes to D-glucose. Cellulose is relatively resistant to oxidizing agents [14].

7.3 Sources of Cellulose

There is a wide range of natural fibers that can be subdivided based on their origins, such as plant fibers, animal fibers, and mineral fibers. All animal fibers (e.g., hair, silk, wool) consist of proteins, while all natural fibers obtained from vegetable or plant sources (e.g., cotton, kapok, jute, flax, ramie, hemp, linen, sisal, abaca, pineapple, coir) are mainly composed of cellulose [2,22]. Classification of natural fibers is shown in Figure 7.3.

Cellulose is the most essential component of all plant fibers. The other main components of plant-based cellulosic fibers are hemicellulose, lignin, pectin, waxes, and water. However, the structure and chemical composition of fibers may vary depending upon the climatic conditions, age, and the extraction process [2,10]. It has been established that properties of fibers such as density, tensile strength, and modulus are strongly related to internal structure as well as chemical composition of the fibers. A single biofiber consists of several cells that are formed out of oriented and crystalline microfibrils

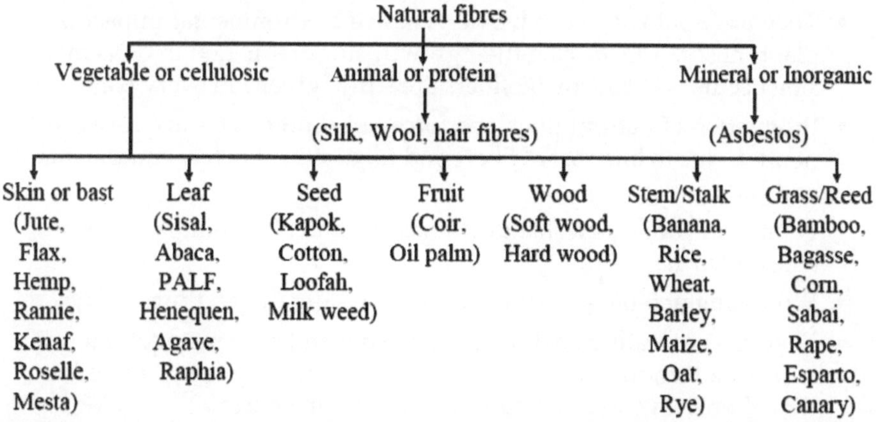

FIGURE 7.3
Classification of natural fibers. (Redrawn from Lyons, W.J., *J. Chem. Phys.*, 9, 377, 1941. With permission.)

based on cellulose and are embedded in an amorphous hemicellulose–lignin matrix of varying compositions [2]. The structure, microfibrillar angle, cell dimensions, defects, cellulosic content in the fiber, and DP of cellulose are the most important variables that determine overall properties of the fibers. Generally, fibers with high cellulosic content, higher DP, and lower microfibrillar angle exhibit higher mechanical strength and modulus [23]. For example, bast fibers like jute, hemp, flax, ramie, etc., having low microfibrillar angles (5°–15°) show very high modulus and low elongation at break.

Apart from plants, some algae, bacteria, and fungi species also produce cellulose [13,24]. Nevertheless, nowadays many man-made cellulosic fibers (e.g., Viscose, Modal, Lyocell, Tencel, Bocell) are designed, based on regenerated cellulose. These man-made cellulose fibers generally offer greater homogeneity in terms of fiber structure, fiber diameter, and mechanical properties when compared with natural fibers [6].

7.4 Pros and Cons of Cellulosic Materials for Making Bio-composites

The growing interest toward bio-composites is mainly due to the following advantages of cellulosic materials as reinforcement [2,5,10,25–29].

- Being lighter in weight, bio-composites exhibit higher specific strength and modulus compared to metal/glass fiber/carbon fiber–reinforced composites.
- They are fully biodegradable.
- They are nontoxic and have no negative environmental impact. As plants absorb CO_2 during their growth, huge reduction in CO_2 emission occurs. That's why biofibers are truly "green" in every way.
- Production of natural fiber–reinforced bio-composites are very economical due to low cost of fibers and requirement of few equipments and little energy.
- Plant fibers are renewable raw materials and also abundantly available on Earth.
- Processing of biofibers are easy and also safer in handling.
- They are generally nonabrasive to mixing and molding equipments, which leads to advantages with regard to easy processing and reduction of process cost of composite materials in general.
- Biofibers possess high electrical resistance, and due to the hollow cellular structure, it provides good thermal and acoustic insulating properties.

Apart from their advantages, these biofibers also have some shortfalls as bio-reinforcements [2,5–6,10]. These are

- Being inherently polar and hydrophilic in nature, the biofibers cannot adhere properly with the hydrophobic polymer matrix. This incompatibility causes ineffective stress transfer throughout the interface of matrix and reinforcement, showing poor mechanical properties.
- Their poor resistance to moisture absorption leads to swelling, which results in the inferior mechanical property and poor dimensional stability of composites.
- Processing temperature is restricted to 200°C due to lower thermal stability of natural fibers above 230°C. Therefore, it limits the number of thermoplastic materials to be considered as matrix material for the preparation of natural fiber–reinforced bio-composites.
- Depending upon different growing conditions, fiber-extraction procedure, maturity of harvested plant, cellulosic fibers have a tendency of nonuniformity in quality and mechanical properties. This variation was observed even between individual plants in the same cultivation. Man-made cellulosic fibers (e.g., viscose, lyocell, modal) are much more uniform than natural biofibers and can be used where uniformity is more important.
- Problematic end-of-life disposal of bio-composites, where nondegrading or slowly degrading petrochemical-derived thermoplastics or bio-based polymers are used as matrix.

Some of these shortcomings have been overcome with modification of biofibers physically or chemically. Physical modification can be done by corona discharge or plasma treatment. Different types of chemical treatment processes have also been developed to modify natural fibers, which include treatment with silane, alkali, enzyme, isocyanate, maleated coupling agents, and others like acetylation, benzoylation, etherification, graft copolymerization, etc. [5]. Lots of research work has been done in this area and is still ongoing. With the development of ACCs, some of the previously specified problems of bio-composites is being avoided due to the use of cellulose both in matrix and reinforcing phase. For example, problem of incompatibility between matrix and reinforcement can be avoided in case of ACC as both components are cellulose-based. Therefore, the need for any chemical or physical modification for improving matrix-reinforcement interfacial bonding could be extremely reduced or even completely eliminated in case of ACCs. On the other hand, being chemically homogeneous, self-reinforced ACCs are easy to recycle and end-of-life disposal is never a problem.

7.5 Basic Concepts of All-Cellulose Composites

A composite material is generally characterized by the useful combination of at least two constituents, termed as matrix and reinforcement—aiming to achieve properties that cannot be obtained using the constituent materials separately. All-cellulose composites represent a relatively new class of self-reinforced bio-composites that are entirely based on cellulosic materials. Unlike other traditional bio-composites, in an ACC, both matrix and reinforcement phase are cellulosic. The matrix phase in ACC is isotropic and the reinforcement may be isotropic or anisotropic in nature.

Pure cellulose can't be melt processed as it degrades melting. However, it can be dissolved partially or completely in a suitable solvent. Different process techniques and solvent systems have been developed to prepare ACCs where cellulosic materials are being used in different forms as raw material. During synthesis of ACC, the cellulosic materials are commonly dissolved in a solvent in the presence of some undissolved cellulosic reinforcement; thereby, after regeneration of dissolved cellulose, it acts as matrix phase. Nowadays, researchers have succeeded in suggesting newer ways to melt process cellulose, for making composites. The concepts of processing of ACCs has been explained in Figure 7.4, showing the process flow.

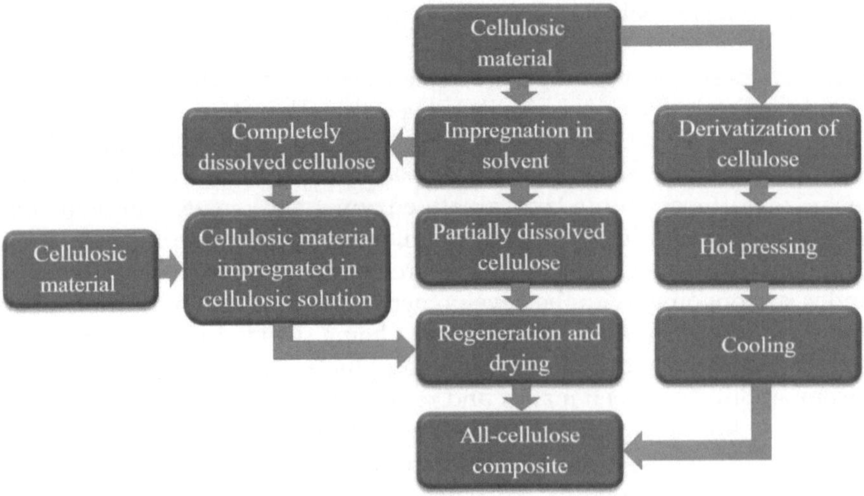

FIGURE 7.4
Process flowcharts of ACC preparation.

7.6 Classification of All-Cellulose Composites

By using different process techniques, different types of ACCs can be produced. ACCs can be classified mainly in three ways, as shown in Figure 7.5.

7.6.1 ACCs Based on Type of Matrix Phase

All-cellulose composites can be prepared through dissolution and regeneration of non-derivatized cellulose or through melting of derivatized cellulose. On the basis of purity of cellulose in matrix phase, ACCs may be of two types:

i. **Non-derivatized ACC:** In this type of ACCs, the cellulosic materials are dissolved in a solvent without forming any cellulose derivative. The dissolved cellulose is regenerated or resolidified afterward, in the presence of undissolved cellulosic material that forms the matrix phase. There are two main pathways in the processing of this type of ACCs: i) introduction of fibrous elements in solvent followed by regeneration or ii) partial dissolution of fibers in solvent followed by regeneration.

ii. **Derivatized ACC:** As native cellulose is not melt processable, some researchers have used cellulose derivative (ester, acetate, carbamate, etc.), which are melt processable. In this type of composite, initially the cellulosic materials are partially converted to thermoplastic or thermoset derivatives of cellulose, which are consolidated by hot pressing or curing to a matrix with the unsubstituted cellulose part (reinforcement). Though researchers have named these as "all-cellulose composites," these are not "all-cellulose composites" in the true sense, as the matrix phase is based on cellulose derivatives. Derivatized ACCs are discussed in detail in Chapter 9.

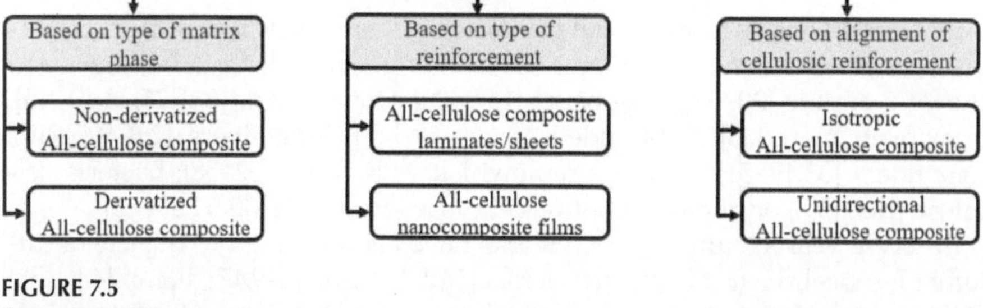

FIGURE 7.5
Classes of ACCs.

7.6.2 ACCs Based on Type of Reinforcement

On the basis of structure or shape of reinforcements, which were used during preparation, ACCs can be classified in to two categories:

i. **All-cellulose composite laminates/thick sheets:** In this class of ACCs, micro/macro-sized reinforcements are used. These ACCs are generated in the form of laminates or thick sheets.
ii. **All-cellulose nanocomposites (films/thin sheets):** In this type of ACCs, nano-reinforcing materials (i.e., nanowhisker, nanofiber, nanofibril, etc.) are used. These are generally prepared in the form of films or thin sheets.

7.6.3 ACCs Based on Alignment of Reinforcements

On the basis of alignment of reinforcements, ACCs may be of two types:

i. **Isotropic ACCs:** In this type of ACCs, the reinforcing phase is distributed evenly in the matrix phase. It shows similar mechanical properties in all directions. Undrawn all-cellulose nanocomposites (ACNCs) mainly fall in this category.
ii. **Unidirectional ACCs:** In this case, the reinforcing materials are aligned in a particular direction. Therefore, in this type of ACCs, mechanical properties are different in transverse and longitudinal directions. All micro/macro-sized ACC sheets/laminates fall in this category. Even all-cellulose nanocomposite films can be transformed to unidirectional ACCs after proper drawing or by aligning nanomaterials in a particular direction by other means.

7.7 Different Forms of Cellulosic Materials for the Preparation of ACCs

Cellulosic materials from annual plants, woods, and agricultural by-products are abundant renewable resources. The starting materials for ACCs may be in the form of pulp [30–37], filter paper [1,38–39], woven fabric [6,17,25,34,40–45], nonwoven matt [26], 3D-braided structure [46], long fibers [2,8–9,47–48], nanofibers [33,49–51], cellulose nanowhiskers (CNWs) [52–58], bacteria cellulose (BC) [59], or microcrystalline cellulose (MCC) [28–30,54,60–63].

In recent years, many research works have been devoted to the use of different lignocellulosic fibers such as flax [17,26], ramie [8–9,47], hemp [48], jute [43], bagasse [51], and so on for preparation of ACCs. If required, lignocellulosic fibers can also be transformed to micro/nano cellulosic fibers by a special process. For example, Ghaderi et al. in their study produced nanofibers from bagasse fibers and used them to produce ACNCs [51]. In a recent study, Shibata et al. [45] reported use of the most popular natural cellulosic fiber, that is, cotton, for the preparation of ACC. The use of regenerated cellulose fibers such as viscose-rayon [6,17,25,34,40], lyocell [3,26,41–42,44], and Bocell [3] in the preparation of ACCs have also been reported in many literatures.

The development of textile technologies has resulted in the formation of composites from woven, knitted, or braided fabric that have superior mechanical properties due to continuous orientation of fibers. The common 2D-weave architectures that are used for composite manufacturing are plain, twill, satin, basket, leno, and mock-leno [2]. Among these, till date mainly different twill- [17,34,41–45] and plain- [40] structured weaves have been used by researchers to prepare ACCs. In addition to woven material, cellulosic fiber–based nonwoven materials can also be used for making ACCs. Gindl-Altmutter et al. used needle-punched nonwoven mats made of flax and lyocell fiber for this purpose [26]. Huber was the first to take an initiative for making ACCs from 3D-braided structure prepared by using rayon yarn [46]. Filter paper, which is based on 100% cellulosic material, can also be used as a preform for preparing ACC [1,38–39].

The mechanical properties of wood fibers depend on the pulping process. Lignin content in wood is relatively higher and varies for different species [10]. For preparation of ACCs from wood pulp, generally a pretreatment is required for removing non-cellulosic materials. Novel high-strength biocomposites based on micro and nanofibrils obtained by mechanically fibrillating pulp from agricultural by-products and wood have been successfully used for reinforcement of a broad variety of polymers [30,64] and to produce ACCs in a few cases [30–31,33].

Bacterial cellulose is one of the most important sources of cellulose. Among different generations of bacteria, the most efficient producers of BC is *Acetobacter xylinum,* which is a gram-negative, acetic acid bacteria [59]. Nanocomposites manufactured from BC fibrils of *Acetobacter xylinum* show exceptionally high mechanical strength, biocompatibility, and optical transparency [65–66]. Hsieh et al. reported that the modulus of a single filament of BC is about 114 GPa [67]. Its excellent mechanical properties come from a strongly hydrogen-bonded network [58,66].

Cellulose nanowhiskers are rod-like nanocrystals, with high potential in mechanical performance and have been successfully used as reinforcing fillers in a series of synthetic and natural polymeric matrices [68–69]. These are manufactured by acid hydrolysis of different natural plant fibers, pulp, bacteria, fungi, and algae. Even by-products from the forest, agriculture, and paper industries may be employed as abundant, inexpensive, and readily available sources for CNWs [33]. During acid hydrolysis with strong mineral acids such as sulfuric acid, MCC of predominantly coarse particulate aggregates are initially produced and further acid hydrolysis yields the desired nanoparticles.

Hepworth et al. reported that nanowhiskers have a Young's modulus and tensile strength of around 130 GPa and 7.5 GPa, respectively [70]. The mechanical strength of the whiskers is thought to arise from their large surface area, the extended-chain conformation, and strong mutual association of cellulose molecules by H-bonding in the crystalline state [53,55].

Depending on these different forms of cellulosic materials, the structure of ACC may vary from nano level (all-cellulose nanocomposite) to macro level (ACC sheets or laminates).

7.8 Manufacturing of Non-Derivatized All-Cellulose Composites

The preparation of non-derivatized ACCs involves three individual steps:

i. Cellulose dissolution,
ii. Regeneration of dissolved cellulose, and
iii. Drying.

7.8.1 Cellulose Dissolution

In this step, a suitable solvent is used for dissolving cellulose fully or partially as per requirements, depending on the process route being followed. The main factors controlling this step are (a) crystallinity of native cellulose, (b) amount of cellulose to be dissolved, (c) time for dissolution, (d) temperature, and (e) solvent type. Crystallinity of native cellulose materials greatly affect the dissolution process. Huber et al. observed that dissolution of linen fibers were slower than that of rayon fibers, due to considerably higher crystallinity of linen fibers compared to rayon [17]. Even the non-cellulosic components of the natural fibers such as hemicelluloses, lignin, and pectin influence the dissolution [12]. Presence of these non-cellulosic components slow down the dissolution of cellulose [17,42]. Generally, higher dissolution time helps in dissolving the cellulose more, but up to a certain extent. On the other hand, dissolution of cellulose in a particular solvent is a function of temperature. For example, in the preparation of ACCs using ionic liquid (IL) and NaOH/urea as solvent, required temperatures are generally about 100°C and –12°C, respectively.

7.8.1.1 Cellulose–Solvent Systems for Manufacturing Non-Derivatized ACCs

Due to the presence of a complex H-bended network structure, pure cellulose cannot be processed by melting [8]. However, cellulose can be processed by dissolving it in many solvents. In many applications, proper dissolution of cellulose is a very important and a very challenging step. Generally, cellulosic solvents are classified in two categories: (i) derivatizing solvents and (ii) non-derivatizing solvents. However, in this module, we will discuss mainly about non-derivatizing solvents. The term "non-derivatizing" denotes the systems capable of dissolving the polymer only by intermolecular interactions and no other derivatives are formed [70]. The solvents may be aqueous or nonaqueous based. The so-called non-derivatizing solvents are of great importance in the preparation of ACCs.

In addition to making the polymer soluble, the resulting cellulosic solution must have certain desirable properties, such as thermal and chemical stability, proper viscoelastic properties, environmental friendliness, easy application, and recovery. Nowadays, the cellulose industry makes extensive use of non-derivatizing solvents. A large number of solvents are available in the market for cellulose dissolution but only a few have been explored up to the semi-industrial scale and can qualify as "sustainable" processes.

Till date, several non-derivatizing solvent systems have been accepted to be capable of dissolving cellulose, such as lithium chloride/N, N-dimethylacetamide (LiCl/DMAc) [3,8–9,28,47,52,54,60–62], calcium and sodium thiocyanate [71], dimethyl sulfoxide (DMSO)/tetrabutylammonium fluoride [72], NH_3/NH_4SCN [73], NaOH/urea [36,53,55], NaOH/urea/ZnO [37,74], polyethylene glycol (PEG)/NaOH [1], molten salt hydrates [75], N-methylmorpholine-N-oxide (NMMO) [48,76], ILs [6,26–27,34,38,41–45,49,63], and many others. The developments in the novel solvent systems for cellulose offer the possibility to prepare ACCs through solution processing [77]. Figure 7.6 shows the percentage usage of different solvents for making ACCs till date. It indicates that mostly LiCl/DMAc, NaOH/urea, and different ILs have been used so far for the preparation of ACCs. From the technical point of view, in comparison to LiCl/DMAc, other solvents like NaOH/urea, NMMO, and ILs are more advantageous as in these cases no activation step is required before dissolution (Figure 7.7).

Nishino et al. were the first to prepare an ACC from pure craft cellulose pulp and unidirectionally aligned ramie fibers using LiCl/DMAc solvent system. This method eliminated the overheating of incorporated fibers during the thermal processing, and thereby the plant fibers were able to persist their high mechanical performance [8].

Another important non-derivatizing solvent system for cellulose is the aqueous solution of NaOH and urea. Zhou et al. observed that solubility of cellulose improved significantly in 6–8 wt% NaOH aqueous solution by adding 2–4 wt% urea [78]. The apparent solubility of cellulose in the NaOH/urea system is strongly dependent on the concentration of solvent, temperature, molecular weight, DP, and crystal types of cellulose. This solvent system is potentially cheap, nonpolluting, effective in very low temperatures, and capable of replacing many conventional solvent systems [36]. Therefore, many researchers have recently started using this solvent for the preparation

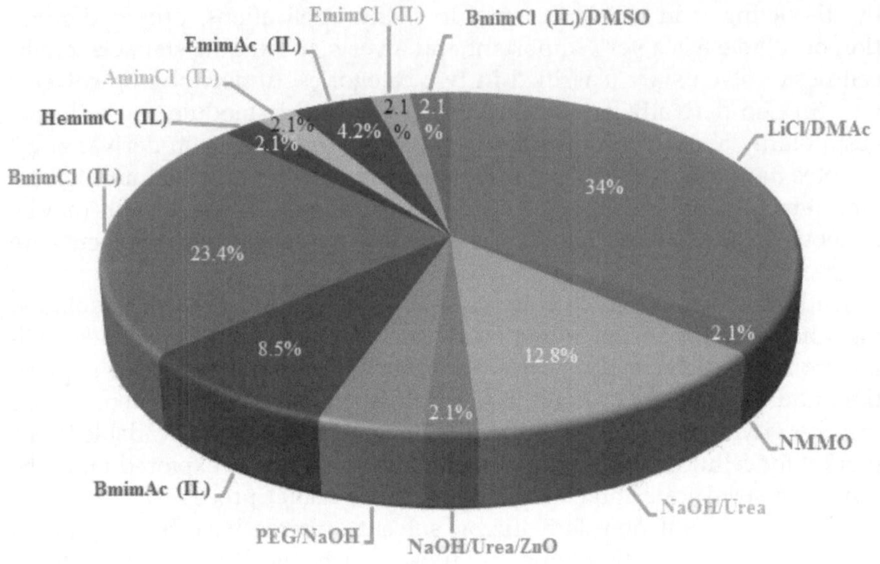

FIGURE 7.6
Share of different non-derivatizing solvents used for preparation of ACCs (percentage values were calculated on the basis of total 47 cases.)

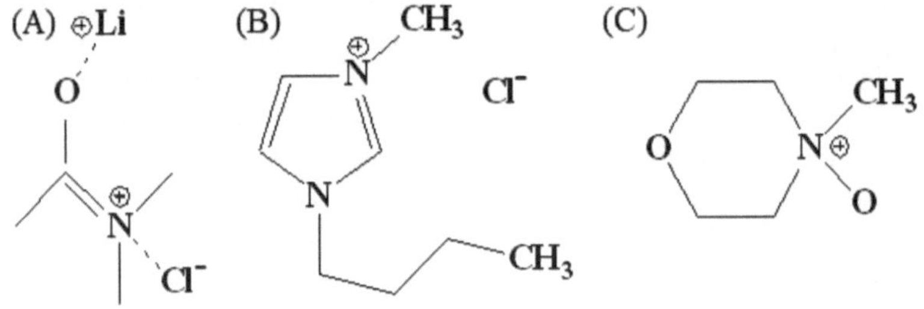

FIGURE 7.7
Structural formula of different solvents: (A) LiCl/DMAc, (B) NMMO, (C) IL (1-butyl-3-methylimidazolium chloride.)

of ACNCs following these basic steps: (a) dissolution, (b) gelation, and (c) regeneration [52–53,55]. The main advantages of this solvent system are – (i) the cellulose-dissolution process is very energy efficient where dissolution takes place at a subzero temperature (–12°C) and (ii) the dissolution process is very fast and gets completed within few minutes [35–37,53].

In recent years, the NMMO-based Lyocell or Tencel process has become a simple physical alternative to the dominating Viscose technology for producing regenerated cellulose fibers, films, membranes, beads without producing any cellulose derivatives and hazardous by-products [76,79]. The main reasons behind high dissolution power of NMMO are: (i) its high polarity and (ii) weakness of N-O bond [24]. The full recovery of this solvent is advantageous for the process [48]. Therefore, NMMO is very useful in the production of non-derivatizing ACCs. However, two major problems regarding NMMO system are tendency of fibrillation in regenerated fibers and requirement of costly investments for safety purposes due to instability of the solvent [79].

Ionic liquids have recently been used extensively in the preparation of ACCs. Ionic liquids are a group of salts, which generally exist as liquids at relatively low temperatures (<100°C). Some of them even exist in liquid form below room temperature and are known as room temperature ionic liquids. The most common examples of these ILs include salts of organic cations, such as 1-alkyl-3-methylimidazolium $[C_n\,mim]^+$, N-alkylpyridinium $[C_nPy]^+$, N,N-2-alkyl imidazole $[R_1R_3Im]^+$, tetraalkylammonium $[NR_4]^+$ or tetraalkylphosphonium $[PR_4]^+$, and anions, such as chloride $[Cl]^-$, bromide $[Br]^-$, thiocyanate $[SCN]^-$, tetrafluoroborate $[BF_4]^-$, hexafluorophosphate $[PF_6]^-$, trifluoroacetate $[CF_3COO]^-$, methanesulfonate $[CH_3SO_3]^-$, trifluoromethane sulfonate $[CF_3SO_3]^-$, and nitrate [80–81]. The cellulose-dissolution efficiency of any IL strongly depends on the alkyl chain length and presence of type of cations and anions. They have many attractive properties over other solvents such as good chemical and thermal stability, non-flammability, very low vapor pressure, easy recovery, and no by-product during cellulose dissolution. In addition to these, having tendency of very low evaporation, no respiratory protection and exhaust systems are required during their usage. This property also enhances their recycling and is one reason for being termed as "green solvents" [12]. However, some ILs, like as 1-butyl-3-methylimidazolium hexafluorophosphate have proved to be toxic, contradicting the green chemistry of ILs [82]. Moreover, conventional ILs, such as imidazolium and pyridinium-based ILs, are very expensive and generally exhibit poor biodegradability [36].

7.8.1.2 Mechanisms of Cellulose Dissolution

Cellulose dissolution is a key aspect for ACC processing. Due to very complex H-bonded hierarchical structure, the mechanisms involved behind dissolution of cellulose are very critical in addition to being interesting. Dissolution kinetics is quite different for crystalline and amorphous zones of cellulosic fibers.

The cell walls of natural cellulosic fibers play a great role in controlling their dissolution. The arrangement and packing of cellulose microfibrils vary in different cell wall structures of cellulosic fibers. Therefore, dissolution of cellulose may initiate the following two steps: (a) heterogeneous swelling of cell walls and (b) dismantling of microfibrils and cellulose chains. This heterogeneous swelling is termed as "ballooning phenomenon" by many researchers. According to these authors, "ballooning phenomenon" is caused by swelling of secondary walls, resulting in the bursting of primary walls [83–84].

There is an enormous number of solvents for cellulose and also have different theories to describe the dissolution mechanism, involving both kinetics and thermodynamics issues. Cellulose is known to be insoluble in water as well as in many organic solvents, but can be dissolved in a number of solvents of intermediate properties, like NMMO, ILs, etc. Cellulose is insoluble in water because of the strong and large number of H-bonding in cellulose [85]. However, it can also be dissolved in water at extreme pHs, particularly if a co-solute of intermediate polarity is added. Actually, cellulose is significantly amphiphilic in nature and it either plays the role of an acid (in a basic medium) or a base (in an acidic medium) [86]. More important than pH, dissolution of cellulose depends on temperature of the system, crystallinity, and DP of cellulose. Typically, a subzero temperature is required for dissolution of cellulose in soda [1], while about 130°C–140°C temperature is required for dissolution in NMMO [48,79]. During dissolution, the portion of cellulose with less crystallinity and low DP has been found to dissolve fast and the portion with higher crystallinity at a slower rate.

Having several hydroxyl groups, cellulose is a polar molecule and thus has good inter and intra-hydrogen–bonding ability [85]. The hydrophobic interaction in cellulose plays an important role in dissolution of cellulose, and thus, elimination of these interactions between cellulose molecules has been found to be very important for its aqueous solubility. It has been described earlier that urea can facilitate cellulose dissolution in aqueous solution of NaOH. Urea has a much lower polarity than water and has a good capability of eliminating hydrophobic association in water. Other substances of intermediate polarity, such as PEG and thiourea have similar effects and are well-known enhancers of solubility of cellulose in an aqueous-based solvent system [87]. Yang et al. observed that solubility of NaOH/urea aqueous solution enhanced significantly with incorporation of small amount of zinc oxide (ZnO), which existed in alkali system as $Zn(OH)_4$. This improvement of dissolution power can be explained by capability of $Zn(OH)_4$ in forming stronger hydrogen bonds with cellulose, compared to hydrated NaOH [37].

For LiCl/DMAc system, many dissolution mechanisms have been proposed by researchers. Generally, the dissolving mechanism is believed to go via an intermediate, involving the interaction of cellulose with Cl^- and an exchange of DMAc in the coordination sphere of lithium by cellulose hydroxyl groups. Accumulation of Cl^- along the cellulose chain produces an anionically charged polymer, $[\text{cellulose-Cl}]^-$ with the macro-cation, $[\text{Li–DMAc}]^+$, as the counter ion (Figure 7.8). This process, where polymer molecules are forced apart due to charge repulsion, is considered as polyelectrolyte effect. In addition to this, the high osmotic pressure creates a continuous influx of solvent resulting in higher disordering of the cellulose-binding forces until the polymer is completely dissolved [72,87–88].

During preparation of ACCs using LiCl/DMAc, initially the dissolution involves an activation step in which the cellulose structure is first swollen via solvent exchange steps using water, acetone, and DMAc. The increased

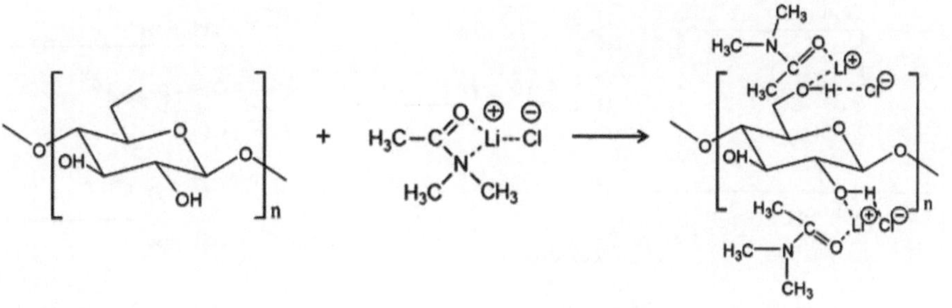

FIGURE 7.8
Mechanism of cellulose dissolution in LiCl/DMAc. (From Medronho, B. and Lindman, B., *Adv. Colloid Interface Sci.*, 222, 502, 2015. With permission.)

molecular mobility of cellulose allows the LiCl/DMAc solvent to penetrate into the cellulose structure more easily [61–62]. It has been reported that the dissolution of cellulose in LiCl/DMAC may take several months, without activation step [24].

The dissolution activity of highly polar solvent like NMMO depends on its N-O group, which is strongly dipolar in character. It is suggested that the oxygen in this group is able to form one or two hydrogen bonds with AGU of cellulose. The proposed dissolution mechanism assumes generation of a soluble complex by cleavage of intermolecular hydrogen bonds of cellulose and formation of new stronger hydrogen bonds between the cellulosic hydroxyl groups and the N-O group of NMMO. The solubility of cellulose in NMMO strongly depends on mixing ratio of cellulose, NMMO, and water [87,89].

In respect of ILs, till date there is no clear understanding on the role of individual ions in dissolution of cellulose. However, it is believed that dissolution of cellulose in IL happens due to interaction between anion and cellulose. Upon interaction, it breaks the intra and intermolecular hydrogen bond present between molecular chains of cellulose. Finally, cellulose gets dissolved in IL with formation of an electron donor–electron acceptor complex [12,81], as shown in Figure 7.9.

Solubility of cellulose in IL depends on DP of cellulose, temperature, chain length of the alkyl group, and type of anions present in the chemical structure of IL. Generally, solubility of cellulose in IL decreases as the alkyl chain length increases [81]. However, many researchers reported that among imidazolium-based ILs, 1-butryl-3-methylimidazolium chloride (BmimCl) shows the maximum solubility for cellulose [90]. At ambient temperature, ILs can only wet cellulose, but cannot dissolve them. However, when heated to 100°C–110°C, cellulose could be dissolved slowly in ILs. Microwave heating can increase rate of dissolution tremendously [79,90].

For synthesis of ACC, it is not necessary for cellulose to always dissolve completely in solvent. In fact, cellulose is dissolved partially in a few processes, leaving some part undissolved. This phenomenon has been discussed later in detail.

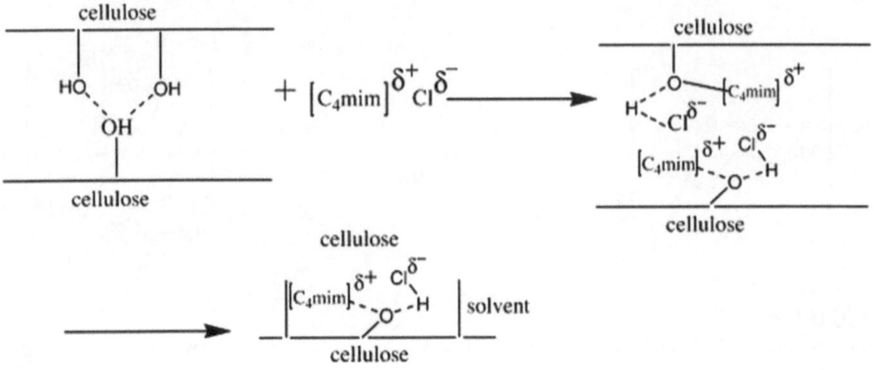

FIGURE 7.9
Dissolution mechanism of cellulose in IL. (From Feng, L. and Chen, Z.L., *J. Mol. Liquids*, 142, 1–5, 2008. With permission.)

7.8.2 Cellulose Regeneration

Regeneration is a step in which dissolved cellulose get transformed to solid cellulose in contact with a non-solvent or coagulant medium. Generally, water, acetone, and different alcohols (ethanol, methanol, etc.) are used as non-solvent or coagulant for cellulose. Regeneration of dissolved cellulose is one of the key aspects in many applications. In this step, when the cellulose solution comes in contact with the non-solvent, a thin skin is initially formed on the surface of dissolved cellulose. Subsequently, the exchange of solvent and non-solvent occurs through weak places/pores present in the skin [91]. The kinetics of regeneration is controlled by counter-diffusion rate of non-solvent and solvent from coagulation bath to solution and from solution to coagulation bath, respectively. This exchange of solvent and non-solvent results in reformation of inter and intramolecular H-bonds in regenerated cellulose [92]. It has been reported that regenerated cellulose is much more hydrophilic than the original one [91]. Additionally, the crystallinity of regenerated cellulose has not approached that of pre-dissolved cellulose. These changes happen due to the change in molecular packing and phase transformation of cellulose after dissolution and regeneration, which has been discussed in detail in the last part of this chapter.

In few solvents like NaOH/urea, a unique gelation behavior is observed. Gelation is an intermediate step between dissolution and regeneration [93]. Gelation behavior of cellulose is a function of concentration of cellulose, temperature, and time or rate of regeneration. In an interesting research, Cai et al. studied the gelation behavior and rheological properties of cellulose in NaOH/Urea aqueous solution [93].

Regeneration is a very important step in the production of ACCs. After regeneration, the regenerated cellulose forms matrix phase and binds the undissolved cellulose fraction (reinforcement). During regeneration, due to hydrophilic nature of cellulose, it swells and after drying creates internal voids in the fiber–matrix interface of ACCs. It ultimately affects the physical and mechanical properties of ACCs. Duchemin et al. observed that rate of regeneration has a great effect on the gel in terms of aggregation structure at the submicron scale and it also controls phase composition. With a slower regeneration rate, higher crystallinity was obtained. When regeneration rate was slow, the dissolved cellulose chain got a higher time to orient and obtain a low-energy configuration resulting in higher crystallinity [94].

7.8.3 Drying

After completion of the regeneration stage, the coagulant is removed by evaporative drying. Subsequently, the chemical similarity of matrix and reinforcement theoretically leads to an "interface-less" composite by enhanced fiber–matrix adhesion. In the drying stage, sometimes the ACCs shrink due to differential shrinkage of matrix and reinforcing phase after removal of coagulant [24,42,95]. This differential shrinkage might cause internal voids or defects inside the ACCs [34,94], which ultimately affects the microstructure as well as mechanical properties of ACCs. However, this effect could be minimized by applying pressure, especially in case of thick laminate-type ACCs [17,34,42].

7.9 Synthesis of ACCs and Different Processing Routes

For preparation of ACCs, different researchers have used different strategies. These are mainly:

i. Impregnation technique,
ii. Partial dissolution technique,
iii. Other approaches.

7.9.1 Impregnation Technique

In this technique, a two-step dissolution process is followed where a particular portion of cellulose is first dissolved in solvent. This cellulosic solution is used to impregnate some undissolved cellulosic materials in the second step. Ultimately, the dissolved cellulose is regenerated in the presence of undissolved cellulose (Figure 7.10). In this case, the cellulosic material for matrix and reinforcement may be the same or different [8,29–30,47–48,54].

Nishino et al. were the first to use this technique where they synthesized ACC by dissolving activated craft pulp in 8% LiCl/DMAc (wt/v) and

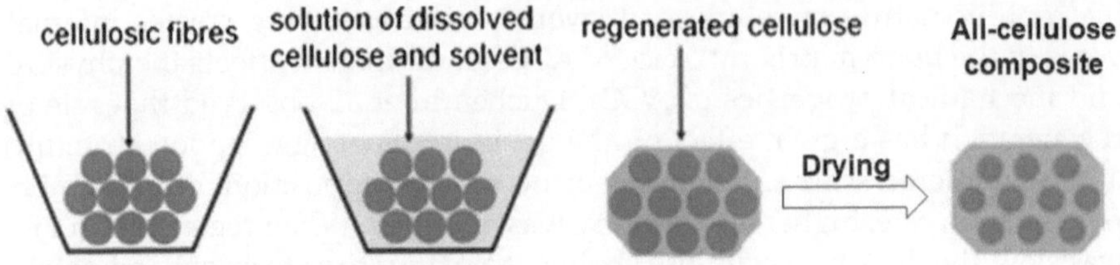

FIGURE 7.10
Schematic of a two-step method for the synthesis of ACCs. (From Huber, T. et al., *J. Mater. Sci.*, 47, 1171, 2012. With permission.)

followed it by regeneration of cellulose in the presence of parallel aligned ramie fibers [8]. Qin et al., in their studies, developed a similar type of ACC by using cellulose solution of different concentration to impregnate unidirectionally aligned ramie fibers [47].

Ouajai et al. modified this process slightly to prepare ACC from hemp fiber using NMMO as solvent. They used a "mechanical blending technique," where a hemp-NMMO solution containing 12% cellulose (wt/v) was prepared and then alkali-treated microfibers were subsequently added in that solution. The composite film was prepared by spreading this suspension on a heated glass plate, then compressed and regenerated allowing proper time in each stages [48]. Zhao et al. also prepared a similar type of composite by mixing de-waxed rice husk (RH) with completely dissolved cellulose (filter paper), followed by casting on Petri dish, coagulation in water, and drying. The thickness of the cellulose/RH ecocomposite films was about 0.2–0.4 mm [77].

Pullawan et al. in their study initially dissolved activated MCC in LiCl/DMAc and then CNWs were added to the dissolved MCC solution to generate CNWs-reinforced nanocomposite films [52,54]. It was similar to the nanocomposite film prepared by Zhao et al., where they used cellulose nanofibrils as reinforcing material in place of CNWs [30]. In a similar approach, Ma et al. prepared green nanocomposite films via addition of nanocrystalline cellulose (NCC) to cellulose solution prepared by dissolving pretreated microcrystalline cellulose in preheated IL, 1-(2-hydroxylethyl)-3-methyl imidazolium chloride ([HeMIM]Cl) [29].

In a recent study, Lourdin et al. used 1-butyl-3-methylimidazolium chloride (BmimCl) and DMSO in a ratio of 1:1 to dissolve MCC. After completion of this step, the required amount of freeze-dried cellulose nanocrystals (CNC) was added to the cellulose solution and kept for 15 days at room temperature. Subsequently, solid all-cellulose nanocomposite film was obtained by film casting, regeneration, and drying [63].

In all this research, the main focus was on the preparation of ACCs being relatively smaller in size and in the form of very thin sheet or film. Schuermann et al. first planned to overcome this by exhibiting the possibility of adapting an existing manufacturing technique to produce larger and thicker ACCs using prepreg style. In this process, cellulose solutions were initially prepared by complete dissolution of cellulose powders and pulp in an IL (1-ethyl-3-methylimidazolium acetate). The rayon textiles (2/2 twill) were then impregnated in the cellulose solutions and stored. Ultimately, four prepreg laminates were hot pressed to consolidate them and washed in water to regenerate the dissolved cellulose forming a matrix phase [34].

7.9.2 Partial Dissolution Technique

In this technique, a one-step dissolution process is used that involves "partial dissolution" of the surface of cellulosic materials. Since the cell wall of natural fibers consists of several layers, the surface layers of fibers can be partially dissolved and transformed into a matrix phase after regeneration. The core part remains undissolved, maintaining its original structure and acts as reinforcement to the composite materials (Figure 7.11). This method has also been described as "surface-selective dissolution" [1,3,9,26,28,59,61–62] by different researchers. In contrast to the first method, the source of cellulosic material is the same for both matrix and reinforcement.

Using this technique, Gindl et al. first prepared an optically transparent all-cellulose nanocomposite (ACNC) where they partly dissolved MCC (cellulose I) and precipitated the dissolved portion to form a matrix around the undissolved portion [28,32].

"Aerogel" is a highly porous material where the liquid in a gel is replaced with one continuous gas phase through freeze or supercritical drying, which was first prepared by Kistler [96]. Following this concept, Duchemim et al. prepared aerogels (or aerocelluloses)–based ACCs by partial dissolution of activated MCC in LiCl/DMAc, followed by a gentle precipitation of dissolved cellulose in the presence of water. During precipitation, a gel was formed initially and it was finally freeze-dried to maintain the openness of the structure [61].

In an interesting study, Nishino et al. used LiCl/DMAc as solvent to produce ACC from commercial filter paper by converting selectively dissolved fiber surface into a matrix phase [39]. A similar type of composite was prepared by Han et al. (2010) by using PEG/NaOH aqueous solution as solvent [1]. Gindl-Altmutter et al., in their study, produced a 0.4-mm

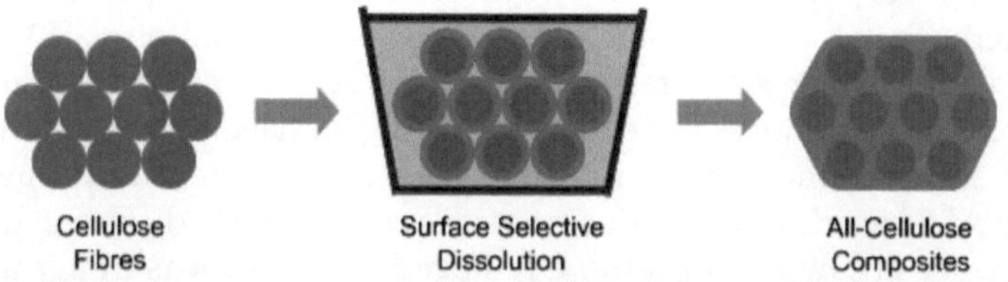

FIGURE 7.11
Schematic diagram of surface-selective dissolution process. (From Soykeabkaew, N. et al., *Compos. Sci. Technol.*, 68, 2201, 2008. With permission.)

thick ACC by partial dissolution of nonwoven matt of flax and lyocell in IL (BmimCl) [26].

Soykeabkaew et al. successfully used longitudinally aligned lyocell and Bocell fibers to prepare ACCs using a surface-selective dissolution method [9]. In another study, Soykeabkaew et al. initially prepared BC sheets for surface-selective dissolution in LiCl/DMAc. By this way all-cellulose nanocomposite (ACNC) films are produced after regeneration and finally drying under compression molding [59]. In a similar effort by Ghaderi et al., the nanofibers from bagasse were successfully incorporated to prepare ACNCs using LiCl/DMAc as solvent [51]. In a study by Yousefi et al. (2011), "nanowelding technique" was used to prepare all-cellulose nanocomposites from ground nanofibers and cellulose microfibers of canola straw using IL (BmimCl) [49].

The majority of studies on ACCs discussed in the earlier sections were focused on producing ACCs in the form of thin films with a thicknesses range of about 0.2–0.5 mm. Huber et al. developed a new technique that was named as "solvent infusion processing (SIP)" to produce thick ACC laminates from rayon textile (2/2 twill), where IL (1-Butyl-3-methyl imidazolium acetate) was used as cellulose solvent. The schematic diagram of the five stages of processing is shown in Figure 7.12. In the first step, which was called as "infusion step," the textile material was placed between two sheets of perforated plastic film (distributing medium) and solvent was infused maintaining a constant differential pressure. In the "dissolution step" the textile layers were hot pressed and in the "regeneration step" water was infused to regenerate dissolved cellulose. Subsequently, the consolidate laminate was washed in water for complete removal of IL. Finally, the laminate was dried by hot pressing. Pressure, temperature, and the position of inlet and outlet were varied to analyze their effect [25,97]. Huber et al. reported a similar approach for making thick ACC laminates from the fabric made of linen and rayon, using hand lay-up technique [17].

Very recently, Dormanns et al. successfully replaced IL by a much cheaper solvent, that is, aqueous NaOH/urea solution to prepare ACC laminates from the same rayon textile by SIP. They were thus able to prepare 25% stronger ACC laminates in addition to the advantage of shorter processing time and 97% reduction in solvent cost [98].

In few of our reported works, we have discussed the effect of process parameters, that is, dissolution time [41,43–44] and pressure [42], on the structure and properties of fabric-based thick ACC laminates produced by using compression molding. In another work, we have discussed the effect of process routes and dissolution time on the structure and properties of lyocell-fabric–based ACC laminates produced by four different methods [44]. Interestingly, in one process we removed the weft yarns from the lyocell fabric with a certain interval before making ACC laminates, which resulted in huge improvement in mechanical properties in longitudinal direction due to a unidirectional effect.

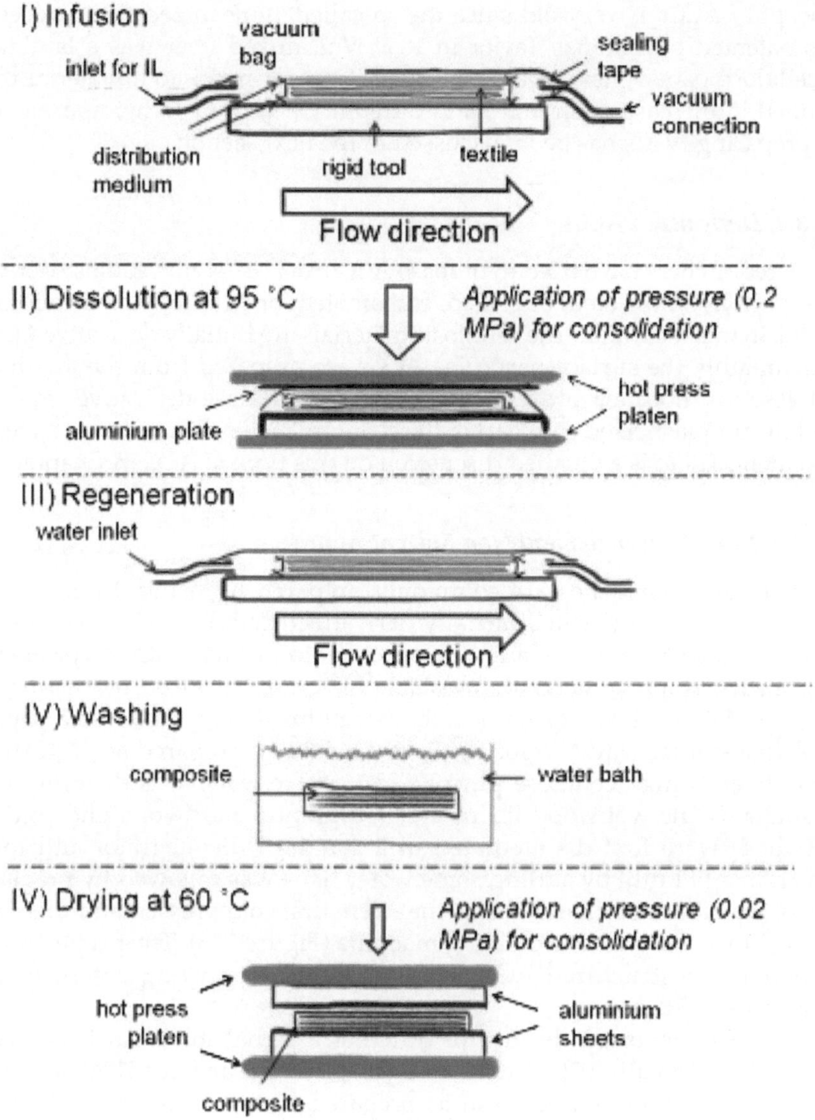

FIGURE 7.12
Schematic of the apparatus and five stages of SIP. (From Huber, T. et al., *Carbohydr. Polym.*, 90, 730, 2012. With permission.)

7.9.3 Other Approaches

The process techniques discussed in earlier sections were focused on preparation of ACCs by dissolution of cellulose in non-derivatizing solvents, followed by regeneration and drying. However, lot of research has been reported on ACCs where the concept is different from this. Actually, the concept of ACCs is very old since the so-called "vulcanized fiber process" was patented by Thomas Taylor in 1859. Vulcanized fiber was a laminate of cellulose paper plies, where paper was transformed into a material that resulted in increase in strength and durability [99]. Some other approaches for preparing ACCs has been discussed in the next section.

7.9.3.1 Derivatized ACCs

Many researchers have tried to prepare ACC from different cellulose derivatives such as benzylated, esterified, carbamated, or oxypropylated cellulose [24,40]. In this technique, the cellulosic materials are initially derivatized partially (mainly the surface part). The ACCs are prepared from the modified cellulose by "hot melt processing," where the cellulose derivatives melt to bind the un-plasticized core part of the cellulose material, which acts as reinforcement. There is a detailed discussion on this type of ACC in Chapter 9.

7.9.3.2 Non-Derivatized and Non-Solvent Approach

All-cellulose composites are commonly prepared by using different cellulose solvents or by using partially derivatized cellulose. However, other approaches were taken by a few researchers to produce ACCs. For example, "Zelfo" is a commercially available material processed by casting or molding, followed by drying of cellulosic pulps (hemp, flax, waste paper) [100]. In a similar way, Nilsson et al., in their study, prepared an ACC from wood fiber of high cellulose purity (97%), where water was the only processing aid. The wet wood fiber pulps (eucalyptus hardwood and conifer softwood) were first disintegrated in a standard disintegrator and then beaten in a PFI mill by adding some water. This was followed by preparation of the laboratory hand-sheets that were first cold pressed and then hot pressed in a compression molding machine (Figure 7.13). The properties of prepared nanostructured ACCs were tailored by applying pressure and temperature [31].

Few works reported the use of different external forces such as magnetic field [33,56,101–102], electric field [103], and shear force [104] to orient CNWs in thin nanocomposite films prepared from an aqueous cellulosic-suspension. Bordel et al. used an electric field to orient nanowhiskers from the suspension of nanowhiskers and ramie fibers in cyclohexane. The suspension was then allowed to evaporate on to thin formvar supporting films [103]. However, orientation of nanowhiskers under a shear force or an electrical field has limited suitability for large-scale production. In contrary, even a weak magnetic field is able to orient the CNWs effectively, and application of a strong magnetic field may be suitable for large-scale production [33]. Li et al. fabricated unidirectionally reinforced nanocomposite paper from CNWs and wood pulp under an externally applied magnetic field [33,56]. The alignment under external force was driven by characteristic negative

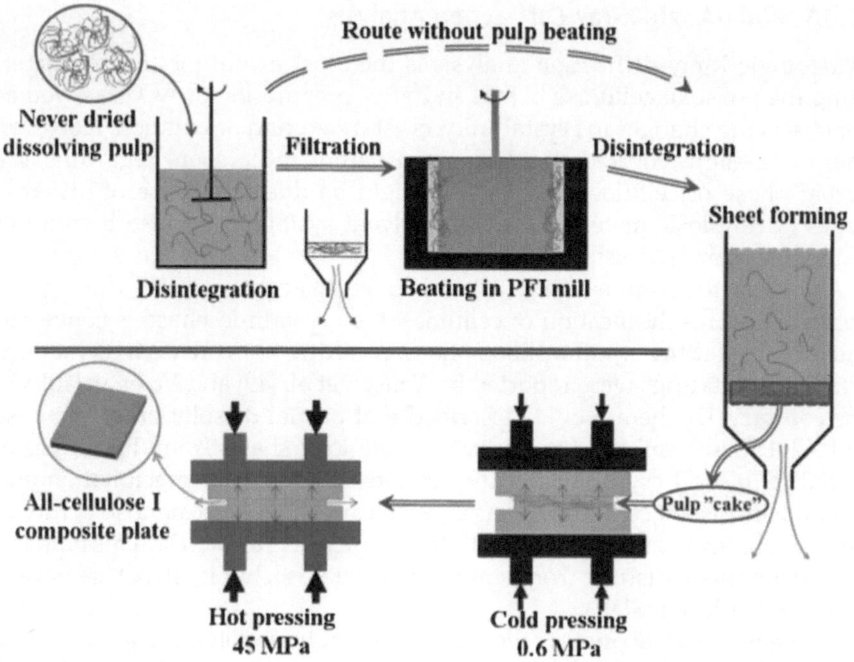

FIGURE 7.13
Schematic representation of the processing route from a never-dried dissolving pulp to an all-cellulose I bio-composite plate. (From Nilsson, H. et al., *Compos. Sci. Technol.*, 70, 1704, 2010. With permission.)

diamagnetic anisotropy of CNWs, which was generated during preparation of cellulose microcrystal [33]. Peces in his PhD work explored the concept of binder-free all-cellulose to avoid chemical substances of high environmental impact (solvents). Here, the intrinsic bonding capability between cellulose fibers by hydrogen-bonded network was utilized to produce ACC of good mechanical performance [105–106].

7.10 Phase Characterization of Cellulose in All-Cellulose Composites

The structural changes of cellulose from its native form to regenerated form in ACCs are characterized and investigated mainly using wide-angle X-ray diffraction (WAXD) [8–9,14,17,28,39–40,49,59–61,77,94,107] and in few cases by using nuclear magnetic resonance (NMR) spectroscopy [31,62,107], Fourier-transform infrared spectroscopy (FTIR) [27,48,55,107], and Raman spectroscopy [47,52,54].

7.10.1 Wide-Angle X-ray Diffraction Analysis

Wide-angle X-ray diffraction analysis is the most useful tool for characterizing the phase of cellulose before and after preparation of ACCs as well as for observing changes in crystallinity, crystal size, and orientation. However, there is a slight contradiction among different researchers regarding the actual phase of cellulose in ACCs. It might be due to the use of different types of cellulosic material, different solvent systems, and also because of using different processing techniques.

According to Nishino et al. [8,39] and Soykeabkaew et al. [9,59], during dissolution and resolidification of cellulose, the crystalline phase is converted into noncrystalline or amorphous phase, resulting in decrease in crystallinity. Similar findings were reported by Yousefi et al. [49] and Vo et al. [40]. On the contrary, Duchemin et al. described that partial dissolution of cellulose in LiCl/DMAc results in the removal of cellulosic sheets from the surface of initial cellulose I crystallites of the activated MCC. After precipitation and drying, the peeled cellulose sheets assemble into a biphasic nanostructure of amorphous and paracrystalline cellulose, which acts as matrix. This biphasic nanostructure is distinct from amorphous cellulose, but its structure is very close to cellulose I [61,94].

Duchemin et al. in another work reported slight depolymerization during the dissolution of cellulose I in IL, and eventually the embedding of highly crystalline undissolved cellulose I crystallites in a poorly crystalline matrix. This transmission was accompanied by a displacement of X-ray diffraction peak (2 0 0) to a lower value, in case of ACC [38].

Gindl et al. [80,60] and Zhao et al. [77] suggested that the dissolution and regeneration of cellulose I lead to a transformation to cellulose II, which is a more stable form of cellulose due to the presence of antiparallel packing of cellulose chains in contrast to parallel packing in cellulose I. Same observations were reported by Han et al. [1] and Pang et al. [107].

From the different WAXD analyses, it is clear that treatment of cellulose with different non-derivatizing solvents leads to decrease in overall crystallinity of cellulose. The overall crystallinity and crystal size in ACCs depends on mainly three factors [94]:

i. Concentration of cellulose in solvent,
ii. Dissolution time, and
iii. Rate of precipitation of cellulose to form a matrix.

It is observed that with increasing dissolution time, both crystallinity and crystal size of cellulose are reduced notably as shown in Figure 7.14, because of higher dissolution of cellulose and formation of a slightly less-ordered paracrystalline or amorphous phase [3,9,28,38–39,60,62,94]. Although, Huber et al. in their study observed that changes in crystallinity between linen

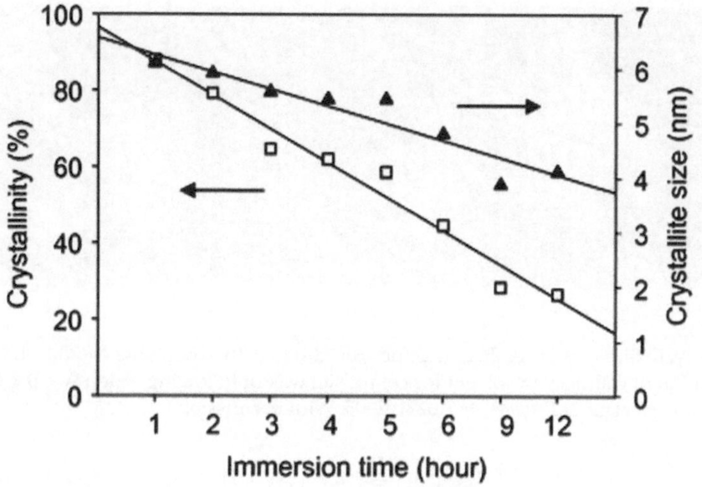

FIGURE 7.14
Effect of immersion time on percentage crystallinity and crystallite size (nm) of cellulose before and after ACCs preparation. (From Soykeabkaew, N. et al., *Compos. Sci. Technol.*, 68, 2201, 2008. With permission.)

fabric and linen laminates is very negligible due to a very high crystallinity of linen fibers [17].

Duchemin et al. observed that a higher crystallinity index value (CrI) could be obtained by slowing the precipitation method than their counterpart obtained with the same dissolution time, but with faster precipitation. The slower precipitation route allows greater time for the cellulose chains to reassemble into a partly ordered paracrystalline state, resulting in higher CrI value [94].

In another work, Duchemim et al. showed that even after activation process, the MCC became slightly disordered and lower crystalline, with crystallinity index decreasing from 87% to 80%. Also, the orientation factor reduced from 0.84 to 0.73 [62].

Figure 7.15 shows the WAXD of ACC films, where the clearly visible Debye–Scherrer rings indicate the presence of cellulose I crystallites. However, the diffraction patterns of ACCs became more diffuse with reducing cellulose concentration, suggesting a decrease in crystallinity and portion of cellulose I by selective dissolving and resolidification process [28].

Gindl et al. [60,108] and Qi et al. [53] used WXRD images to measure the orientation of drawn and undrawn ACC films, using the equation of Harman's orientation factor (*f*). The Herman's orientation factor (*f*) was calculated from the azimuthal profile, assuming a cylindrical symmetry according to the following equations [53,108].

$$f = \frac{1}{2}\left(3\langle \cos^2 \gamma \rangle - 1\right) \tag{7.1}$$

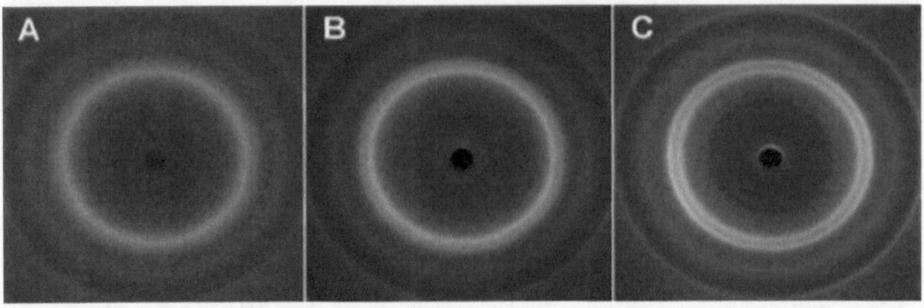

FIGURE 7.15
WAXD of ACC films, types A, B, and C, produced by partly dissolving MCC in LiCl/DMAc (concentration of cellulose in solvent for composite was in following order: A < B < C). (From Gindl, W. and Keckes, J., *Polymer*, 46, 10221, 2005. With permission.)

$$\langle \cos^2 \gamma \rangle = 1 - 2\langle \cos^2 \varnothing \rangle \tag{7.2}$$

$$\langle \cos^2 \varnothing \rangle = \frac{\int_0^{\pi/2} I(\varnothing) sin\varnothing \cos^2 \varnothing \, d\varnothing}{\int_0^{\pi/2} I(\varnothing) sin\varnothing \, d\varnothing} \tag{7.3}$$

Where ∅ represents the azimuthal angle and I(∅) is the intensity along the Debye–Scherrer ring. Maximum orientation parallel to the reference direction is found when $f = 1$. Whereas, $f = 0$ indicates random orientation [108]. Gindl et al. reported that after a wet drawing of prepared ACC film with a draw ratio of 1.5, the orientation factor changed from 0 to 0.0325, resulting in huge improvement in mechanical properties of ACC films [108].

Figure 7.16 shows a typical WXRD spectra for MCC, regenerated cellulose, and ACCs with varying content of cellulose I and cellulose II, produced by dissolving different amount of cellulose in solvent. The crystalline part of MCC powder corresponding to cellulose I showed highest scattering intensity at 22.7° (2 0 0). For regenerated cellulose, this peak was shifted to a lower value at 20.4° [28]. In some other studies, similar findings were reported by Duchemin et al. [38,62].

In Figure 7.16, "Composite A," where the ratio of cellulose I and cellulose II was 24/76, was very similar to regenerated cellulose, except a very small shoulder at scattering angle of 22.7° indicating the presence of cellulose I. This shoulder became more prominent in "Composite B" (ratio of cellulose I and cellulose II was 43/57) and finally developed into a clear intensity peak in "Composite C" (ratio of cellulose I and cellulose II was 59/41). The ratio of scattering intensity at 22.7° vs. intensity at 20.4°, that is, $I_{22.7}/I_{20.4}$ was an indicative of cellulose I vs. cellulose II content. This ratio was maximum for composite C due to lowest cellulose II content, indicating maximum crystallinity than other composites [28].

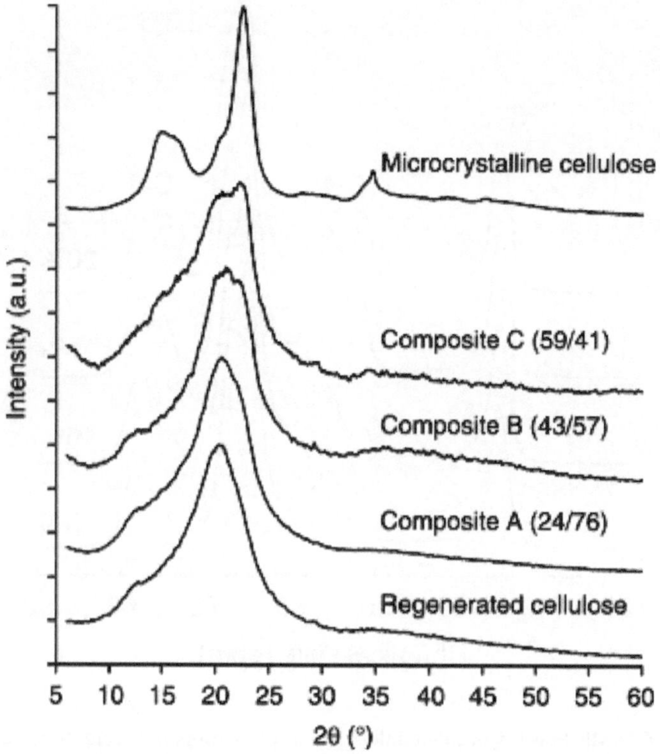

FIGURE 7.16
X-ray powder diffractogram for MCC, regenerated cellulose, and ACCs with varying cellulose I/cellulose II ratio. (From Gindl, W. and Keckes, J., *Polymer*, 46, 10221, 2005. With permission.)

In an interesting study, Gindl observed that during mechanical testing of ACCs, the cellulose crystallites are irreversibly oriented with their crystallographic C-axis and parallel to the force direction, resulting in increase of orientation factor of cellulose in ACCs. During tensile straining of ACCs containing both cellulose I and cellulose II, the increase in orientation for cellulose I part was higher compared to that of the cellulose II part [60].

7.10.2 CP/MAS^{13}C NMR Spectra Analysis

Solid-state ^{13}C NMR Spectra is one of the most powerful tools available for organic structure determination. It was used in few reported works [31,62,107] for characterizing molecular packing of cellulose or supramolecular nanostructure of cellulose.

Figure 7.17 shows typical NMR spectra of ACCs produced with varying cellulose concentration. In all spectra, two partly resolved C-4 signals were observed at 89 ppm and 84 ppm, which were assigned to crystallite-interior chains of cellulose I and crystallite-surface glucosyl residues, respectively

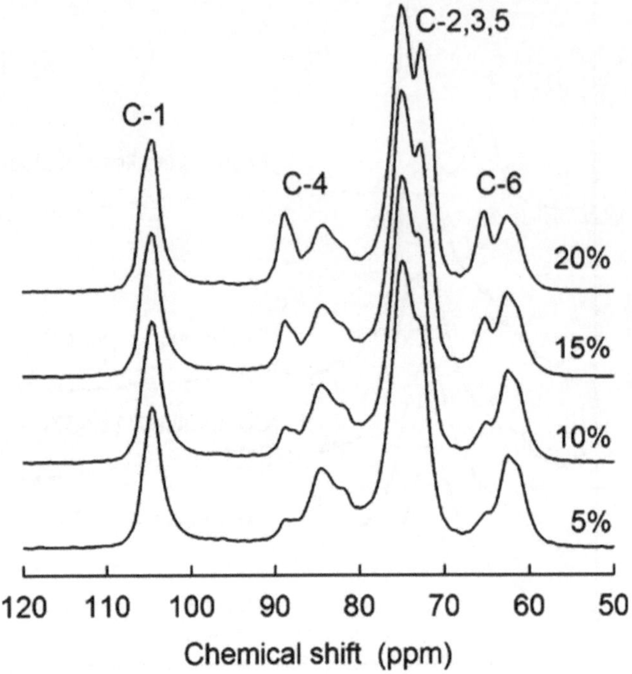

FIGURE 7.17
Solid-state ^{13}C NMR Spectra of all-cellulose composites prepared from activated cellulose mixed with solvent (LiCl/DMAc) at different weight % values (as shown above each trace) and left for 8 h of dissolution. (From Benoit, J.C.Z.D., et al., *Cellulose* 14, 311, 2007. With permission.)

[62,109]. Duchemim et al. reported that area under peaks at 89 ppm decreased with decreasing cellulose concentration (c) in the solvent (Figure 7.17), indicating decrease in crystallinity [62]. This result justifies the X-ray diffraction results for ACCs.

Newman et al. and Duchemin et al. reported that activated MCC exhibits slight change in^{13}C NMR spectrum showing a broader peak in the sub-spectrum of crystalline cellulose I. It indicates the change in molecular packing and transformation of cellulose I to a less-ordered form, justifying WAXD results [62,109].

According to Horii et al., the peak at 79–81 ppm is characteristic of the random coil conformation of dissolved cellulose [110] or due to regenerated amorphous cellulose [109]. Whereas, the chemical shift in a range of 84–90 ppm are characteristic of the liner chain conformation in crystalline cellulose [109–110].

Pang et al. observed that C-6 peak of the regenerated cellulose film shifts to 62.8 ppm as a single peak, as compared to the double peaks for native cellulose at 65.4 ppm. This would suggest that "trans-gauche" conformation of the C6-OH group for crystalline part of cellulose I changed into a "trans-gauche" conformation of cellulose II in regenerated films [107]. Pang et al.

also noticed that in comparison to the native cellulose, the C-4 peak for the regenerated films shifted to a lower value, that is, from 89.2 to 87.9, indicating huge decrease in crystallinity of cellulose due to the dissolution and regeneration process [107]. This observation supports the findings of WAXD spectra for ACCs.

7.10.3 FTIR Spectra Analysis

This characterization technique was utilized in few reported works [27,48,55,107] to obtain direct information on chemical changes in cellulose during dissolution and regeneration.

Figure 7.18 shows a typical FTIR spectrum in the spectra range 1800–600 cm^{-1} comparing the phase of cellulose in NaOH-treated hemp fiber, a typical Tencel fiber, and regenerated hemp cellulose film [48]. The 8% NaOH-treated fiber showed characteristic of cellulose I without a carbonyl peak at 1740 cm^{-1}, suggesting removal of non-cellulosic components (lignin, hemicellulose, and

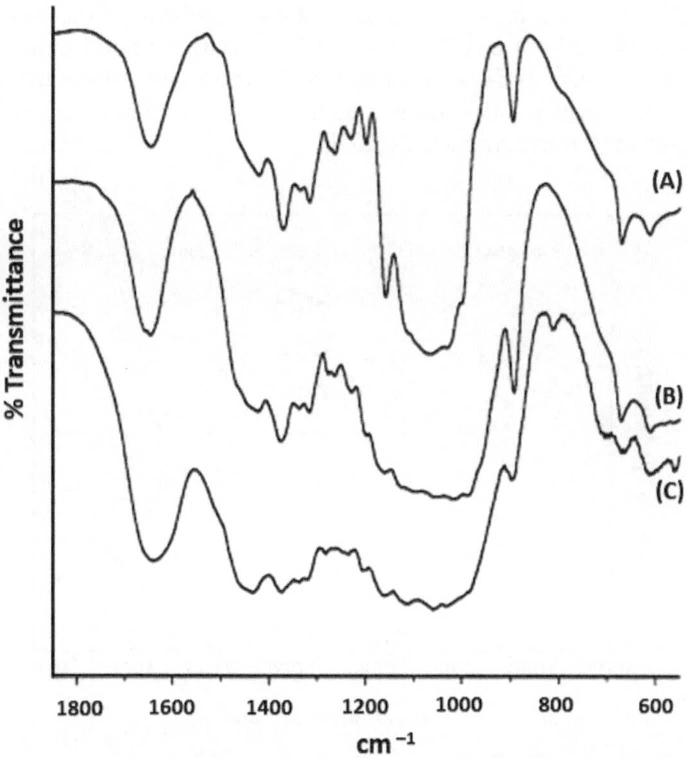

FIGURE 7.18
FTIR spectrum of: (A) regenerated hemp film, (B) Tencel fiber and (C) 8% NaOH-treated hemp fibers. (From Ouajai, S. and Shanks, R. A. *Compos. Sci.Technol.* 69, 2119, 2009. With permission.)

pectin) by alkali [43,48]. The band at 1643 cm^{-1} could be due to the water in the amorphous region of cellulosic fibers [43,107,111]. The absorbance of the band at 1430 cm^{-1} corresponded to a crystalline absorption and was closely related to the portion of cellulose I structure [38]. The peaks at 1375, 1335, 1315, and 1055 cm^{-1} were attributed to CH stretching, O–H bending vibration, CH_2 wagging, and C–O stretching in cellulose, respectively [107,111].

The cellulose I having a parallel-closed packing, the vibration of β-1, 4-glycosidic linkage at 893 cm^{-1} was limited, whereas for cellulose II, intensity of that vibration increased, due to its antiparallel chain alignment and loosely packed structure [48,59,112]. Therefore, the rise in the intensity of the band at 893 cm^{-1} in case of regenerated cellulose and Tencel suggested the transformation from cellulose I to cellulose II [48,55].

Wang et al. in their study used FTIR spectra (Figure 7.19) to characterize phase of cellulose in neat CNWs powder (cellulose I) and regenerated cellulose gel having different CNWs content. It was observed that when the CNWs concentration was low, the absorbance of the bands at 1430 and 710 cm^{-1} were very weak, indicating that CNWs were dispersed and embedded in the regenerated cellulose matrix. However, with increasing CNWs concentration, gradually it formed the continued phase, resulting in increased intensity of the peaks at 1430 and 710 cm^{-1} [55]. Pang et al. reported that compared to native cellulose, no new peaks appeared for the regenerated samples, indicating that no derivatization occurs during dissolution and regeneration of cellulose for manufacturing ACCs [107].

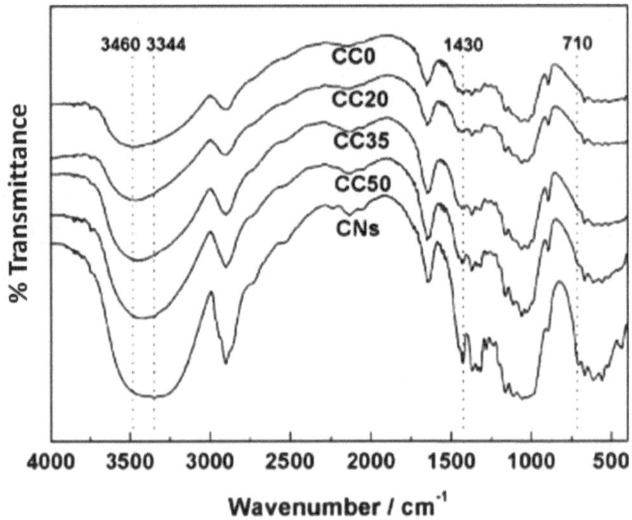

FIGURE 7.19
FTIR spectra of regenerated cellulose gels having different CNW contents and neat CNW powder. The codes CC0, CC20, CC35, and CC50 correspond to CN content of 0, 20, 35, and 50 wt%, respectively. (From Wang, Y. and Chen, L., *Carbohydr. Polym.*, 83, 1937, 2011. With permission.)

7.10.4 Raman Spectra

In few studies, Raman spectroscopy has been used to phase-characterize cellulose in ACCs [47,52,54]. Figure 7.20 shows typical Raman Spectra for pure MCC matrix, pure CNWs, and nanocomposites, with CNWs embedded in MCC matrix [10]. The most prominent peak in the Raman Spectra was found at 1095 cm^{-1} for all samples, which are due to C–O stretch mode along the cellulose backbone [113] or due to the stretching mode involving the glycosidic linkage (C–O–C) [114]. In case of CNWs, the absence of peak at 895^{-1} cm proved the presence of only cellulose I [52,54], which was first reported by Atalla et al. [115]. The Raman spectrum for the pure matrix material (Figure 7.20a) exhibited both peaks at 1095^{-1} cm and 895^{-1} cm, demonstrating the presence of chemically regenerated cellulose II. Similarly, the CNWs-embedded nanocomposites showed both peaks, indicating presence of both cellulose I and cellulose II. However, the peaks located at 1095 cm^{-1} and 895 cm^{-1} for pure matrix exhibited a strain-induced shift toward a lower wavenumber with addition of CNWs. The Raman peaks for nanocomposite not only shifted, but also asymmetrically broadened, indicating a nonuniform distribution of stress over the structure [54].

Qin et al. in their study used Raman Spectra to investigate molecular orientation of aligned ramie fiber–based mercerized and unmercerized ACCs. The difference in the characteristic spectra of these two ACCs was caused mainly by a change to a more oriented cellulose conformation (cellulose II) from cellulose I after mercerization [47].

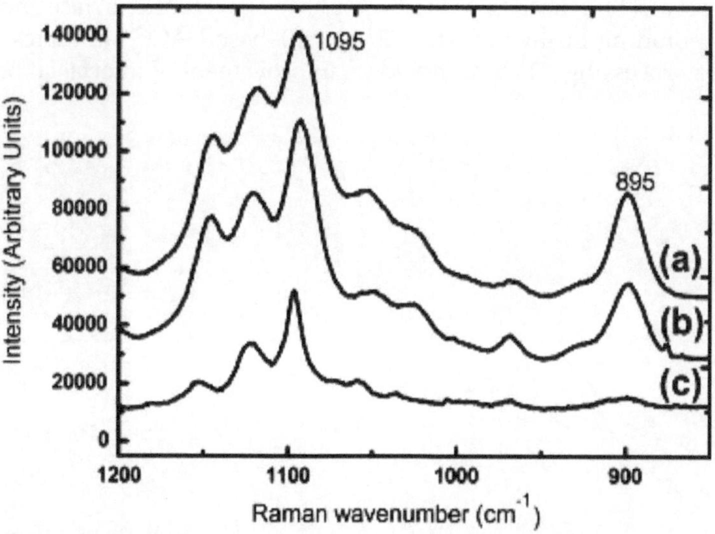

FIGURE 7.20
Raman Spectra for (a) activated MCC dissolved in LiCl/DMAc, (b) nanocomposite of activated MCC dissolved in LiCl/DMAc with CNWs (1.0 wt%), and (c) pure CNWs. (From Pullawan, T. et al., *Compos. Sci. Technol.*, 70, 2325, 2010. With permission.)

7.11 Microstructural Analysis of Different ACCs

The final properties of any composite material strongly depends on its morphology. Therefore, change in morphology of ACCs should be analyzed to understand its properties. Different microscopic techniques such as optical microscopy, scanning electron microscopy (SEM), transmission electron microscopy (TEM), and atomic force microscopy (AFM) can be utilized for analyzing microstructure of ACCs. In many literatures, the microstructure, internal voids, interfacial binding between matrix and reinforcement, dispersion of nano-reinforcing materials in matrix as well as the surface morphology of ACCs were analyzed mainly by using SEM [6,8–9,14,34,47,77,59,61,97,116], and in few cases, by using TEM [29,94], AFM [33,53,107,116], and OM [41,48]. The microstructures of ACCs mainly depends on different process parameters such as dissolution time, temperature, pressure, external forces, etc. Moreover, final microstructure of ACCs may change depending on solvent and the source/type of cellulosic material used in the reinforcement and matrix phase.

In general, with longer immersion time, more cellulose is dissolved and larger noncrystalline domains are formed, leaving smaller cellulose crystallites (fiber fraction) as undissolved parts. It has led to the formation of more homogenous structure with reduction of microvoids [9], as shown in the SEM micrographs (Figure 7.21). Qin et al. used SEM micrographs to observe cross-section of ACCs before and after mercerization, conforming an improved structure after mercerization (Figure 7.22) [47].

Huber et al. (2012) used SEM micrographs to analyze the surface and cross-sectional binding of rayon textile (2/2 twill)–based ACC laminates, before and after processing. They observed an improvement of interfacial bonding

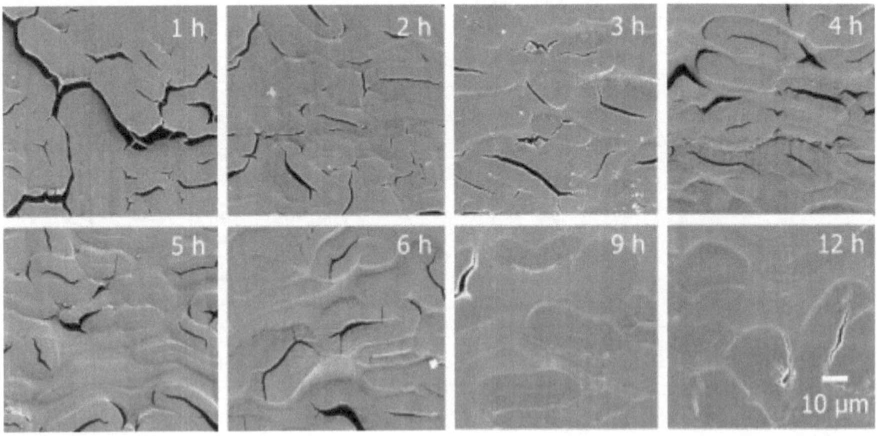

FIGURE 7.21
SEM micrographs of the cross-section of ACCs produced with varying immersion time. (From Soykeabkaew, N. et al., *Compos. Sci. Technol.* 68, 2201, 2008. With permission.)

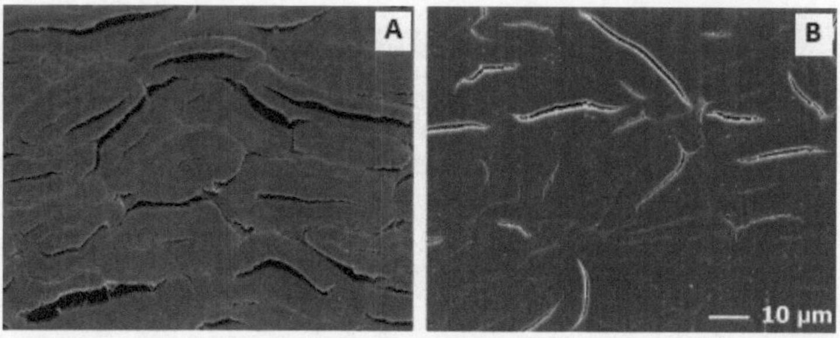

FIGURE 7.22
SEM micrographs for cross-section of (A) unmercerized and (B) mercerized ACCs prepared with 4% cellulose concentration in solution. (From Qin, C. et al., *Carbohydr. Polym.*, 71, 458, 2008. With permission.)

by applying pressure and also by increasing the dissolution time [17,97]. In a biodegradability study, Kalka et al. observed the surface of ACC laminates under SEM before and after soil-burial test. In the SEM micrograph of soil-burial tested sample, some clear voids were observed, which might have formed due to bio-degradation by organisms [6].

The structure of ACNCs depends on the size of the reinforcing part (undissolved cellulose) [49]. Additionally, successful transfer of the nanomaterial's properties into the nanocomposites depends on the proper dispersion of nano-reinforcing material in the matrix. Figure 7.23 (A & B) shows the typical TEM and AFM micrographs, respectively, representing the dispersion of NCC and CNWs, respectively, in the cellulose matrix [29,33].

SEM micrographs in different reported works revealed that cross-section as well as surface morphology are relatively smooth for ACNCs produced with lower nanowhiskers compared to the ACNCs produced with higher CWs content [30]. Ma et al. observed that pure regenerated cellulose film (Com-N0) or Com-N5 exhibited a smooth surface (Figure 7.24), suggesting

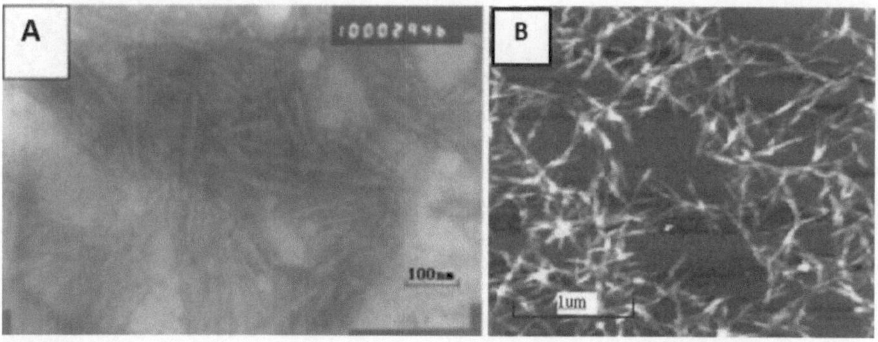

FIGURE 7.23
(A) TEM image showing dispersion of NCC dried from aqueous suspension [29], (B) AFM image showing CNWs dried from aqueous suspension on a glass slide. (From Li, D. et al., *Polym. Bulletin*, 65, 635, 2010. With permission.)

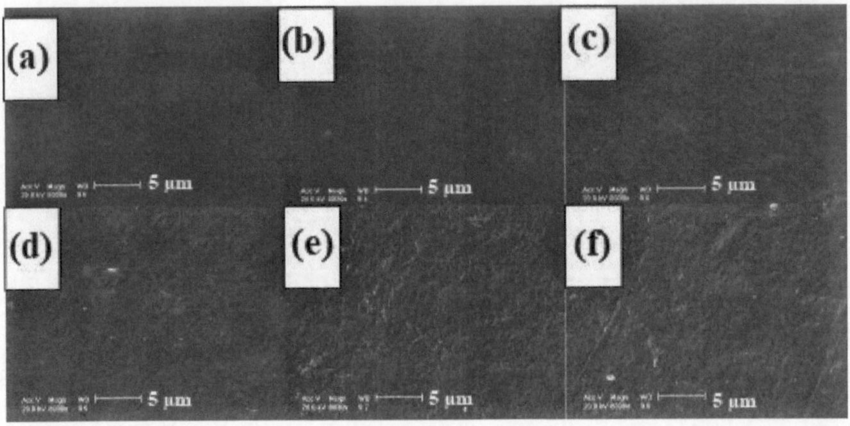

FIGURE 7.24
SEM images of surface of ACNC films: (a) Com-N0 (b) Com-N5 (c) Com-N10 (d) Com-N15 (e) Com-N20 (f) Com-N25 (where the numbers indicates the wt % of NCC). (From Ma, H. et al., *Carbohydr. Polym.*, 84, 383, 2011. With permission.)

that the cellulose nanoparticles were properly embedded into the regenerated cellulose matrix because of strong interaction between cellulose I and cellulose II. However, when the content of NCC gradually increased from 10% to 25%, the surface of nanocomposite films became rougher, indicating the possibility of phase separation [29].

The aspect ratio of the prepared nano-reinforcing materials can also be measured using these characterization techniques. Ma et al. used SEM to measure aspect ratio of the prepared needle-like nanowhiskers [29], while Qi et al. used AFM images for this purpose [53]. In a study by Li et al., the SEM micrographs were used to investigate the effect of applied magnetic field, to align the CNWs in the perpendicular direction to the magnetic field [33].

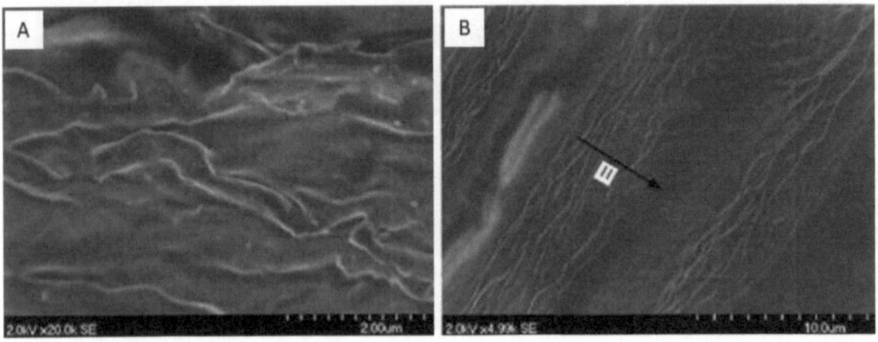

FIGURE 7.25
FE-SEM micrographs of ACNCs fabricated (A) in the absence of magnetic field and (B) with application of external magnetic field. (From Li, D. et al., *Polym. Bulletin*, 65, 635, 2010. With permission.)

The microstructures of ACNCs fabricated under magnetic field are shown in Figure 7.25B, where the CNWs are oriented in a particular direction. On the contrary, in case of nanocomposites that were prepared in the absence of a magnetic field, the CNWs were not oriented (Figure 7.25A).

7.12 Conclusion

Nowadays, due to the growing environmental awareness, the demand of environmentally friendly materials has significantly risen. From this point of view, fully cellulose-based ACCs have great importance as biodegradable "green composite." The ACCs, other than having similar or superior mechanical properties compared to other bio-composites, have overcome matrix-reinforcement adhesion. Cellulosic materials can be incorporated in many forms with different solvent systems to produce ACCs. However, till date ACCs have been mainly confined to the laboratory level and have not been explored much on the industrial scale. However, considering the benefits of ACCs, it is expected that ACCs will get more importance in near future.

Though the concept of ACC is new, many process techniques have been explored by different researchers to synthesize ACC from nano to macro form. Bacterial cellulose or nano-cellulosic fibers to macro-sized cellulosic fiber-based fabrics have been used to obtain required structure of ACCs. During processing, the dissolution and regeneration processes alter crystalline phase of cellulose in ACCs resulting in change of crystallinity, crystal size, and orientation. Depending on process conditions, the microstructure of ACCs can also be varied. The final microstructure has a great role in controlling the properties of ACCs, which will be thoroughly discussed in the next chapter.

References

1. Han, D., and L. Yan. Preparation of all-cellulose composite by selective dissolving of cellulose surface in PEG/NaOH aqueous solution. *Carbohydrate Polymers* 79, no. 3 (2010): 614–619.
2. John, M. J., and S. Thomas. Biofibres and biocomposites. *Carbohydrate Polymers* 71, no. 3 (2008): 343–364.
3. Soykeabkaew, N., T. Nishino, and T. Peijs. All-cellulose composites of regenerated cellulose fibres by surface selective dissolution. *Composites Part A: Applied Science and Manufacturing* 40, no. 4 (2009): 321–328.
4. Sujito, S., and J. K. Pandey, Mechanical properties of "green" composites based on poly-lactic acid resin and short single bamboo fibres. In *Proceedings of 18th International Conference on composite materials*, Jeju Island, Korea, 2011.

5. Kabir, M. M., H. Wang, K. T. Lau, and F. Cardona. Chemical treatments on plant-based natural fibre reinforced polymer composites: An overview. *Composites Part B: Engineering* 43, no. 7 (2012): 2883–2892.
6. Kalka, S., T. Huber, J. Steinberg, K. Baronian, J. Müssig, and M. P. Staiger. Biodegradability of all-cellulose composite laminates. *Composites Part A: Applied Science and Manufacturing* 59(2014): 37–44.
7. Madsen, B., and H. Lilholt. Physical and mechanical properties of unidirectional plant fibre composites: An evaluation of the influence of porosity. *Composites Science and Technology* 63, no. 9 (2003): 1265–1272.
8. Nishino, T., I. Matsuda, and K. Hirao. All-cellulose composite. *Macromolecules* 37, no. 20 (2004): 7683–7687.
9. Soykeabkaew, N., N. Arimoto, T. Nishino, and T. Peijs. All-cellulose composites by surface selective dissolution of aligned ligno-cellulosic fibres. *Composites Science and Technology* 68, no. 10–11 (2008): 2201–2207.
10. Bledzki, A. K., and J. Gassan. Composites reinforced with cellulose based fibres. *Progress in Polymer Science* 24, no. 2 (1999): 221–274.
11. Nevell, T.P and S. H. Zeronian. *Cellulose Chemistry and Its Applications.* Chichester, UK: Ellis Horwood Ltd, 1985.
12. Pinkert, A., K. N. Marsh, S. Pang, and M. P. Staiger. Ionic liquids and their interaction with cellulose. *Chemical Reviews* 109, no. 12 (2009): 6712–6728.
13. Klemm, D., B. Heublein, H. P. Fink, and A. Bohn. Cellulose: Fascinating biopolymer and sustainable raw material. *Angewandte Chemie International Edition* 44, no. 22 (2005): 3358–3393.
14. Kalia, S., B. S. Kaith, and I. Kaur, *Cellulose Fibers: Bio-and Nano-Polymer Composites: Green Chemistry and Technology.* New York: Springer Science & Business Media, 2011.
15. Finkenstadt, V. L., and R. P. Millane. Crystal structure of Valonia cellulose Iβ. *Macromolecules* 31, no. 22 (1998): 7776–7783.
16. Kaplan, D. L. Introduction to biopolymers from renewable resources. In *Biopolymers from Renewable Resources.* In Macromolecular Systems—Materials Approach, ed. D. L. Kaplan. Berlin, Heidelberg: Springer, 1998.
17. Huber, T., S. Pang, and M. P. Staiger. All-cellulose composite laminates. *Composites Part A: Applied Science and Manufacturing* 43, no. 10 (2012): 1738–1745.
18. Nishino, T., K. Takano, and K. Nakamae. Elastic modulus of the crystalline regions of cellulose polymorphs. *Journal of Polymer Science Part B: Polymer Physics* 33, no. 11 (1995): 1647–1651.
19. Gassan, J., A. Chate, and A. K. Bledzki. Calculation of elastic properties of natural fibers. *Journal of Materials Science* 36, no. 15 (2001): 3715–3720.
20. Chen, W., G. C. Lickfield, and C. Q. Yang. Molecular modeling of cellulose in amorphous state. Part I: model building and plastic deformation study. *Polymer* 45, no. 3 (2004): 1063–1071.
21. Lyons, W. J. Crystal density of native cellulose. *The Journal of Chemical Physics* 9, no. 4 (1941): 377–378.
22. Jawaid, M. H. P. S., and H. A. Khalil. Cellulosic/synthetic fibre reinforced polymer hybrid composites: A review. *Carbohydrate Polymers* 86, no. 1 (2011): 1–18.
23. Jayaraman, K. Manufacturing sisal–polypropylene composites with minimum fibre degradation. *Composites Science and Technology* 63, no. 3–4 (2003): 367–374.

24. Huber, T., J. Müssig, O. Curnow, S. Pang, S. Bickerton, and M. P. Staiger. A critical review of all-cellulose composites. *Journal of Materials Science* 47, no. 3 (2012): 1171–1186.
25. Huber, T., S. Bickerton, J. Müssig, S. Pang, and M. P. Staiger. Flexural and impact properties of all-cellulose composite laminates. *Composites Science and Technology* 88(2013): 92–98.
26. Gindl-Altmutter, W., J. Keckes, J. Plackner, F. Liebner, K. Englund, and M. P. Laborie. All-cellulose composites prepared from flax and lyocell fibres compared to epoxy–matrix composites. *Composites Science and Technology* 72, no. 11 (2012): 1304–1309.
27. Wu, R. L., X. L. Wang, F. Li, H. Z. Li, and Y. Z. Wang. Green composite films prepared from cellulose, starch and lignin in room-temperature ionic liquid. *Bioresource Technology* 100, no. 9 (2009): 2569–2574.
28. Gindl, W., and J. Keckes. All-cellulose nanocomposite. *Polymer* 46, no. 23 (2005): 10221–10225.
29. Ma, H., B. Zhou, H. S. Li, Y. Q. Li, and S. Y. Ou. Green composite films composed of nanocrystalline cellulose and a cellulose matrix regenerated from functionalized ionic liquid solution. *Carbohydrate Polymers* 84, no. 1 (2011): 383–389.
30. Zhao, J., X. He, Y. Wang, W. Zhang, X. Zhang, X. Zhang, Y. Deng, and C. Lu. Reinforcement of all-cellulose nanocomposite films using native cellulose nanofibrils. *Carbohydrate Polymers* 104(2014): 143–150.
31. Nilsson, H., S. Galland, P. T. Larsson, E. K. Gamstedt, T. Nishino, L. A. Berglund, and T. Iversen. A non-solvent approach for high-stiffness all-cellulose biocomposites based on pure wood cellulose. *Composites Science and Technology* 70, no. 12 (2010): 1704–1712.
32. Gindl, W., T. Schöberl, and J. Keckes. Structure and properties of a pulp fibre-reinforced composite with regenerated cellulose matrix. *Applied Physics A* 83, no. 1 (2006): 19–22.
33. Li, D., Z. Liu, M. Al-Haik, M. Tehrani, F. Murray, R. Tannenbaum, and H. Garmestani. Magnetic alignment of cellulose nanowhiskers in an all-cellulose composite. *Polymer Bulletin* 65, no. 6 (2010): 635–642.
34. Schuermann, H., T. Huber, and M. P. Staiger. Prepreg style fabrication of all cellulose composites. In *Proceedings of 19th international conference on composite materials, Canada*, pp. 5626–5634. 2013.
35. Yang, Q., A. Lue, and L. Zhang. Reinforcement of ramie fibers on regenerated cellulose films. *Composites Science and Technology* 70, no. 16 (2010): 2319–2324.
36. Piltonen, P., N. C. Hildebrandt, B. Westerlind, J. P. Valkama, T. Tervahartiala, and M. Illikainen. Green and efficient method for preparing all-cellulose composites with NaOH/urea solvent. *Composites Science and Technology* 135(2016): 153–158.
37. Yang, Q., H. Qi, A. Lue, K. Hu, G. Cheng, and L. Zhang. Role of sodium zincate on cellulose dissolution in NaOH/urea aqueous solution at low temperature. *Carbohydrate Polymers* 83, no. 3 (2011): 1185–1191.
38. Duchemin, B. J. C., A. P. Mathew, and K. Oksman. All-cellulose composites by partial dissolution in the ionic liquid 1-butyl-3-methylimidazolium chloride. *Composites Part A: Applied Science and Manufacturing* 40, no. 12 (2009): 2031–2037.
39. Nishino, T., and N. Arimoto. All-cellulose composite prepared by selective dissolving of fiber surface. *Biomacromolecules* 8, no. 9 (2007): 2712–2716.

40. Vo, L. T. T., B. Široká, A. P. Manian, H. Duelli, B. MacNaughtan, M. F. Noisternig, U. J. Griesser, and T. Bechtold. All-cellulose composites from woven fabrics. *Composites Science and Technology* 78(2013): 30–40.
41. Adak, B., and S. Mukhopadhyay. Effect of the dissolution time on the structure and properties of lyocell-fabric-based all-cellulose composite laminates. *Journal of Applied Polymer Science* 133, no. 19 (2016).
42. Adak, B., and S. Mukhopadhyay. Effect of pressure on structure and properties of lyocell fabric-based all-cellulose composite laminates. *The Journal of the Textile Institute* 108, no. 6 (2017): 1010–1017.
43. Adak, B., and S. Mukhopadhyay. Jute based all-cellulose composite laminates. *Indian Journal of Fibre & Textile Research (IJFTR)* 41, no. 4 (2016): 380–384.
44. Adak, B., and S. Mukhopadhyay. A comparative study on lyocell-fabric based all-cellulose composite laminates produced by different processes. *Cellulose* 24, no. 2 (2017): 835–849.
45. Shibata, M., N. Teramoto, T. Nakamura, and Y. Saitoh. All-cellulose and all-wood composites by partial dissolution of cotton fabric and wood in ionic liquid. *Carbohydrate polymers* 98, no. 2 (2013): 1532–1539.
46. Huber, T. Processing of all cellulose composites via an ionic liquid route. PhD diss., University of Canterbury, New Zealand, 2012.
47. Qin, C., N. Soykeabkaew, N. Xiuyuan, and T. Peijs. The effect of fibre volume fraction and mercerization on the properties of all-cellulose composites. *Carbohydrate Polymers* 71, no. 3 (2008): 458–467.
48. Ouajai, S., and R. A. Shanks. Preparation, structure and mechanical properties of all-hemp cellulose biocomposites. *Composites Science and Technology* 69, no. 13 (2009): 2119–2126.
49. Yousefi, H., T. Nishino, M. Faezipour, G. Ebrahimi, and A. Shakeri. Direct fabrication of all-cellulose nanocomposite from cellulose microfibers using ionic liquid-based nanowelding. *Biomacromolecules* 12, no. 11 (2011): 4080–4085.
50. Yousefi, H., T. Nishino, A. Shakeri, M. Faezipour, G. Ebrahimi, and M. Kotera. Water-repellent all-cellulose nanocomposite using silane coupling treatment. *Journal of Adhesion Science and Technology* 27, no. 12 (2013): 1324–1334.
51. Ghaderi, M., M. Mousavi, H. Yousefi, and M. Labbafi. All-cellulose nanocomposite film made from bagasse cellulose nanofibers for food packaging application. *Carbohydrate Polymers* 104(2014): 59–65.
52. Pullawan, T., A. N. Wilkinson, L. N. Zhang, and S. J. Eichhorn. Deformation micromechanics of all-cellulose nanocomposites: comparing matrix and reinforcing components. *Carbohydrate Polymers* 100(2014): 31–39.
53. Qi, H., J. Cai, L. Zhang, and S. Kuga. Properties of films composed of cellulose nanowhiskers and a cellulose matrix regenerated from alkali/urea solution. *Biomacromolecules* 10, no. 6 (2009): 1597–1602.
54. Pullawan, T., A. N. Wilkinson, and S. J. Eichhorn. Discrimination of matrix–fibre interactions in all-cellulose nanocomposites. *Composites Science and Technology* 70, no. 16 (2010): 2325–2330.
55. Wang, Y., and L. Chen. Impacts of nanowhisker on formation kinetics and properties of all-cellulose composite gels. *Carbohydrate Polymers* 83, no. 4 (2011): 1937–1946.
56. Li, D., X. Sun, and M. A. Khaleel. Materials design of all-cellulose composite using microstructure based finite element analysis. *Journal of Engineering Materials and Technology* 134, no. 1 (2012): 010911.

57. Bondeson, D., P. Syre, and K. O. Niska. All cellulose nanocomposites produced by extrusion. *Journal of Biobased Materials and Bioenergy* 1, no. 3 (2007): 367–371.
58. Pullawan, T., A. N. Wilkinson, and S. J. Eichhorn. Orientation and deformation of wet-stretched all-cellulose nanocomposites. *Journal of Materials Science* 48, no. 22 (2013): 7847–7855.
59. Soykeabkaew, N., C. Sian, S. Gea, T. Nishino, and T. Peijs. All-cellulose nanocomposites by surface selective dissolution of bacterial cellulose. *Cellulose* 16, no. 3 (2009): 435–444.
60. Gindl, W., K. J. Martinschitz, P. Boesecke, and J. Keckes. Structural changes during tensile testing of an all-cellulose composite by in situ synchrotron X-ray diffraction. *Composites Science and Technology* 66, no. 15 (2006): 2639–2647.
61. Duchemin, B. J. C., M. P. Staiger, N. Tucker, and R. H. Newman. Aerocellulose based on all-cellulose composites. *Journal of Applied Polymer Science* 115, no. 1 (2010): 216–221.
62. Benoît, J. C. Z. D., R. H. Newman, and M. P. Staiger. Phase transformations in microcrystalline cellulose due to partial dissolution. *Cellulose* 14, no. 4 (2007): 311–320.
63. Lourdin, D., J. Peixinho, J. Bréard, B. Cathala, E. Leroy, and B. Duchemin. Concentration driven cocrystallisation and percolation in all-cellulose nanocomposites. *Cellulose* 23, no. 1 (2016): 529–543.
64. Nakagaito, A. N., and H. Yano. Novel high-strength biocomposites based on microfibrillated cellulose having nano-order-unit web-like network structure. *Applied Physics A* 80, no. 1 (2005): 155–159.
65. Yano, H., J. Sugiyama, A. N. Nakagaito, M. Nogi, T. Matsuura, M. Hikita, and K. Handa. Optically transparent composites reinforced with networks of bacterial nanofibers. *Advanced Materials* 17, no. 2 (2005): 153–155.
66. Nakagaito, A. N., S. Iwamoto, and H. Yano. Bacterial cellulose: The ultimate nano-scalar cellulose morphology for the production of high-strength composites. *Applied Physics A* 80, no. 1 (2005): 93–97.
67. Hsieh, Y. C., H. Yano, M. Nogi, and S. J. Eichhorn. An estimation of the Young's modulus of bacterial cellulose filaments. *Cellulose* 15, no. 4 (2008): 507–513.
68. Chen, Y., C. Liu, P. R. Chang, X. Cao, and D. P. Anderson. Bionanocomposites based on pea starch and cellulose nanowhiskers hydrolyzed from pea hull fibre: effect of hydrolysis time. *Carbohydrate Polymers* 76, no. 4 (2009): 607–615.
69. Lu, Y., L. Weng, and X. Cao. Morphological, thermal and mechanical properties of ramie crystallites—reinforced plasticized starch biocomposites. *Carbohydrate Polymers* 63, no. 2 (2006): 198–204.
70. Hepworth, D. G., and D. M. Bruce. A method of calculating the mechanical properties of nanoscopic plant cell wall components from tissue properties. *Journal of Materials Science* 35, no. 23 (2000): 5861–5865.
71. Hattori, M., Y. Shimaya, and M. Saito. Solubility and dissolved cellulose in aqueous calcium-and sodium-thiocyanate solution. *Polymer Journal* 30, no. 1 (1998): 49.
72. Ostlund, Å., D. Lundberg, L. Nordstierna, K. Holmberg, and M. Nydén. Dissolution and gelation of cellulose in TBAF/DMSO solutions: The roles of fluoride ions and water. *Biomacromolecules* 10, no. 9 (2009): 2401–2407.
73. Cuculo, J. A., C. B. Smith, U. Sangwatanaroj, E. O. Stejskal, and S. S. Sankar. A study on the mechanism of dissolution of the cellulose/NH3/NH4SCN system. II. *Journal of Polymer Science Part A: Polymer Chemistry* 32, no. 2 (1994): 241–247.

74. Jiao, L., J. Ma, and H. Dai. Preparation and characterization of self-reinforced antibacterial and oil-resistant paper using a NaOH/Urea/ZnO solution. *PloS One* 10, no. 10 (2015): e0140603.
75. Sen, S., J. D. Martin, and D. S. Argyropoulos. Review of cellulose non-derivatizing solvent interactions with emphasis on activity in inorganic molten salt hydrates. *ACS Sustainable Chemistry & Engineering* 1, no. 8 (2013): 858–870.
76. Fink, H. P., P. Weigel, H. J. Purz, and J. Ganster. Structure formation of regenerated cellulose materials from NMMO-solutions. *Progress in Polymer Science* 26, no. 9 (2001): 1473–1524.
77. Zhao, Q., R. C. M. Yam, B. Zhang, Y. Yang, X. Cheng, and R. K. Y. Li. Novel all-cellulose ecocomposites prepared in ionic liquids. *Cellulose* 16, no. 2 (2009): 217–226.
78. Zhou, J., and L. Zhang. Solubility of cellulose in NaOH/urea aqueous solution. *Polymer Journal* 32, no. 10 (2000): 866.
79. Lewin, M. *Hand Book of Fibre Chemistry*. 3rd edition, Boca Raton, London, New York: CRC Press, Taylor & Francis group, 2007.
80. Seddon, K. R., A. Stark, and M. J. Torres. Influence of chloride, water, and organic solvents on the physical properties of ionic liquids. *Pure and Applied Chemistry* 72, no. 12 (2000): 2275–2287.
81. Feng, L., and Z. L. Chen. Research progress on dissolution and functional modification of cellulose in ionic liquids. *Journal of Molecular Liquids* 142, no. 1–3 (2008): 1–5.
82. Swatloski, R. P., J. D. Holbrey, and R. D. Rogers. Ionic liquids are not always green: hydrolysis of 1-butyl-3-methylimidazolium hexafluorophosphate. *Green Chemistry* 5, no. 4 (2003): 361–363.
83. Budtova, T., and P. Navard. Cellulose in NaOH–water based solvents: a review. *Cellulose* 23, no. 1 (2016): 5–55.
84. Zhang, S., W. C. Wang, F. X. Li, and J. Y. Yu. Swelling and dissolution of cellulose in NaOH aqueous solvent systems. *Cellulose Chemistry and Technology* 47, no. 9–10 (2013): 671–679.
85. Medronho, B., A. Romano, M. G. Miguel, L. Stigsson, and B. Lindman. Rationalizing cellulose (in) solubility: reviewing basic physicochemical aspects and role of hydrophobic interactions. *Cellulose* 19, no. 3 (2012): 581–587.
86. Lindman, B., B. Medronho, L. Alves, C. Costa, H. Edlund, and M. Norgren. The relevance of structural features of cellulose and its interactions to dissolution, regeneration, gelation and plasticization phenomena. *Physical Chemistry Chemical Physics* 19, no. 35 (2017): 23704–23718.
87. Medronho, B., and B. Lindman. Competing forces during cellulose dissolution: from solvents to mechanisms. *Current Opinion in Colloid & Interface Science* 19, no. 1 (2014): 32–40.
88. Medronho, B., and B. Lindman. Brief overview on cellulose dissolution/regeneration interactions and mechanisms. *Advances in Colloid and Interface Science* 222(2015): 502–508.
89. Franks, N. E., and J. K. Varga. Process for making a shapeable cellulose and shaped cellulose products. U.S. Patent 4,196,282, issued April 1, 1980.
90. Swatloski, R. P., S. K. Spear, J. D. Holbrey, and R. D. Rogers. Dissolution of cellose with ionic liquids. *Journal of the American Chemical Society* 124, no. 18 (2002): 4974–4975.

91. Guillen, G. R., Y. Pan, M. Li, and E. M. V. Hoek. Preparation and characterization of membranes formed by nonsolvent induced phase separation: A review. *Industrial & Engineering Chemistry Research* 50, no. 7 (2011): 3798–3817.
92. Qi, H. Novel Regenerated Cellulosic Materials. In *Novel Functional Materials Based on Cellulose*, pp. 25–43. Cham: Springer, 2017.
93. Cai, J., and L. Zhang. Unique gelation behavior of cellulose in NaOH/urea aqueous solution. *Biomacromolecules* 7, no. 1 (2006): 183–189.
94. Duchemin, B. J. C., R. H. Newman, and M. P. Staiger. Structure–property relationship of all-cellulose composites. *Composites Science and Technology* 69, no. 7–8 (2009): 1225–1230.
95. Sirviö, J. A., M. Visanko, and N. C. Hildebrandt. Rapid preparation of all-cellulose composites by solvent welding based on the use of aqueous solvent. *European Polymer Journal* 97(2017): 292–298.
96. Kistler, S. S. Coherent expanded aerogels and jellies. *Nature* 127, no. 3211 (1931): 741.
97. Huber, T., S. Bickerton, J. Müssig, S. Pang, and M. P. Staiger. Solvent infusion processing of all-cellulose composite materials. *Carbohydrate Polymers* 90, no. 1 (2012): 730–733.
98. Dormanns, J. W., J. Schuermann, J. Müssig, B. J. C. Duchemin, and M. P. Staiger. Solvent infusion processing of all-cellulose composite laminates using an aqueous NaOH/urea solvent system. *Composites Part A: Applied Science and Manufacturing* 82(2016): 130–140.
99. Brown, W. F. Vulcanized fibre-an old material with a new relevancy. In *Electrical Insulation Conference and Electrical Manufacturing & Coil Winding Conference, 1999. Proceedings*, pp. 309–312. IEEE, 1999.
100. Svoboda, M. A., R. W. Lang, R. Bramsteidl, M. Ernegg, and W. Stadlbauer. Zelfo—An Engineering Material Fully Based on Renewable Resources. *Molecular Crystals and Liquid Crystals Science and Technology. Section A. Molecular Crystals and Liquid Crystals* 353, no. 1 (2000): 47–58.
101. Sugiyama, J., H. Chanzy, and G. Maret. Orientation of cellulose microcrystals by strong magnetic fields. *Macromolecules* 25, no. 16 (1992): 4232–4234.
102. Stephen, E., A. Wilkinson, and T. Pullawan. Orientation of Cellulose Nanofibers Using Magnetic Fields and Wet-Stretching. In *Proceedings of The Fiber Society 2012 Fall Meeting and Technical Conference, Boston, USA*, November 7–9, 2012.
103. Bordel, D., J. L. Putaux, and L. Heux. Orientation of native cellulose in an electric field. *Langmuir* 22, no. 11 (2006): 4899–4901.
104. Ebeling, T., M. Paillet, R. Borsali, O. Diat, A. Dufresne, J. Y. Cavaille, and H. Chanzy. Shear-induced orientation phenomena in suspensions of cellulose microcrystals, revealed by small angle X-ray scattering. *Langmuir* 15, no. 19 (1999): 6123–6126.
105. Goutianos, S., R. Arévalo, B. F. Sørensen, and T. Peijs. Effect of processing conditions on fracture resistance and cohesive laws of binderfree all-cellulose composites. *Applied Composite Materials* 21, no. 6 (2014): 805–825.
106. Peces, R. A. Preparation and Characterisation of Binder-Free All-Cellulose Composites. PhD diss., Queen Mary University of London, 2014.
107. Pang, J. H., X. Liu, M. Wu, Y. Y. Wu, X. M. Zhang, and R. C. Sun. Fabrication and characterization of regenerated cellulose films using different ionic liquids. *Journal of Spectroscopy* 2014(2014).

108. Gindl, W., and J. Keckes. Drawing of self-reinforced cellulose films. *Journal of Applied Polymer Science* 103, no. 4 (2007): 2703–2708.
109. Newman, R. H., and J. A. Hemmingson. Carbon-13 NMR distinction between categories of molecular order and disorder in cellulose. *Cellulose* 2, no. 2 (1995): 95–110.
110. Horii, F., A. Hirai, and R. Kitamaru. Cross-polarization-magic angle spinning carbon-13 NMR approach to the structural analysis of cellulose. 1987.
111. Carrillo, F., X. Colom, J. J. Sunol, and J. Saurina. Structural FTIR analysis and thermal characterisation of lyocell and viscose-type fibres. *European Polymer Journal* 40, no. 9 (2004): 2229–2234.
112. Nelson, M. L., and R. T. O'Connor. Relation of certain infrared bands to cellulose crystallinity and crystal lattice type. Part II. A new infrared ratio for estimation of crystallinity in celluloses I and II. *Journal of Applied Polymer Science* 8, no. 3 (1964): 1325–1341.
113. Wiley, J. H., and R. H. Atalla. Band assignments in the Raman spectra of celluloses. *Carbohydrate Research* 160(1987): 113–129.
114. Gierlinger, N., M. Schwanninger, A. Reinecke, and I. Burgert. Molecular changes during tensile deformation of single wood fibers followed by Raman microscopy. *Biomacromolecules* 7, no. 7 (2006): 2077–2081.
115. Atalla, R. H., and B. E. Dimick. Raman-spectral evidence for differences between the conformations of cellulose I and cellulose II. *Carbohydrate Research* 39, no. 1 (1975): C1–C3.
116. Adak, B., and S. Mukhopadhyay. All-cellulose composite laminates with low moisture and water sensitivity. *Polymer* 141(2018): 79–85.

8

Properties of Non-Derivatized All-Cellulose Composites

8.1 Introduction

All-cellulose composites are one of the emerging categories of biocomposites having versatile properties. As both the matrix and the reinforcing phase of ACCs are cellulosic materials, a strong interfacial adhesion could be expected between them. Very high fiber volume fraction has been achieved with ACCs. In spite of very low fraction of matrix phase in most of the ACCs, wetting of fiber into matrix is not an issue, due to chemical similarity of these two phases. The good compatibility assists in very efficient stress transfer between matrix and reinforcement of ACCs, resulting in high values of strength and modulus [1–3].

The evaluation of mechanical properties could provide important information about internal microstructure of composite materials as these two are strongly interrelated. Mechanical properties of fiber-reinforced composites depends on many factors—intrinsic properties of fibers and polymers, geometry and orientation of reinforcement, types of matrix, matrix-reinforcement adhesion, and fiber volume fraction [4–6]. The mechanical properties of ACCs depend on inherent properties of reinforcing cellulose materials, dissolution time, and regeneration conditions. There are two contradictory effects of dissolution time in controlling the mechanical properties of ACCs. These are: (i) with increasing dissolution time, dissolution of more cellulose causes a reduction in basic strength of the cellulosic material, whereas (ii) increase in amount of dissolved cellulose results in a better fiber–matrix adhesion, which leads to an improvement of mechanical properties [7,8]. The alignment or orientation of reinforcing materials also has a great effect on mechanical properties of ACCs [9]. On the basis of alignment of reinforcing materials, the composites may have isotropic or unidirectional character.

In addition to mechanical properties, some other properties are also important considering different applications of ACCs. For example, the viscoelastic properties and thermal properties of composites are important to understand their behavior when these composites are used against some heat source or thermal loading.

Till date, research has reported different combination of fibers, matrices, and solvent systems to get a wide range of properties for ACCs. This chapter has covered mainly the mechanical properties, thermal properties, viscoelastic properties, and other miscellaneous properties of non-derivatized ACCs.

8.2 Mechanical Properties

8.2.1 Factors Affecting Mechanical Properties of ACCs

There are many factors that control mechanical properties (tensile, flexural, and impact properties) of ACCs. Among these, the main factors are:

- Intrinsic mechanical properties of cellulosic raw materials
- Types, sources, qualities, and many other characteristics like crystallinity, molecular weight, degree of polymerization, composition, etc. of cellulosic raw material
- Process techniques
- Process conditions such as temperature, pressure, processing time, concentration and viscosity of cellulosic solution, regeneration and drying conditions
- Solvent system and solubility of raw material in it
- Adhesion level between matrix and reinforcement
- Fiber volume fraction in the composite
- Microstructure and internal void content in the composite
- Orientation of reinforcement in matrix phase

8.2.2 Mechanics of ACCs

The mechanics of any composite material deals with forces or stresses, strain, and deformation under application of mechanical or thermal loads [6]. When tensile load is applied to a composite material, it is distributed to the fiber and matrix by a shearing mechanism between fibers and matrix. However, since the matrix phase generally has a lower modulus and much higher longitudinal strain than adjacent reinforcement or fibers, the reinforcement bears much higher load compared to matrix.

ACCs can be classified into two categories on the basis of mechanics—isotropic and unidirectional. In ACCs, as both reinforcement and matrix phase are chemically identical (both are cellulosic material), after preparation of ACCs it is very difficult to draw a clear distinction among their constituents. As a result, the mechanics of ACCs are more complex than that of other conventional bio-composites. However, due to the chemical similarity of fiber and matrix phase in ACCs, generally an excellent bonding is achieved, which causes effective distribution of stresses across the fiber–matrix interface. The inter and intramolecular hydrogen bonding between matrix and reinforcing materials also plays a significant role in the mechanics of ACCs [3].

8.2.3 Rule of Mixtures: General and Modified Equations

For quantitative analysis of mechanical properties of composites, often different modeling is used. In materials science, the "rule of mixtures" is used to predict various properties of a composite material that depends on the relative amounts and properties of the individual constituents [10]. According to the rule of mixture, general equations for modulus and strength of unidirectional composite are:

$$E_C = V_f E_f + (1 - V_f) E_m \tag{8.1}$$

$$\sigma_C = V_f \sigma_f + (1 - V_f) \sigma_m \tag{8.2}$$

Where, E_c and σ_c are the modulus and strength of the composite; V_f is the volume fraction of fiber; E_f and E_m are modulus of fiber and matrix, respectively; σ_f, σ_m are strength of fiber and matrix, respectively.

Some researchers used different modified model equations of the rule of mixtures to measure the strength and modulus of unidirectional ACCs. A few of these will be highlighted here. The modified Cox–Krenchel equation for composite stiffness can be written as:

$$E_C = \eta_{fE} \eta_0 V_f E_f + (1 - V_f) E_m \tag{8.3}$$

Similarly, for modeling of strength, Kelly and Tyson extended the equation of rule of mixtures as given:

$$\sigma_C = \eta_{f\sigma} \eta_0 V_f \sigma_f + (1 - V_f) \sigma_m \tag{8.4}$$

Where η_{fE}, $\eta_{f\sigma}$ are the fiber efficiency factors for composite modulus and composite strength considering stress transfer from the matrix to the fibers; η_0 is the fiber-orientation factor (η_0 = 1 for a unidirectional composite) [1–2,4–5,7].

8.2.4 Tensile Properties of Unidirectional ACCs

In case of unidirectional ACCs, the reinforcing material is aligned in a particular direction. Therefore, it shows better properties in that particular direction compared to other directions. For these type of ACCs, the tensile properties mainly depend on:

i. types of reinforcing material,
ii. fiber volume fraction, and
iii. dissolution time.

Soykeabkaew et al. in their study observed that longitudinal strength and Young's modulus of ramie fiber–based ACCs (produced by surface-selective dissolution method) increased with increasing dissolution time up to 2h. It was due to the formation of an efficient matrix by dissolving the outer layer of fiber, which could bind core fibers effectively, leading to good interfacial bonding and stress transfer in the prepared ACCs [9]. However, for higher dissolution time, the remaining fiber volume fraction in the resultant composites reduced, resulting in deteriorating mechanical properties of ACCs (Figure 8.1) [9].

A similar type of finding was reported by Adak et al. [8] for Lyocell-based ACC laminates. Nishino et al. produced ACC by impregnating unidirectionally aligned ramie fibers in dissolved kraft pulps and studied for a longer dissolution time (up to 72h) [11]. The optimum tensile strength and modulus obtained for the ACCs was found when impregnated for 24h.

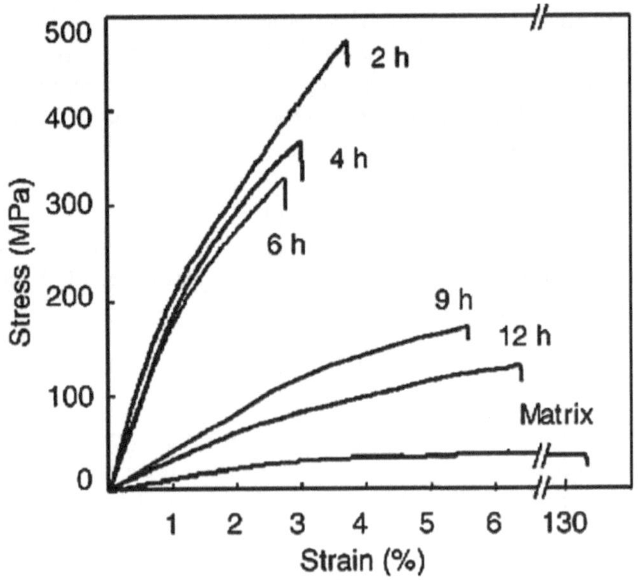

FIGURE 8.1
Stress–strain curve for ramie fiber–based ACCs (produced by surface-selective dissolution method) and pure cellulose matrix as a function of dissolution time. (From Soykeabkaew, N. et al., *Compos. Sci. Technol.*, 68, 2201, 2008. With permission.)

Whereas for 72-h impregnated ACCs, both the tensile strength and Young's modulus decreased due to the over dissolving of fibers during impregnation [11].

In an interesting study, Qin et al. investigated the effect of fiber volume fraction by impregnating aligned ramie fibers in 1%–7% (wt/v) cellulosic solutions of ramie fibers dissolving in LiCl/DMAc [1]. Initially, an increasing trend was observed in the tensile strength of the ACCs as cellulose concentration increased up to a level of 4%, and above this concentration, a subsequent reduction in tensile strength was recorded (Figure 8.2). The optimal cellulose concentration of 4% synchronized with the concentration that led to lowest fiber volume fraction (85%) among all composites. It happened due to formation of an adequate matrix phase in this concentration (4%), which resulted in good adhesion and better stress transfer between fiber and matrix. In contrary, for a cellulose concentration above and below 4%, inferior properties were observed due to the presence of insufficient matrix phase for formation of optimal composites. At lower concentration (<4%), enough matrix phase was not formed, and at higher concentration (>4%), the fibers were poorly wetted out due to high matrix viscosity, resulting in ineffective stress-transfer between the matrix and fibers in both cases, which leaded to inferior properties [1].

Qin et al. used an additional processing step of "mercerization," that is, treatment with 9% NaOH to improve the properties of prepared ACCs. After mercerization, the tensile strength of the ACCs were enhanced by 15%–95% (Figure 8.2b). The composites produced with 4% cellulose solution showed best tensile properties. During mercerization, alkali solution penetrated through the composites and swelling of fibers occurred, which filled the internal voids and cracks significantly [1]. It also removed hemicellulose that allowed cellulose microfibrils to rearrange themselves with better chain orientation and packing, which resulted in an increase in tensile strength of fibers as well as composites [1,12,13].

Huber et al. showed that application of a higher pressure during dissolution process caused a reduction of internal voids that exhibited a substantial improvement in interlaminar bonding, which led to more compact composite structure. Thus, a higher tensile strength and modulus were achieved for the ACC laminates produced with application of a higher and uniform pressure [14]. In a similar study, better mechanical properties were reported for the Lyocell fabric–based ACC laminates produced by compression molding, applying an optimally higher pressure (1 MPa) [15]. Shibata et al. also reported that application of higher pressure caused a significant increase in Young's modulus and tensile strength of cotton fabric–based ACCs produced by partial dissolution technique using ionic liquid (BmimCl) [16].

In another work, Huber et al. compared mechanical properties of flax (linen) and rayon fabric–based ACCs. The composition of linen or flax fiber is cellulose I 70%–75%, hemicelluloses 15%, pectin 10%–15%, lignin 2%, and

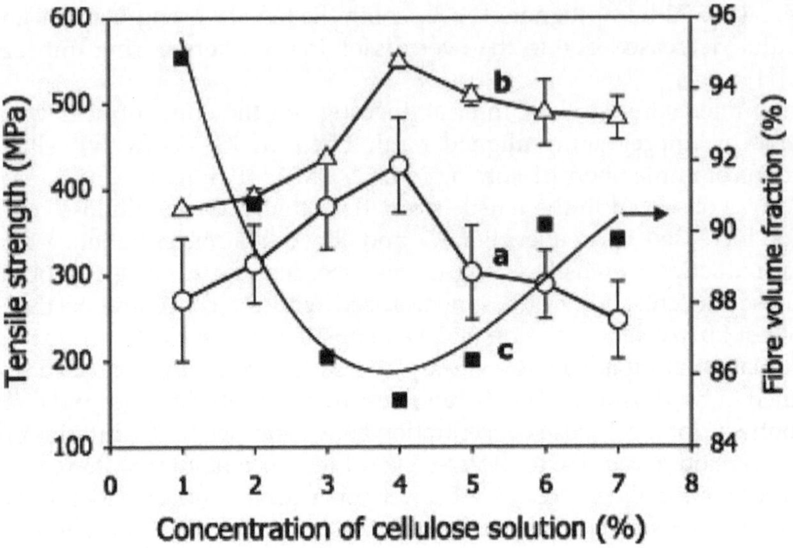

FIGURE 8.2
Tensile strength of (a) "O" unmercerized and (b) "Δ"mercerized composites and (c) "■" fiber volume fraction (%) of the prepared ACCs as an effect of cellulose solution concentration (%). (From Qin, C. et al., *Carbohydr. Polym.*, 71, 458, 2008. With permission.)

waxes 2%. Whereas, rayon is a regenerated cellulose (RC) fiber containing mainly cellulose II. Although flax fibers are considered as stiff, the flax fiber–based laminates showed very low stiffness (only 0.86 GPa) due to the absence of sufficient matrix phase, leading to poor interfacial adhesion [17]. In contrast, the rayon laminates had almost twice the tensile strength and thrice the Young's modulus compared to linen laminates, though rayon fibers were less strong and stiff compared to flax fibers. It was a consequence of better fiber–matrix adhesion and decrease in void content in case of rayon-based composites, which came from basic structural differences of rayon and flax fibers [17].

Soykeabkaew et al. in their study observed that mechanical properties of Lyocell-based composites reduced considerably compared to native Lyocell fibers for longer dissolution time (>10 min), due to highly developed skin-core structure of Lyocell fibers (Figure 8.3a). Once the highly oriented skin of these fibers dissolved to form a matrix phase, the remaining less-oriented weaker core reinforced the composites, resulting in reduction of strength and modulus [2].

In comparison to this, the Bocell ACCs required much longer immersion time than the Lyocell ACCs due to exceptionally high crystallinity of the Bocell fibers and a little skin-core effect in their structure. It exhibited far better mechanical properties in Bocell ACCs (Figure 8.3b). The Bocell ACC produced with an optimum immersion time (1.5 h) was the strongest ACC reported ever. It had an average tensile strength of 910 MPa and Young's

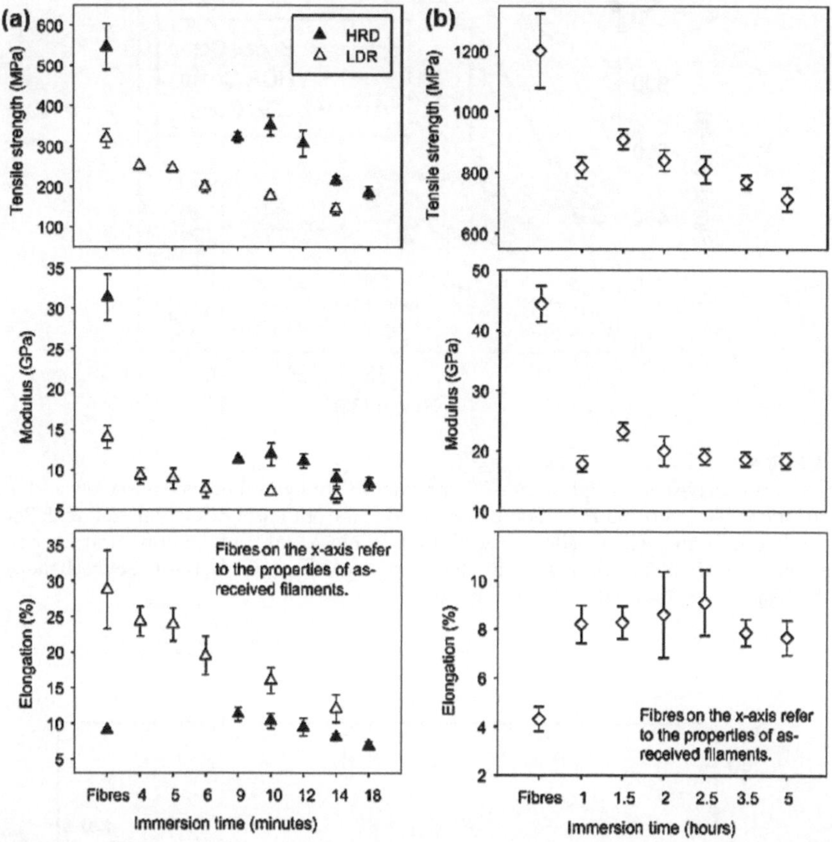

FIGURE 8.3
Effect of immersion time on tensile properties of ACCs produced from low and high draw ratio Lyocell fiber (a) and Bocell fiber (b) (From Soykeabkaew, N. et al., *Compos. Part A*, 40, 321, 2009. With permission.)

modulus of 23 GPa, with 8% elongation at break [2]. Figure 8.4 suggests that the tensile strength and modulus of this Bocell composite were very high compared to the Lyocell composite and any other traditional unidirectional natural fiber–reinforced composites.

In a very recent study, Piltonen et al. reported a 14-fold increase in tensile strength of ACCs produced from sulfite-treated pulp fibers treated with NaOH/urea only for 30 s, as shown in Figure 8.5 [18].

For most of the unidirectional ACCs, both the tensile strength and Young's modulus in transverse direction were poor compared to that in the longitudinal direction [9,11,19]. Nishino et al. reported that the transverse tensile strength for ramie fiber–based unidirectional ACC was only 12 Mpa and the percentage strain to failure was about 5%, while in longitudinal direction these values were 480 MPa and 4%, respectively [11].

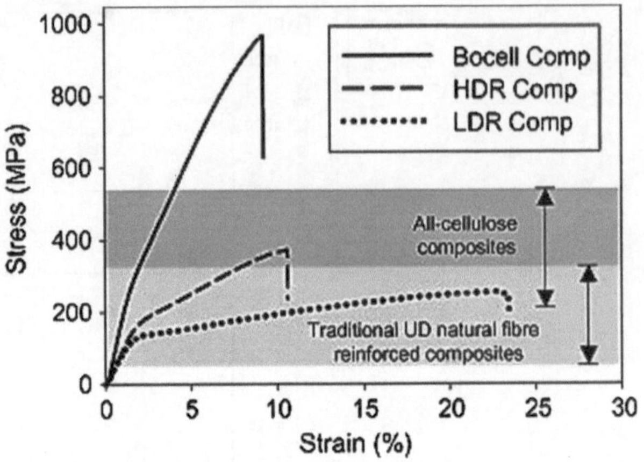

FIGURE 8.4
Stress–strain curves of the optimum ACCs prepared with Lyocell fibers having low and high draw ratio (LDR Comp and HDR Comp) compared to the optimum ACCs prepared with Bocell fibers (Bocell Comp). The tensile strength of all-cellulose and traditional natural fiber composite have shown by dark gray area and light gray area, respectively. (From Soykeabkaew, N. et al., *Compos. Part A*, 40, 321, 2009. With permission.)

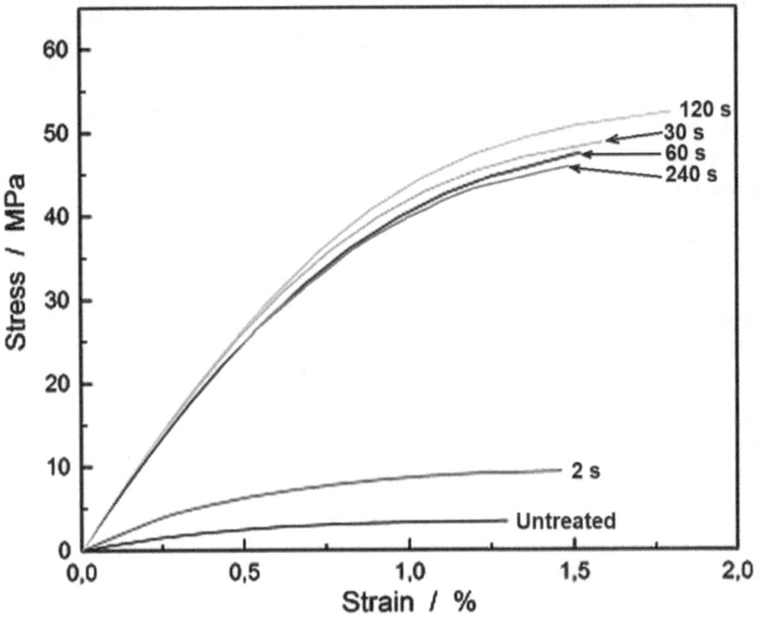

FIGURE 8.5
Averaged stress–strain curves of ACCs produced using sulfite-dissolving pulp fiber sheets and NaOH/Urea. (From Piltonen, P. et al., *Compos. Sci. Technol.*, 135, 153, 2016. With permission.)

Soykeabkaew et al. observed that transverse strength of longitudinally aligned ramie fiber–based ACCs showed an opposite trend compared to the longitudinal tensile strength with respect to effect of dissolution time. As per their investigation, with increasing immersion time up to 12h, transverse tensile strength of ACCs increased about 30% compared to the ACCs produced with immersion time of 6h, while longitudinal tensile strength decreased by about 60% for the same. As the immersion time increases, the matrix phase also increases and the interface becomes more homogenous, leading to efficient interfacial adhesion in those composites [9].

8.2.5 Tensile Properties of Isotropic ACCs

The tensile properties of isotropic ACCs are uniform in all directions, and properties are not much affected by the orientation of reinforcements. In most cases, for these type of composites, the reinforcements are present in nanoform and hence are also called as all-cellulose nanocomposites (ACNCs). Tensile properties of these composites mainly depend on:

i. Amount and type of reinforcing material [cellulose nanowhiskers (CNWs), nanocrystalline cellulose (NCC), cellulose nanofibers (CNFs), etc.],
ii. Aspect ratio and orientation of the nano-reinforcing phase,
iii. Amount of reinforcement,
iv. Dissolution time,
v. Rate of regeneration, and
vi. After-treatments such as wet drawing.

It has been seen in different literatures that in the absence of any nano-reinforcing material, the tensile properties of the regenerated ACC films were not so good. However, with addition of only 1 vol% CNWs to the MCC matrix, Pullawan et al. observed a massive increase in crystallinity, tensile strength, and Young's modulus by about 85%, 41%, and 153%, respectively [7].

Ma et al. reported that both tensile strength and Young's modulus of nanocomposites increased as the content of NCC was increased up to 10wt% in the cellulose matrix phase (Figure 8.6). It was due to a better compatibility and better stress transfer between matrix and NCC in prepared nanocomposites [20]. Similar findings were reported by Qi et al. [21], Zhao et al. [22], and Pullawan et al. [3] in other studies. However, with the addition of a higher concentration (>10wt%) of NCC or CNWs, almost no further increment in tensile strength was observed [20,22].

Some researchers have reported a reduction in tensile strength at a CNWs concentration about 20% and above. This reduction in tensile strength of nanocomposites might be due to phenomenon of phase separation as a result

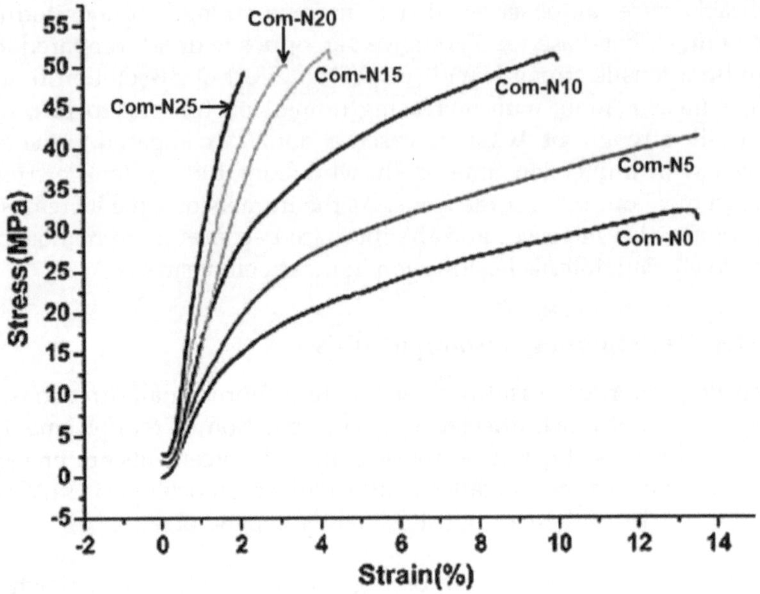

FIGURE 8.6
Stress–strain curves of all-cellulose nanocomposite films (with varying concentration of NCC from 5–25 wt%) and pure RC film (NCC content 0%). (From Ma, H. et al., *Carbohydr. Polym.*, 84, 383, 2011. With permission.)

of aggregation of nano-reinforcing material at a higher loading. However, modulus or stiffness still increased (with increasing content of CNWs/NCC) due to strong interactions of cellulose matrix with NCC or CNWs through hydrogen bonds [3,21,23]. The strain to failure of those composites gradually decreased as the content of NCC increased, leading to an embrittlement of the material. For example, Ma et al. found that strain to failure of ACNCs decreased from 13.89% to 2.11% when the content of CNCs increased from 0% to 25% [20].

In an interesting study Li et al. reported that aspect ratio and orientation of CNWs had a great role in controlling the tensile properties of ACNCs. They also compared the tensile properties of ACNCs and the micro paper sheet directly prepared from pulp. The nanocomposite possessed far superior mechanical properties than the micro paper sheet as shown in Figure 8.7. With implementation of magnetic field during preparation of nanocomposite, the CNWs were oriented with little effect on the orientation of the cellulose pulp fibers, leading to an anisotropy in the composites. It was observed that the tensile strength and modulus of the ACNCs along the direction perpendicular to the magnetic field was much higher than that parallel to the magnetic field, where the values of tensile strength and modulus of ACNC produced without magnetic field were intermediate. Moreover, the elastic modulus of the composites also increased with increasing aspect ratio of CNWs [24].

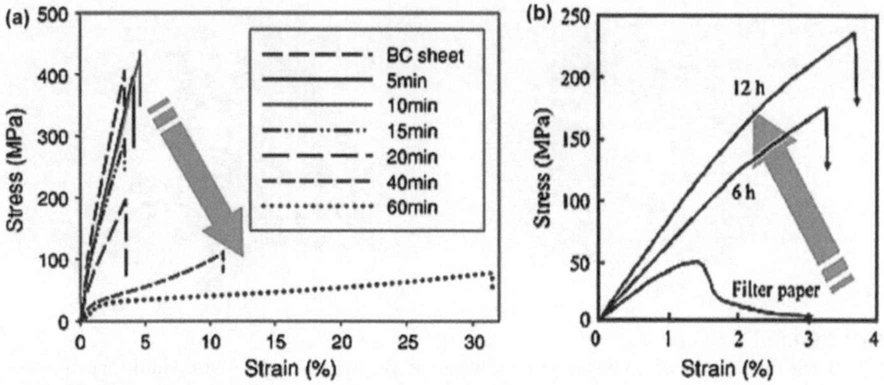

FIGURE 8.7
(a) Stress–strain curves of a BC-based sheet and ACCs prepared with nanosize BC at various immersion times; (b) ACCs prepared with microsize cellulose fibers of filter paper. (From Soykeabkaew, N. et al., Cellulose, 16, 435, 2009; Nishino, T. and Arimoto, N., *Biomacromolecules*, 8, 2712, 2007. With permission.)

Soykeabkaew et al. in their investigation observed that with increasing immersion time the tensile strength and modulus of bacteria cellulose (BC)–based nanocomposite gradually reduced after an optimum dissolution time (10 min) [25], as shown in Figure 8.7a. A similar finding was reported by Ghaderi et al. for bagasse nanofiber–based ACNCs [26]. Interestingly, Nishino et al. [27] and Han et al. [28] reported an opposite trend for ACCs prepared from filter paper, showing a continuous improvement of tensile strength and modulus with increasing immersion time.

Unlike BC nanocomposites, filter paper–based ACCs showed a tremendous improvement in tensile strength (from 50 MPa to 211 MPa) with increasing immersion time, as shown in Figure 8.7b [27]. This was due to the difference in internal network structure of these two different types of composites. In comparison to BC, filter paper consists of a much weaker, microsized cellulose fiber network with larger voids and less H-bonded fiber–fiber interaction, which has been schematically shown in Figure 8.8. Thus, after partial dissolution of cellulose, voids were filled and an improved network was obtained, leading to increase in tensile strength with increasing immersion time. In contrast, BC sheets already had a high level of interfiber bonding by hydrogen bonding as a result of nanosized cellulose ribbons [29]. Moreover, Yamanaka et al. reported that BC fibrils possess "three-way branching points" [30], which might be another reason for their high initial strength. Hence, after an optimum dissolution time (10 min), no further improvements in tensile strength were possible for BC nanocomposites. In fact, tensile strength of BC-based ACCs gradually reduced with increasing dissolution time after 10 min, due to the loss of original BC fiber structure by disruption of hydrogen-bonded network. The findings indicated that there is an optimum processing window for synthesizing high-performance ACNCs [25].

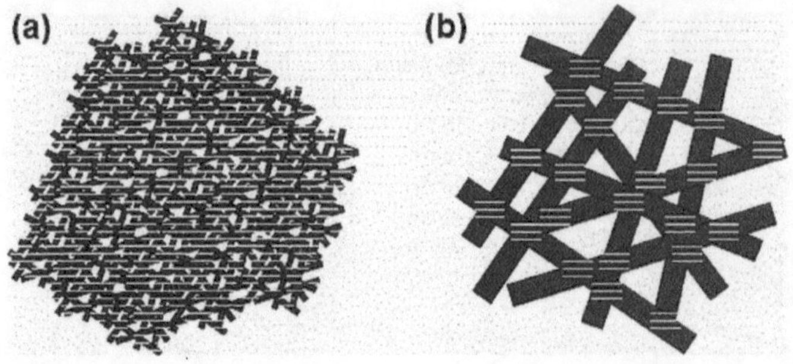

FIGURE 8.8
Schematic illustration of extensive H-bonding in (a) BC (nanosized network) and (b) cellulose-based filter paper (microsized network). (From Soykeabkaew, N. et al., *Cellulose*, 16, 435, 2009. With permission.)

In an interesting study, Duchemin et al. observed that composites obtained by slow precipitation exhibited highly improved mechanical properties relative to their counterparts obtained via the fast precipitation route, as shown in Figure 8.9. This was due to the formation of partially ordered paracrystalline cellulose when the precipitation was slow and fully amorphous cellulose when precipitation was fast [23].

Few studies have been undertaken to improve tensile properties of prepared ACCs by incorporating after-treatments. Qi et al. reported that wet drawing of ACC resulted in increase of tensile strength by increasing orientation of composite films. In the self-reinforced undrawn films there was random orientation of nanowhiskers, due to the absence of any orientating force during drying process. During wet drawing, crystallinity of prepared ACC might not have changed, but orientation was found to increase

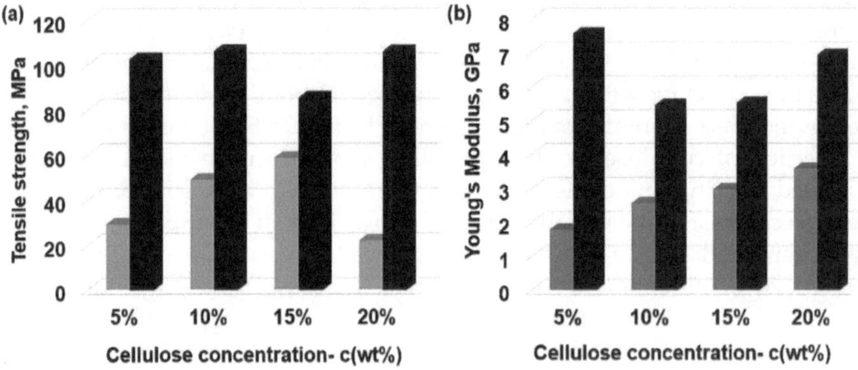

FIGURE 8.9
Tensile strength and Young's modulus of novel ACC films prepared by fast precipitation (gray bar) and slow precipitation (black bar), with varying concentrations of MCC and allowing a dissolution time up to 48h. These graphs have been plotted taking data from the literature. (From Duchemin, B.J.C. et al., *Compos. Sci. Technol.*, 69, 1225, 2009. With permission.)

in the direction of applied force. Drying of drawn samples afterward kept the crystallites oriented in the same position, showing an anisotropy in the sample with improved longitudinal tensile strength [21]. Similar studies were reported by Gindl et al. [31–33]. By wet drawing with a draw ratio of 1.5 and after subsequent drying, the tensile strength of ACC film reached up to 428 MPa and modulus of elasticity up to 33.5 GPa, with an increase of about 112% and 238%, respectively [32]. In a similar study, Pullawan et al. reported that a higher strain (3%) causes better orientation of CNWs in cellulosic matrix resulting in better tensile properties of ACNC, compared to the ACNC produced with low strain (1%) [34].

Pang et al. in their study discussed effects of different solvents in the production of regenerated films and also studied their properties. According to their study, the cellulose film regenerated from 1-ethyl-3-methylimidazolium chloride (EmimCl) showed the best tensile strength (119 MPa) [35]. The tensile strength for the RC films obtained from 1-alkyl-3-methylimidazolium chloride (AmimCl) was lowest in comparison with all samples [35]. However, it was still slightly higher than largely used commercial polyolefin (PE or PP) films having tensile strength in a range of 20–40 MPa [36].

Tables 8.1 and 8.2 shows tensile properties of unidirectional ACCs. Whereas the tensile properties of isotropic ACCs have been enlisted in Tables 8.3 and 8.4. Figure 8.10 shows a graphical representation of tensile strength and Young's modulus values of unidirectional and isotropic ACCs. From the trend lines it is clear that extent of tensile strength and Young's modulus is relatively higher in case of unidirectional ACCs compared to that of isotropic ACCs. This might be because of orientation of reinforcing material in a particular direction in case of unidirectional/anisotropic ACCs. However, isotropic/random ACCs also can show slight anisotropy with improved tensile properties by an additional wet-drawing process [21,32].

TABLE 8.1

Tensile Properties of Unidirectional ACCs Produced by Two-Step Dissolution Method

Sl. No	Cellulosic matrix	Cellulosic reinforcement	Solvent	Fiber vol. fraction (%)	Tensile strength (MPa)	Young's modulus (GPa)	Strain to failure (%)	Reference
1	Craft pulp	Ramie fiber	LiCl/DMAc	80	480/12^{T}	–	4.0/5^{T}	Nishino et al. [11]
2	Ramie fiber	Ramie fiber	LiCl/DMAc	85	410/540^{M}	25	4	Qin et al. [1]
3	Cellulose powder	2/2 twill weave (rayon)	IL (EmimCl)	80	105	2.4	16	Schuermann et al. [37]

TABLE 8.2

Tensile Properties of Unidirectional ACCs Produced by One-Step Dissolution (Surface-Selective Dissolution) Method

Sl. No.	Cellulose Source for Matrix and Reinforcement	Solvent	Fiber vol. Fraction (%)	Tensile Strength (MPa)	Young's Modulus (GPa)	Strain to Failure (%)	Reference
1	Ramie fiber	LiCl/DMAc	84 vol%	460/29[T]	28/2.5[T]	3.7/4.5[T]	Soykeabkaew et al. [9]
2	LDR Lyocell fiber	LiCl/DMAc	72 vol%	250	9	24	Soykeabkaew et al. [2]
3	HDR Lyocell fiber	LiCl/DMAc	73 vol%	350	12	10	Soykeabkaew et al. [2]
4	Bocell fiber	LiCl/DMAc	88 vol%	910	23	8.2	Soykeabkaew et al. [2]
5	Plain linen fabric	IL (BmimAc)	–	46	0.86	–	Huber et al. [17]
6	2/2 twill weave (rayon)	IL (BmimAc)	–	70	2.45	–	Huber et al. [17]
7	2/2 twill weave (rayon)	IL (BmimAc)	>70 vol%	91	4	–	Huber et al. [14]
8	3/1 twill weave (Lyocell)	IL (BmimCl)	–	44.24	1.78	20	Adak et al. [15]
9	3/1 twill weave (Lyocell)	IL (BmimCl)	–	99.5	4.3	8.5	Adak et al. [19]
10	Herringbone twill weave (Jute)	IL (BmimCl)	–	26	0.3	25	Adak et al. [38]
11	Sulfite-dissolving pulp fibers	NaOH/ urea	–	52.4	5.8	1.8	Piltonen et al. [18]

[T], longitudinal/transverse; [M], before/after mercerization; LDR, low draw ratio; HDR, high draw ratio; IL, ionic liquid; EmimCl, 1-ethyl-3-methylimidazolium chloride; BmimAc, 1-butyl-3-methylimidazolium acetate; BmimCl, 1-butyl-3-methylimidazolium chloride; LiCl, lithium chloride; DMAc, dimethylacetamide.

TABLE 8.3

Tensile Properties of Isotropic ACCs Produced by Two-Step Dissolution Method

Sl.No	Cellulose Source for Matrix	Cellulose Source for Reinforcement	Solvent	Fiber Fraction in (vol% or wt%)	Tensile Strength (MPa)	Young's Modulus (GPa)	Strain to Failure (%)	Reference
1	MCC	CNWs	LiCl/DMAc	1 vol%	128.4	4.8	–	Pullawan et al. [7]
2	MCC	CNFs	LiCl/DMAc	20 wt%	99.9	4.16	6.93	Zhao et al. [22]
3	Cotton linter pulps	CNWs	LiCl/DMAc	15 vol%	142.5	8	3.4	Pullawan et al. [3]
4	Cotton linter pulps	Tunicate CNWs	LiCl/DMAc	15 vol%	165.4	11.8	3.2	Pullawan et al. [3]
5	MCC	Tunicate CNWs	LiCl/DMAc	15 v/v%	154.6/170.2^D	10.5/13.6^D	3.5/ 1.5^D	Pullawan et al. [34]
6	Cotton linter pulps	CNWs	NaOH/Urea	15 vol%	127.4	7.2	4.3	Pullawan et al. [3]
7	Cotton linter pulps	Tunicate CNWs	NaOH/Urea	15 vol%	137.1	9.8	4.1	Pullawan et al. [3]
8	Cotton linter pulps	CNWs	NaOH/Urea	10 wt%	124/157^D	5.1	–	Qi et al. [21]
9	Cellulose powder	Hemp fiber	NMMO	40 wt%	28.9	1.82	20.8	Ouajai et al. [39]
10	Filter paper	Rice husk	IL(BmimCl)	60 wt%	56.5	2.92	2.24	Zhao et al. [40]
11	Filter paper	Rice husk	IL(BmimCl)	40 wt%	57.5	4.89	5.76	Zhao et al. [40]
12	MCC	NCC	IL[Hemim]Cl	25 wt%	49.24	3.63	2.11	Ma et al. [20]
13	Cotton linter pulps	Alkali-treated short ramie fiber	NaOH/Urea	15 wt%	124.3	5.2	4	Yang et al. [41]

D, before/after wet drawing; MCC, microcrystalline cellulose; CNWs, cellulose nanowhiskers; CNFs, cellulose nanofibers; IL, ionic liquid; BmimCl, 1-butyl-3-methylimidazolium chloride; [Hemim]Cl, 1-hexyl-3-methyl-imidazolium-chloride; LiCl, lithium chloride; DMAc, dimethylacetamide; NMMO, N-methylmorpholine N-oxide.

TABLE 8.4

Tensile Properties of Isotropic ACCs Produced by One-Step Dissolution (Surface-Selective Dissolution Method)

Sl. No.	Cellulose Source for Matrix and Reinforcement	Solvent	Fiber vol Fraction (%)	Tensile Strength (MPa)	Young's Modulus (GPa)	Strain to Failure (%)	Reference
1	MCC	LiCl/DMAc	–	242.8	13.1	8.6	Gindl et al. [42]
2	MCC	LiCl/DMAc	–	105.7	6.9	3.3	Duchemin et al. [23]
3	MCC	LiCl/DMAc	–	202/428^D	9.9/33.5^D	16.1/2.3^D	Gindl et al. [32]
4	Beech pulp fiber	LiCl/DMAc	80 vol%	154	12.2	2.4	Gindl et al. [43]
5	Filter paper	LiCl/DMAc	16 vol%	211	8.2	3.8	Nishino et al. [27]
6	BC	LiCl/DMAc	–	411	18	4.3	Soykeabkaew et al. [25]
7	Filter paper	PEG/NaOH	–	74.7	–	9.26	Han et al. [28]
8	MFC paper	IL (BmimCl)	–	124.1	10.8	2	Duchemin et al. [44]
9	Filter paper	IL (BmimCl)	–	91.8	5.75	3.76	Duchemin et al. [44]
10	Cellulose micro fiber	IL (BmimCl)	–	208	20	9.8	Yousefi et al. [45]
11	Ground nano fiber	IL (BmimCl)	–	197	19.5	12	Yousefi et al. [45]
12	Nonwoven matt of flax	IL (BmimCl)	–	34	4.6	0.8	Altmutter et al. [46]
13	Nonwoven mat of Lyocell	IL (BmimCl)	–	78	7.2	5.2	Altmutter et al. [46]
14	Cotton linters	IL (BmimCl)	–	106	–	8.5	Pang et al. [35]
15	Cotton linters	IL (EmimAc)	–	83	–	7.3	Pang et al. [35]
16	Cotton linters	IL (EmimCl)	–	119	–	8.8	Pang et al. [35]
17	Cotton linters	IL (AmimCl)	–	52	–	6.7	Pang et al. [35]

D, before/after wet drawing; MCC, microcrystalline cellulose; MFC, microfibrillated cellulose; IL, ionic liquid; BmimCl, 1-butyl-3-methylimidazolium chloride; EmimCl, 1-ethyl-3-methylimidazolium chloride; AmimCl, 1-alkyl-3-methylimidazolium chloride; LiCl, lithium chloride; DMAc, dimethylacetamide.

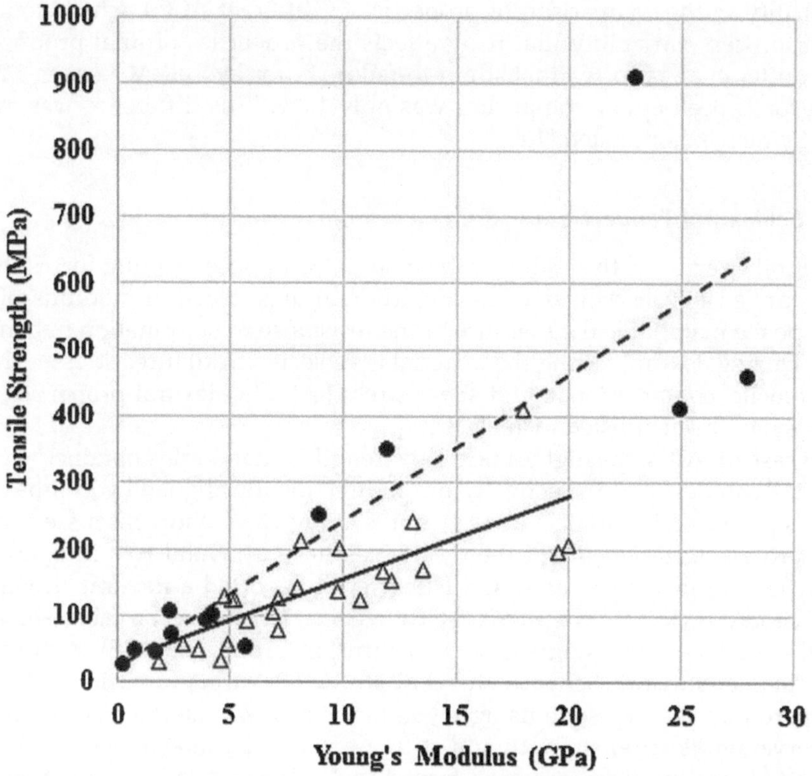

FIGURE 8.10
Tensile strength vs Young's modulus for unidirectional ACCs ("●" and dashed line) and isotropic ACCs ("Δ" and solid trend line), taking data from Tables 8.1–8.4.

The strain to failure of ACCs mainly depends on

i. Types of fibers used for reinforcement,
ii. Volume fraction of matrix, and
iii. Processing techniques and variables.

Rayon and Lyocell fibers have very high elongation break. Therefore, composites made of these fibers have also reported very high strain-to-failure values. For example, reported values of strain to failure for Lyocell (LDR)-based ACCs were 12%–24% [2] and for rayon fabric–based prepreg ACCs were 16%–32% [37]. On the contrary, ligno-cellulosic ramie fiber had a low strain to failure, about 1.2%–4% [47]. Therefore, a lower range of strain to failure (about 3%–6.5% [9]) was reasonable for ramie fiber–reinforced unidirectional ACCs. Interestingly, an unexpected strain to failure was obtained (up to 26%) by Ouajai et al. for hemp fiber–reinforced isotropic ACCs [39], though hemp fibers have a low elongation at break (1%–3.5% [47]). This is the difficulty in the comparison of properties of different ACCs, which comes from process variability that truly affects the properties of final products. Altmutter et al. reported that strain-to-failure % for Lyocell ACCs was 5.2% and for Lyocell-epoxy composites was only 1.2%. This difference was evidently matrix dominated [46].

8.2.6 Flexural Properties of ACCs

Flexural strength is the ability of the material to oppose bending forces that are applied perpendicular to its longitudinal axis. Flexural modulus of a composite material is the measure of the resistance to deformation in bending. During flexural testing the material is subjected to all three stresses that are tensile, compressional, and shear stress [48]. The flexural properties of ACCs have been enlisted in Table 8.5.

In case of ACCs, the highest flexural strength and modulus obtained were 178.MPa and 11 GPa, respectively, for unidirectionally aligned Lyocell-based ACCs produced by Adak et al. [19]. In this study, the authors discussed that the process techniques and their variables have a pivotal role in controlling flexural properties of ACCs. Huber et al. reported a flexural strength and modulus, 135.24 MPa and 3.72 GPa, respectively, for rayon fabric–based ACCs produced by using the solvent infusion processing (SIP) technique [49]. In another work, Schuermann et al. studied flexural properties of ACCs prepared by prepreg style using rayon fabric as raw material. As per their observation, flexural strength and stiffness values became almost doubled as the processing time increased from 1 to 4 h. It was due to a reduction in void content with increasing processing time, leading to improved bonding, which produced a significant increase in flexural properties. The highest flexural strength and modulus values achieved were about 42 MPa and 2 GPa, respectively [37], which were lower than that of the ACCs prepared by Adak et al. [19].

Duchemin et al. studied flexural properties of novel aerogels (or aerocellulose) based on ACCs that were prepared by partial dissolution of microcrystalline cellulose (MCC) in 8 wt % LiCl/DMAc solution. As per their observation, both flexural strength and flexural modulus of ACCs increased initially with increasing cellulose concentration (c), obtaining maxima at "c" equal to 10–15 wt%. At a low cellulose concentration due to higher dissolution of MCC fibrils, a more homogenous paracrystalline phase was formed after regeneration. However, as cellulose concentration increased up to about 10%, crystallinity increased due to higher portion of undissolved cellulose, leading to improved mechanical properties with an optimal matrix-to-reinforcement ratio. With further increase of cellulose concentration ($c > 15\%$), due to the presence of a very high proportion of undissolved cellulose and highly crystalline cellulose I, the ductility of aerocellulose was reduced [50].

TABLE 8.5
Flexural Properties of ACCs

Sl No.	Cellulose for Reinforcement	Cellulose Source for Matrix	Composite Type	Solvent	Fibre vol Fraction (%)	Flexural Strength (Mpa)	Flexural Modulus (GPa)	Reference
1	MCC	Partially dissolved cellulose	Film (IT)	LiCl/DMAc	12	8.1	0.28	Duchemim et al. [50]
2	2/2 twill weave (rayon)	Partially dissolved cellulose	Laminate (UD)	IL (BmimAc)	92	135.24	3.72	Huber et al. [49]
3	2/2 twill weave (rayon)	Cellulose powder	Laminate (UD)	IL (EmimCl)	80	44	2	Schuermann et al. [37]
4	3/1 twill weave (Lyocell)	Partially dissolved cellulose	Laminate (UD)	IL (BmimCl)	–	53.96	1.2	Adak et al. [8]
5	3/1 twill weave (Lyocell)	Partially dissolved cellulose	Laminate (UD)	IL (BmimCl)	–	48.95	0.96	Adak et al. [15]
6	3/1 twill weave (Lyocell)	Partially dissolved cellulose	Laminate (UD)	IL (BmimCl)	–	178.3	11	Adak et al. [19]

IT, isotropic, UD, unidirectional, IL, ionic liquid; BmimCl, 1-butyl-3-methylimidazolium chloride; BmimAc, 1-butyl-3-methylimidazolium acetate; BmimCl, 1-butyl-3-methylimidazolium chloride; LiCl, lithium chloride; DMAc, dimethylacetamide.

Goutianos et al. discussed the effect of processing conditions on the fracture resistance and cohesive laws of binder-free ACCs. The composites were prepared by a mechanical refinement process that allowed the defibrillation and formation of intermolecular bonds among cellulose molecules during the drying process. The fracture resistance was determined by sandwiching the specimen in a double cantilever beam (DCB) fixer, as shown in Figure 8.11a. The path-independent *J* integral along a path following the external boundaries of the specimen was used to calculate the fracture resistance. A good agreement was made between the experimental flexural stress–strain data and curve-fitting data (Figure 8.11b). The fracture resistance and peak cohesive stress both increased with increasing fibrillation time due to the presence of more H-bonded or Van der Waals network structure in ACCs [51].

8.2.7 Impact Properties of ACCs

The impact property of a material is its capacity to absorb and dissipate energies under impact or shock loading [52]. It is directly proportional to the overall toughness of the material [53]. The composite materials sometimes are subjected to impact loading and in response to that they may be damaged during their service life.

Many bio-composites have been found to be very sensitive to impact loading. Therefore, their performance should be checked before use. There are mainly two types of impact testing methods – (1) high-velocity impact test and (2) low-velocity impact test [54]. Charpy impact testing and Izod impact testing fall into the category of low-velocity impact testing, whereas ballistics impact testing falls into the category of high-velocity impact testing. During impact testing, very high contact forces are applied over a small area for a little duration. Generally, in case of low-velocity impact test, materials may

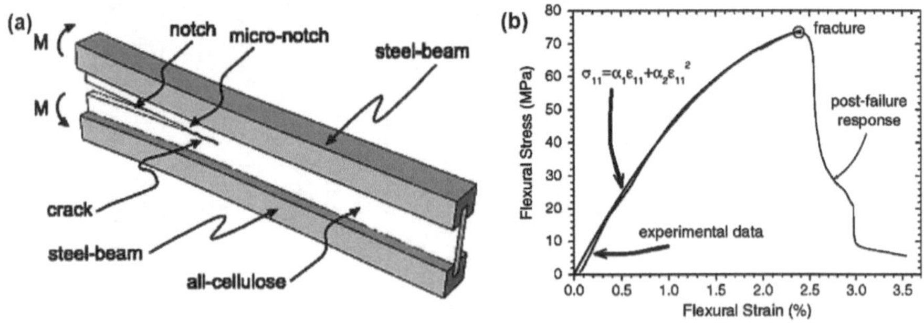

FIGURE 8.11
(a) Specimen loaded with pure bending moments in a DCB; (b) Experimental stress–strain curve (three-point bending) for an ACC produced with 5-h refining time. The experimental data up to failure are fitted with the equation shown in the plot, where, $\sigma_{11}$is the normal stress in the direction parallel to the specimen, α_1 is Young's modulus of ACC when $\varepsilon_{11} \rightarrow 0$. (From Goutianos, S. et al., *Appl. Compos. Mater.*, 21, 805, 2014. With permission.)

not always be fully damaged, but in case of high-velocity impact test, materials are completely destroyed by the striker. During impact testing with increasing impact loading, the energy is dissipated by de-bonding, fiber and/or matrix fracture, and fiber pull-out, causing complete or partial failure of composites [49,55].

The interfacial bonding between fiber and matrix plays an important role in controlling impact strength of fiber-reinforced composites. Due to good interfacial adhesion between matrix and reinforcement in ACCs, it is expected that ACCs would show good impact properties. The average unnotched Charpy impact strength of the ACCs prepared from rayon textile (2/2 twill weave) by SIP was 41.54 ± 4.44 kJ/m^2, as reported by Huber et al. [49]. Bax et al. in their study observed that the impact strength of a natural fiber–reinforced composite can be enhanced by the addition of high strain fibers (Cordenka) [55]. Similarly, the high elongation of rayon fibers was the main reason behind the high impact strength of the ACCs prepared by Huber et al. [49].

First time, Rubio-López et al. predicted the low-velocity impact behavior of woven fabric–based ACC laminates using a finite element model (FEM). A good appropriation was established between experimental data and theoretical or model-predicted values of absorb energy and force displacement [56].

8.2.8 Peel Strength of ACC Laminates

Peel strength is the measure of interlaminar adhesion strength of a laminate. In case of ACC laminates, due to chemical similarity of matrix and reinforcing phase, a high interlaminar strength is expected. However, the process parameters, fiber volume fraction, and dissolution efficiency of solvent have a vital role in controlling peel strength of ACC laminates [19]. Adak et al. reported a twelvefold increase in peel strength of Lyocell fabric–based ACC laminates by increasing dissolution time from 30 min to 4 h [8], as shown in Figure 8.12. Similarly, application of lower pressure during composite manufacturing and drying of ACC laminates leads to inferior peel strength. An optimum pressure is required to obtain a good interlaminar adhesion strength in ACC laminates [15].

8.2.9 Fracture Behavior of ACCs

The fracture behavior of tensile/flexural/impact tested ACC samples have been reported in many literatures [8,11,15,20,23,27,43,45,49,50,57]. SEM micrographs are generally used to observe fractured surface of composites and to understand interfacial adhesion characteristics between fiber and matrix. Mainly two types of fractures have been observed in the testing of mechanical properties of composites; these are – (1) brittle fracture and (2) ductile fracture. Among these, ductile fracture is generally preferred for most of

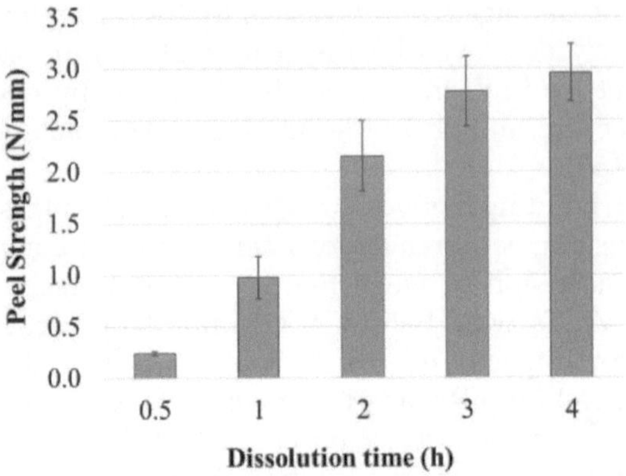

FIGURE 8.12
Change in the peel strength of the Lyocell fabric–based ACCs produced by varying dissolution times. (From Adak, B. and Mukhopadhyay, S. *J. Appl. Polym. Sci.*, 133, 2016. With permission.)

the engineering applications. However, depending on the level of interfacial adhesion, the following fracture modes may be observed in ACCs:

i. Pull-out of fiber from the matrix,
ii. Fiber failure (e.g., breakage or buckling),
iii. Breakage of yarn,
iv. Splitting of matrix connecting individual fibers or fiber bundles,
v. Delamination of laminate layers due to interlaminar stresses [49].

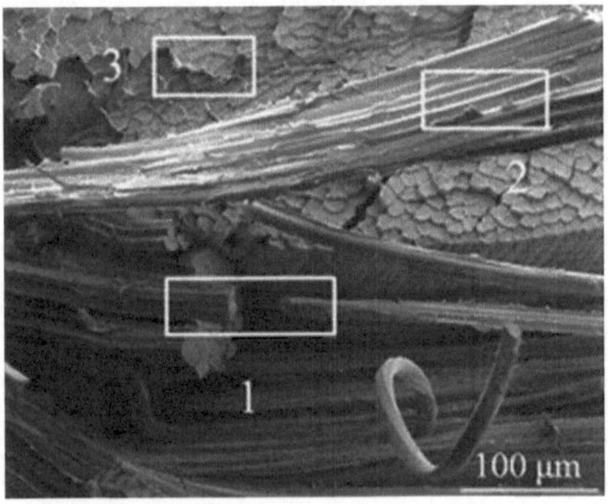

FIGURE 8.13
SEM micrographs of the fractured surface of puncture impact tested ACC laminate showing fractured individual rayon fibers (1), fiber delamination within the yarn (2), and fractured yarns (3). (From Huber, T. et al., *Compos. Sci. Technol.*, 88, 92, 2013. With permission.)

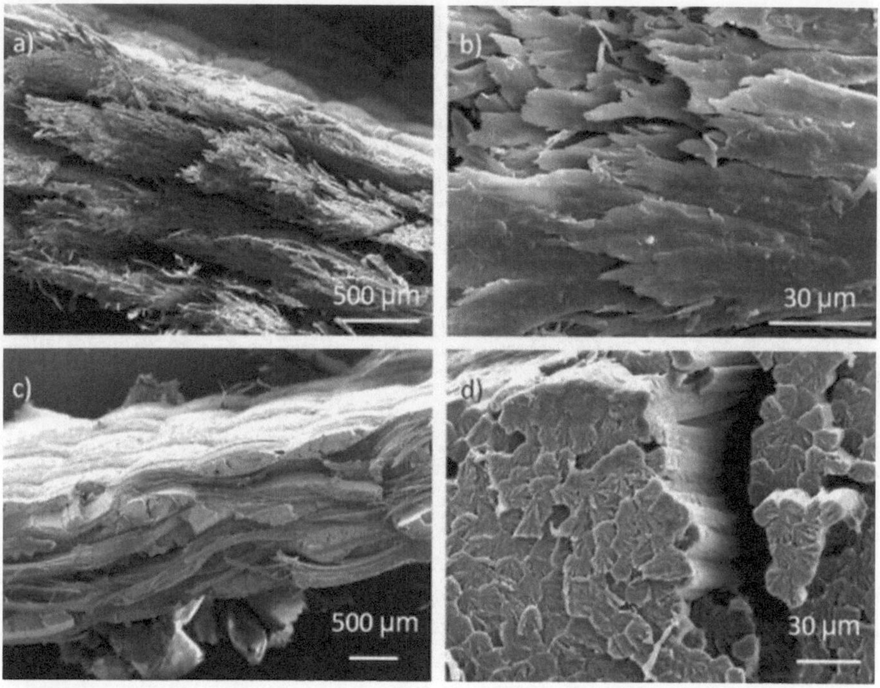

FIGURE 8.14
SEM micrographs showing fractured surfaces of linen ACCs (a and b) and rayon ACCs (c & d). (From Huber, T. et al., *Compos. Part A*, 43, 1738, 2012. With permission.)

Figure 8.13 represents different fractured modes (fiber failure, matrix failure, and delamination) of a rayon-based ACC after puncture impact test [49].

Huber et al. compared the cross-sections of fractured surface for linen and rayon laminates (Figure 8.14). The fractured surface of linen laminate exhibited delamination of laminated layers and separation of fiber bundles within a single layer, showing lack of bonding between adjacent fibers. However, the fractured behavior of the rayon laminate was totally different compared to that of the linen laminate, with very little separation of individual fibers under load, indicating better adhesion between fibers [17].

Schuermann et al. observed that for a higher processing time (2h), the reinforcing fibers adhered better, resulting in an extensive failure of fibers and a comparatively less fiber pull-out during fracture. Whereas, for a lower hot-pressing time, as some parts of composites fiber and matrix were not effectively bound, more fiber pull-out was observed in poorly bonded regions [37].

Nishino et al. found that in ACCs produced from un-pretreated ramie fibers, the adhesion between fiber–matrix was not good. It resulted in fiber pull-out from the matrix and a clear boundary was observed between these two constituents in the fractured surface (Figure 8.15a). By contrast, for the

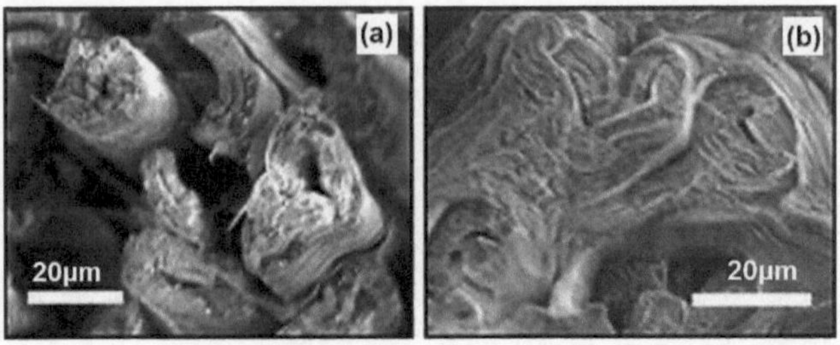

FIGURE 8.15
Fractured surface of ACCs (24-h impregnated) prepared with (a) un-pretreated ramie fibers and (b) pretreated ramie fibers. (From Nishino, T. et al., *Macromolecules*, 37, 7683, 2004. With permission.)

ACCs produced with pretreated ramie fibers, no clear boundary was visible between fiber and matrix, indicating better fiber–matrix adhesion (Figure 8.15b) [11].

A similar finding was reported by Gindl et al. where they used SEM micrographs of fractured surface to reveal the important microstructural difference between the cellulose-epoxy composite and ACC. In case of cellulose-epoxy composite, fiber and matrix was clearly identified, where a smoother surface was observed for epoxy. In contrary to this, in the fractured surface of ACC, no clear boundaries between fiber and matrix were observed. It indicated much better compatibility and interfacial adhesion between these two constituents [43].

Zhao et al. observed that the fractured surface of rice husk (RH)–based all-cellulose eco-composite became smoother when it was produced from prolonged pretreated RH. On the other hand, with a higher pretreatment temperature (130°C) of RH, the filler/matrix interface of the ACC almost diminished and exhibited high level of homogeneity in the fractured surface. It suggests an indication of high-level fiber–matrix adhesion [40].

Duchemin et al. observed that fractured surface of ACC gradually became rougher when the cellulose (MCC) concentration increased from 5% to 20%. The composite prepared with lower cellulose concentration (c) (5%–10%) and fast precipitation was fractured to expose lamellae with smooth surface (Figure 8.16a). In this case, there was a formation of more homogeneous material because of thorough dissolution of MCC and regeneration of a paracrystalline phase with decreased longitudinal order. However, when the cellulose concentration was gradually increased up to 20%, the fracture surfaces appeared rougher, indicating pull-out of fibrils during fracture (Figure 8.16b) [23]. Similar observations were reported by Ma et al. [20] and by Duchemin et al. [50].

Goutianos et al. developed an innovative way to observe the in situ crack growth for binder-free flax-based ACCs, by incorporating an environmental scanning electron microscope inside the chamber containing DCB fixer [51]. Adak et al. used optical microscopy images to analyze the fractured surface

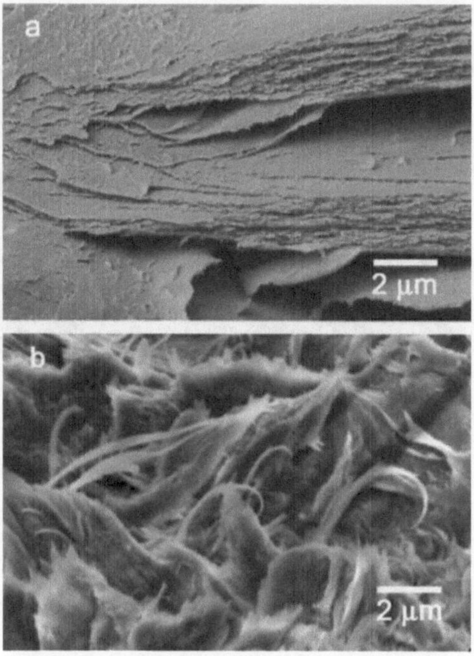

FIGURE 8.16
Fractured surface of ACCs made with 1-h dissolution time and with varying cellulose concentrations (a - 5%, b - 20%), followed by fast precipitation. (From Duchemin, B.J.C. et al., *Compos. Sci. Technol.*, 69, 1225, 2009. With permission.)

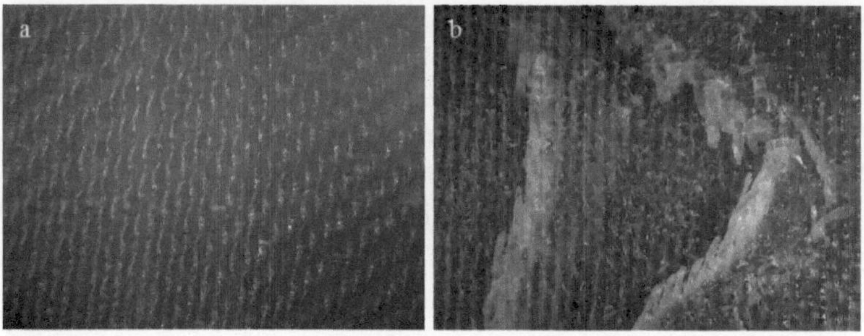

FIGURE 8.17
Optical microscopy image of internal surface of ACC laminates after T-peel testing: (a) smooth internal surface of ACC-1 (produced by applying 0.25 MPa pressure) and (b) fractured internal surface of ACC-3 (produced by applying 1 MPa pressure). (From Adak, B. and Mukhopadhyay, S., *J. Text. Inst.*, 108, 1010, 2017. With permission.)

of T-peel strength-tested samples. The tearing of internal laminas before separating out of layers and a rough surface due to this (Figure 8.17b) is an indication of good interlayer bonding. A relatively smooth fracture surface (Figure 8.17a) indicates that interlayer bonding is inferior [15].

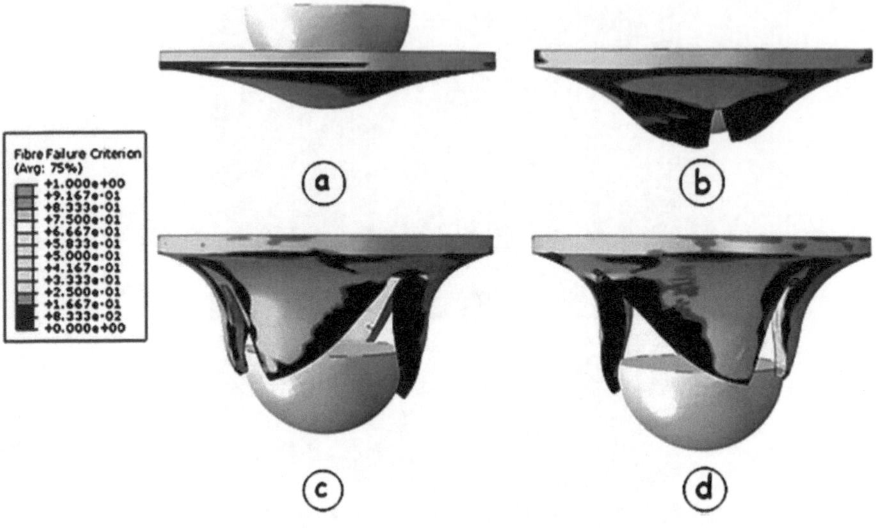

FIGURE 8.18
Failure modes of ACC plates under different impact energies. (a) Impact energy 4.77 J. (b) Impact energy 19.08 J. (c) Impact energy 42.93 J. (d) Impact energy 83.71 J. (From Rubio-Lopez, A. et al., *Compos. Struct.*, 122, 139, 2015. With permission.)

Rubio-López et al. established an FEM model to analyze the failure mechanism of ACC plates under different impact energies. This model was able to reproduce the force history and different damage modes observed in experiments. When impact energy was very low, the behavior of the plate was elastic, showing impact energy equal to zero. Subsequently, impact energy increased with increasing velocity and a crack was initiated. At very high velocities (higher than 30 J), impact energy was quite higher than the energy absorbing capability, resulting in no change in energy of absorption. The main failure modes were fiber breakage and subsequent crack propagation in principle direction, which led to a four-petal pyramid shape (Figure 8.18) [56].

8.3 Viscoelastic and Thermal Properties of ACCs

8.3.1 Dynamic Mechanical Analysis

Dynamic mechanical analysis (DMA) has been reported in many works to investigate the viscoelastic properties as well as thermal properties of ACCs [2,11,27,40,46,58,59]. Viscoelasticity is the property of materials that exhibits both elastic and viscous characteristics when undergoing deformation. Most of the ACCs are viscoelastic in nature. In DMA, the storage modulus (E′), which is a property of elastic part of a material, measures dynamic rigidity of the material under cyclic deformation, while loss modulus (E″) measures the losses due to the viscose or amorphous component [60].

Nishino et al. observed that thermomechanical performance of ACCs (based on ramie fiber [11] or filter paper [27]) was better than many other polymer-based composites. It was observed that although storage modulus of ACCs decreased with increasing temperature, the drop was limited, and a high storage modulus was maintained up to around 250°C [11,27], as shown in Figure 8.19. At a temperature above 250°C cellulose generally starts to degrade, causing a drop in storage modulus. Nishino et al. reported that storage modulus of the ACCs based on unidirectionally aligned ramies were about 45 GPa at 25°C, which was even higher than conventional glass fiber–reinforced composites [11]. The elastic modulus of crystalline regions of cellulose I is independent to temperature, which contributes to excellent thermal as well as mechanical properties of the ACCs [27] and the drop of storage modulus with increasing temperature is initiated by softening as well as degradation of cellulose matrix [40].

Soykeabkaew et al. found that Bocell ACCs exhibit undoubtedly higher storage modulus value than Lyocell ACCs. It was attributed to the less

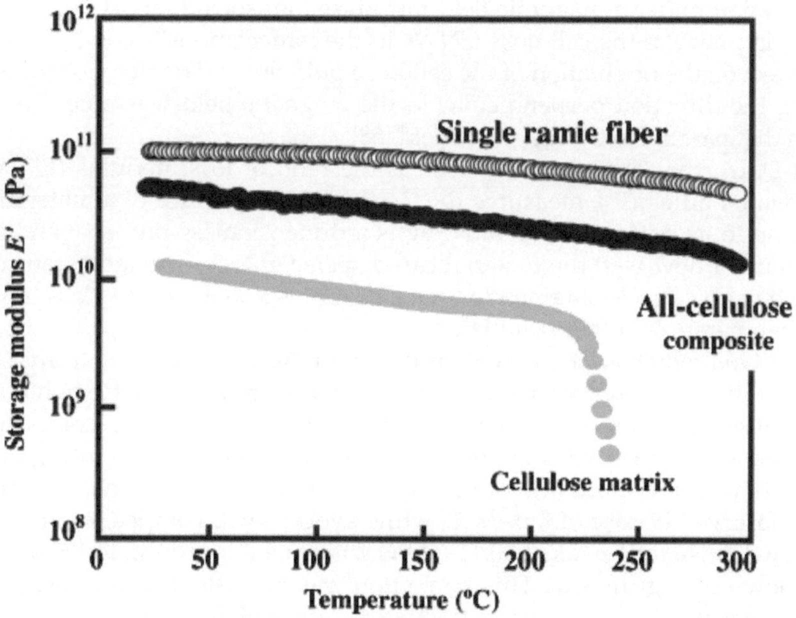

FIGURE 8.19
Temperature dependence of dynamic storage modulus E' of ramie fiber–based ACC, the single ramie fiber and the cellulose matrix. (From Nishino, T. et al., *Macromolecules*, 37, 7683, 2004. With permission.)

skin-core effect of Bocell fibers and formation of a more homogenous structure, resulting in better interfacial bonding [2].

Dynamic mechanical properties of ACNCs depend on three crucial factors:

i. Dispersion of filler or reinforcement in matrix
ii. Alignment of nano-reinforcement in matrix
iii. Reinforcement-matrix interfacial bonding.

As seen by Zhao et al., storage modulus of the ACCs increased with increasing RH content from 20 wt% to 40 wt%, but storage modulus reduced if wt% increased to 60. This thing could be explained by poor dispersion and aggregation of RH fillers in the cellulose matrix above the critical value of 50 wt% [40]. They also reported that storage modulus of prepared ACCs could be increased by increasing both temperature and time during pretreatment of RH in IL, as these resulted in better interfacial bonding between matrix and reinforcement (RH) [40].

Li et al. reported that in the presence of CNWs and application of a magnetic force, the storage modulus of ACNCs can reach up to 4884 MPa. For paper sheet (without CNWs) and the ACNCs fabricated without magnetic field, the storage modulus were only 652 MPa and 3955 MPa, respectively. Fabrication under a magnetic field introduced anisotropy in ACNCs, causing alignment of the cellulose CNWs in the nanocomposites, with little or no effect on the orientation of the cellulose pulp fibers. The storage modulus along the direction perpendicular to the magnetic field was much higher than that parallel to the magnetic field [58].

In DMA, the "tanδ factor," which is the ratio of loss modulus (E") and storage modulus (E'), measures the damping performance of a material in relation to its stiffness. The damping is a dimensionless property and is a measure of how well the material can disperse energy. The temperature at which peak value is obtained for "tanδ factor" is known as the glass transition temperature of that material.

The tanδ factor is not very significant for ACCs as no clear transition temperature is observed in case of cellulose such as synthetic polymers. Altmutter et al. reported that Lyocell and flax-based epoxy composites showed a clear tanδ peak at approximately 80°C–90°C, representing glass transition temperature (Tg) of the used epoxy [46]. However, no such peak was observed in case of flax-ACC, while Lyocell-ACC composite exhibited a very diffuse tanδ peak at 70°C–220°C with a center around at 170°C [46], as shown in Figure 8.20. This relaxation was consistent in that temperature range with the glass transition temperature of dry cellulose, as previously reported by Szcześniak et al. [61]. Here in this case, the viscoelastic response of Lyocell ACC might have been due to its low crystallinity (only 17%) [46].

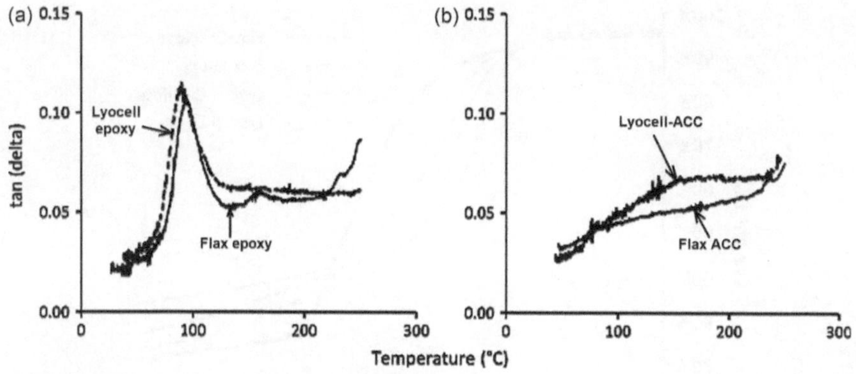

FIGURE 8.20
Tanδ graphs obtained from DMA analysis of flax and Lyocell composites with epoxy matrix (a) and ACC (b), respectively. (From Gindl-Altmutter, W. et al., *Compos. Sci. Technol.*, 72, 1304, 2012. With permission.)

8.3.2 Thermogravimetric Analysis

Thermal stability of composites can be checked by thermogravimetric analysis (TGA), where weight loss corresponding to temperature is noted. The thermal stability of any composite material is dependent on:

i. Chemical composition
ii. Crystallinity
iii. Crystal size

In a TGA, Altmutter et al. showed that both Lyocell and flax-based ACCs were more thermally stable than their epoxy-based counterparts (Figure 8.21). The highest thermal stability was found in case of flax-based ACC due to more crystallinity in flax-based composites than Lyocell-based composites [46].

In an interesting study, Qin et al. analyzed the thermogravimetry (TGA), derivative thermogravimetry (DTGA), and second-derivative thermogravimetry (2DTGA) curves for unmercerized and mercerized ACCs (Figure 8.22). For both mercerized and unmercerized ACCs (prepared with 4% cellulose concentration), there was an initial weight loss between 60°C and 100°C for removal of water. Subsequently, for unmercerized composite sample, two decomposition peaks were observed at 313°C and 323°C, which were due to degradation of hemicellulose (14.2% weight loss) and α-cellulose (57.3% weight loss), respectively. Whereas, for mercerized composite sample, only one peak was observed at 305°C for degradation of α-cellulose only, showing a weight loss of about 58% [1]. The missing peak for hemicellulose indicated that during mercerization all hemicellulose was removed, resulting in increasing residual char formation and an inferior thermal stability than unmercerized ACC [1,62–63].

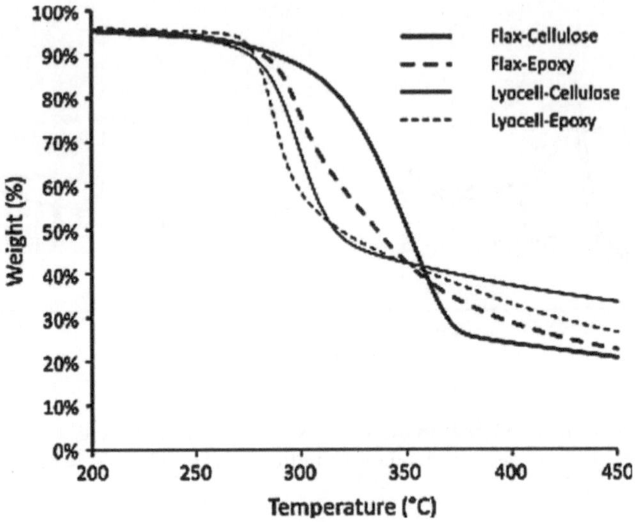

FIGURE 8.21
Thermogravimetric analysis of flax and Lyocell composites with epoxy matrix compared to ACCs. (From Gindl-Altmutter, W. et al., *Compos. Sci. Technol.*, 72, 1304, 2012. With permission.)

Pang et al. observed that cotton linter–based RC films produced by using different ionic liquids showed higher decomposition rate and lower decomposition temperature than virgin cellulose. This observation was probably due to lower crystallinity and smaller crystal size in regenerated ACC films, which accelerated their degradation process. The highest thermal stability was observed in case of ACC film produced using 1-Ethyl-3-methylimidazolium chloride (EmimCl) [35].

Similarly, Ghaderi et al. reported that the thermal stability of bagasse nanofiber–based ACNC films was gradually reduced with increasing partial dissolution time (Figure 8.23). The reason suggested by the researchers was due to a decrease in crystallinity and crystal size of the ACNC with increasing dissolution time, resulting in lower thermal stability [26].

8.3.3 Thermal Expansion Coefficient

One big advantage of an ACC is its very low thermal expansion coefficient. Nishino et al. showed that with increasing temperature, the matrix-cellulose gradually expanded, showing a linear thermal expansion coefficient (α) of 1.4×10^{-5} K^{-1}. The ramie fiber–based ACC, on the other hand, showed almost no thermal expansion or contraction (Figure 8.24). The thermal expansion coefficient (α) of ramie ACCs (1.7×10^{-7} K^{-1}) was very much lower than metals [11].

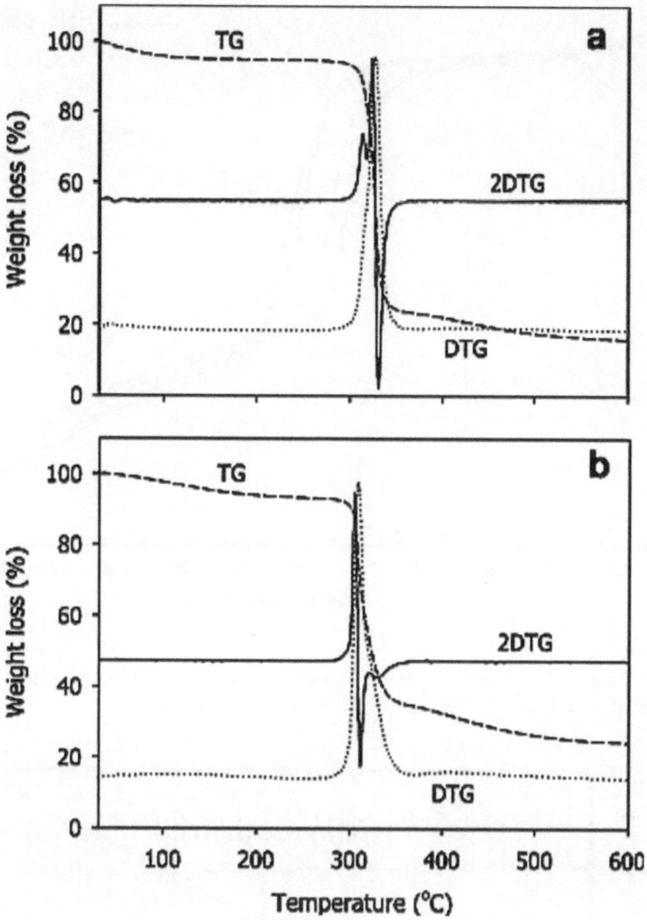

FIGURE 8.22
Thermogravimetric analyses of (a) unmercerized and (b) mercerized composites prepared with 4% cellulose solution. (From Qin, C. et al., *Carbohydr. Polym.*, 71, 458, 2008. With permission.)

8.4 Optical Transparency of ACCs

Among different ACCs, optical transparency is mainly observed in some ACNC films and also in some specially prepared ACC sheets. The main factors that control optical transparency of ACCs are:

i. Dissolution time (during composite preparation),
ii. Composite thickness,
iii. Matrix-reinforcement interface, and
iv. Wt% of nano-reinforcement embedded in matrix (in ACNC).

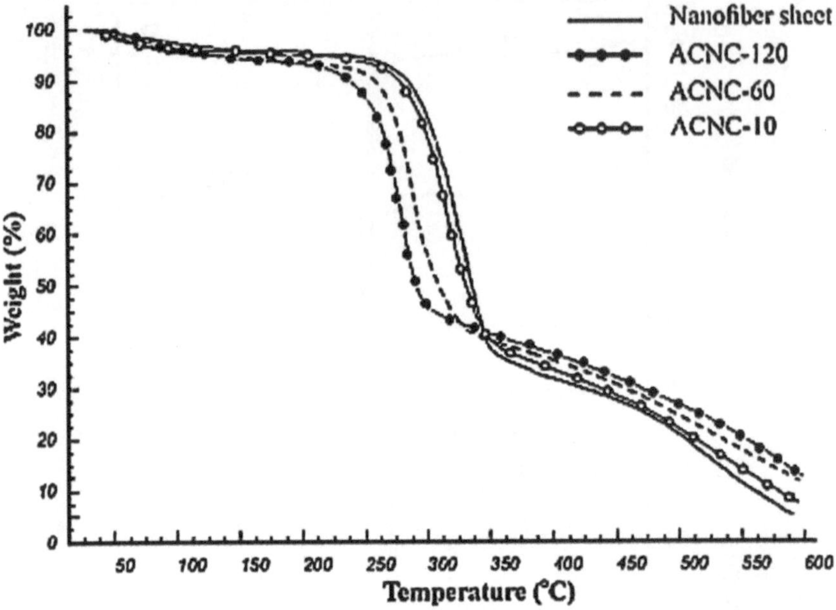

FIGURE 8.23
TGA curves for nanofiber sheet and ACNCs produced with varying dissolution time. (From Ghaderi, M. et al., *Carbohydr. Polym.*, 104, 59, 2014. With permission.)

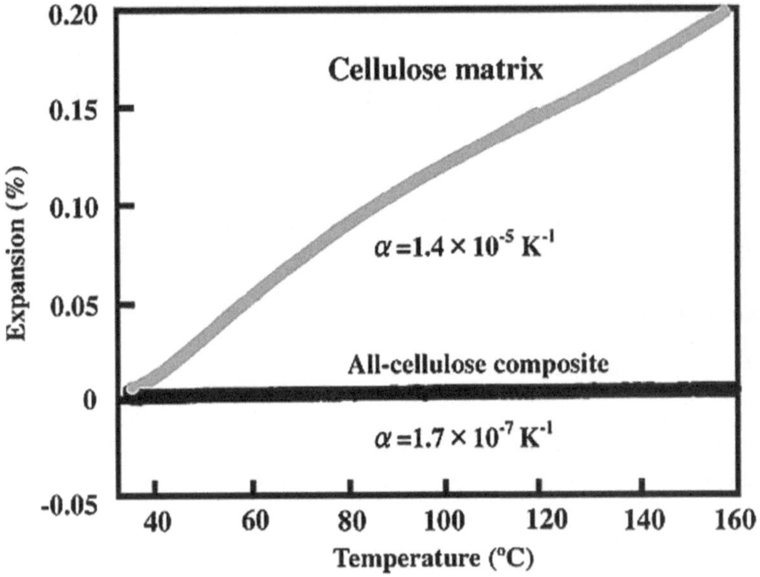

FIGURE 8.24
Thermal expansion behavior of cellulosic matrix and ACCs. (From Nishino, T. et al., *Macromolecules*, 37, 7683, 2004. With permission.)

Optical transparency is also a useful criterion to access the miscibility of the composite elements [20–22,41]. Qi et al. reported that cellulose nanocomposite films embedded with 5–10 wt% CNWs exhibited good optical transparency, indicating good miscibility. However, with increase in the CNWs content from 0 to 25 wt%, optical transmittance (T_r) value decrease from 87% to 49% (Figure 8.25). The decrease in optical transparency (at higher CNWs loading) might be due to the aggregation of CNWs and more scattering of light caused by larger particles [21]. Similar thing was reported by Ma et al [20] and Zhao et al [22]. Interestingly, Yang et al. observed that even the ramie fiber–reinforced ACCs became transparent because of certain compatibility between matrix and reinforcement. However, the transparency decreased from 86.9% to 16.2% (at 800 nm), with increasing ramie fiber content from 0% to 25% [41].

Nishino et al. [27] and Han et al. [28] in their research compared optical transparency of filter paper and the ACCs produced from it, with varying immersion times (Figure 8.26). Filter paper was opaque and white, but after treating in a solvent (LiCl/DMAc [27] or PEG/NaOH [28]) for a longer time, became transparent, and the transparency increased with increasing immersion time up to 12 h. Here, with increasing dissolution time, the selectively dissolved and resolidified fiber surface exhibited good interface and excellent bonding, resulting in high mechanical performance and optical transparency [27,28].

In a study by Yousefi et al., transparency of ACNCs prepared from micro/nano fibers was observed to increase about 250-fold, in comparison

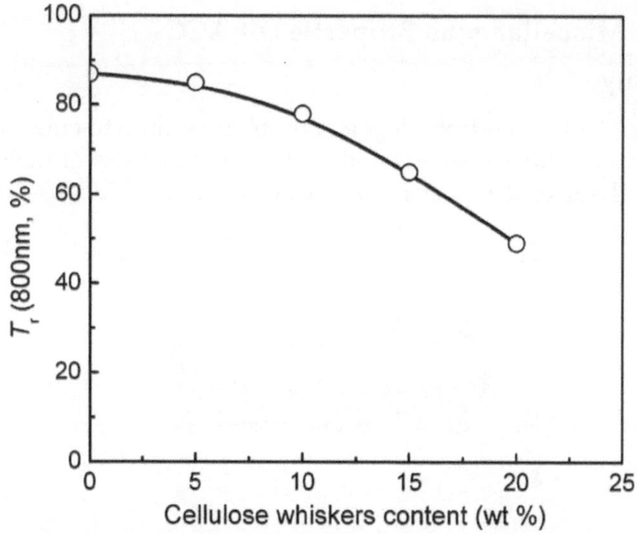

FIGURE 8.25
Optical transmittance (T_r) of ACC films and pure RC films at 800 nm. (From Qi, H. et al., *Biomacromolecules*, 10, 1597, 2009. With permission.)

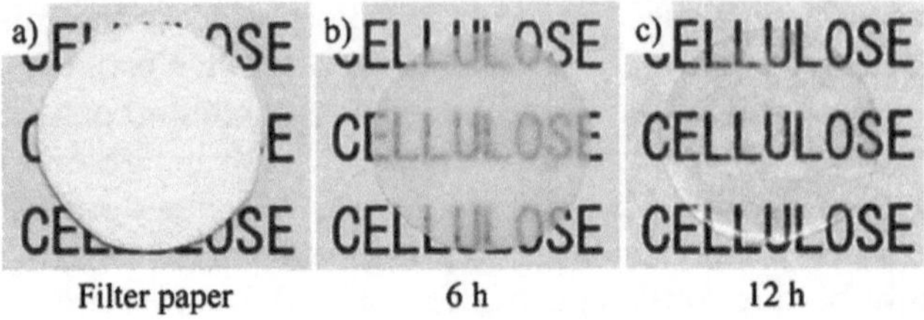

FIGURE 8.26
Photographs showing optical transmittance of filter paper (a) and ACCs (immersion times): (b) 6h and (c) 12h. (From Nishino, T. and Arimoto, N., *Biomacromolecules*, 8, 2712, 2007. With permission.)

to the microfiber sheet [45]. Qin et al. observed that after mercerization of a unidirectionally aligned ramie fiber–based ACC, it showed better optical transparency, as shown in Figure 8.27. It was due to improved interfacial interaction, resulting less light scattering at the composite interface [1].

In a recent study, Adak et al. reported that Lyocell-based fabric without weft yarns became transparent after the preparation of ACC laminates by partial dissolution in ionic liquid (BmimCl). The transparency of the ACC laminates increased when dissolution time increased from 1 to 2h, as shown in Figure 8.28 [19].

8.5 Other Miscellaneous Properties of ACCs

8.5.1 Density

The density of ACCs entirely depends on fiber volume fraction or amount of undissolved cellulose content and also on compactness of the composite materials. Nilsson et al. reported that the density of the ACCs based on wood

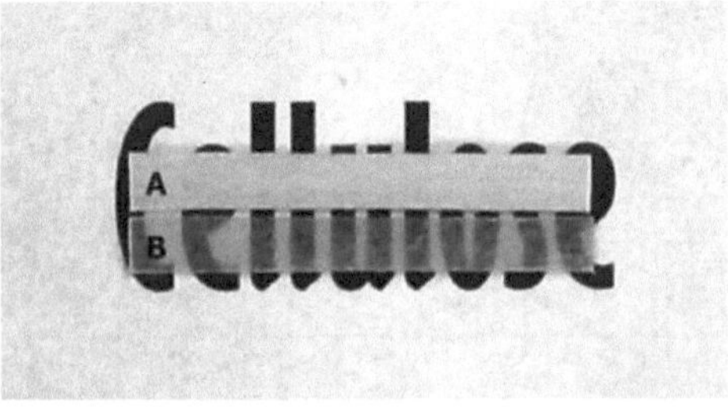

FIGURE 8.27
Photographs showing optical transmittance of (A) un-mercerized and (B) mercerized ACCs prepared with 4% cellulose solution. (From Qin, C. et al., *Carbohydr. Polym.*, 71, 458, 2008. With permission.)

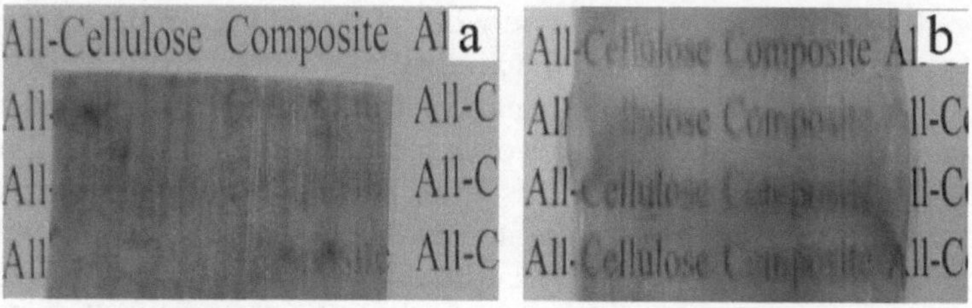

FIGURE 8.28
Photographs showing transparency of the ACC laminates produced from a modified Lyocell-based preform with 1h (a), 2h (b) dissolution time. (From Adak, B. and Mukhopadhyay, S., *Cellulose*, 24, 835, 2017. With permission.)

pulps varied from 1.25 to 1.35 g/cm^3 and the corresponding porosity varied from 10%–17%, assuming a density of 1.5 g/cm^3 for cellulose [56]. Duchemin et al. observed that density of aerocellulose-based ACCs were in a range of 120–350 Kg/m^3. With increasing concentration of cellulose (MCC) in solvent, the density of composite also increased gradually, but no further increase was noticed after a certain concentration (above 15%) [50].

In case of ACC laminates, process parameters (dissolution time, pressure, and temperature) and process techniques have a great role in controlling their density and porosity [8,15,19]. As density and porosity are inversely related, significant reduction in porosity were observed with higher dissolution time [8] and pressure [15], due to the formation of more compact and denser composite structure.

8.5.2 Thickness

The ACCs reported in most of the literatures were invariably in the form of thin films having thickness less than 1mm [17]. First time, Huber et al. reported manufacturing of thicker ACCs from rayon textiles (2/2 twill) having thickness around 2mm. Application of pressure during dissolution and drying stage had a great role in controlling thickness and compactness of ACCs [14,15,37]. Even thickness of ACC laminates was found to vary depending on dissolution time [8], number of lamina used in the preparation of laminate, and the process techniques [19].

8.5.3 Fluid Permeability and Barrier Property

The permeability test measures the transport of different fluids such as air/gas/water vapor/water through the faces, that is, perpendicular to the plane of the material. Nanocellulosic reinforcements such as nanofibrils (CNFs) and cellulose nanocrystals (CNCs) are strong gas barrier material having a great potential to reduce gas permeability through polymeric films [64]. These cellulosic nanomaterials have large specific surface area, high aspect ratio, and the diameter generally varies in the range of 2–50 nm. The capability of forming strong hydrogen-bonded network makes it very hard for gas molecules to pass through, resulting in excellent gas barrier property [65].

The aspect ratio of these cellulosic nano-reinforcements is very high. Hence when these are dispersed as well as oriented properly in a cellulosic matrix, the tortuous path length for gas diffusion increases significantly (Figure 8.29). It results in a significant improvement in gas barrier property of all-cellulosic nanocomposite films. Moreover, moisture content, density, crystallinity, and thickness also strongly affect the gas permeability through cellulosic films [66]. The biodegradable and gas barrier ACNC films have a huge demand for many applications, especially for food packaging application.

Yang et al. fabricated high gas barrier and transparent regenerated cellulosic films from different cellulose sources (MCC powder, cotton linters, and softwood bleached kraft pulp) using NaOH/urea as solvent with varying dissolution and regeneration conditions. Depending on the process condition, the oxygen permeability of prepared films varied widely from 0.003 to 0.03 mL μm m^{-2} day^{-1} kPa^{-1} (at 0% relative humidity). The cellulosic film prepared from 6 wt% cellulose solution by regeneration with acetone (at 0°C) showed lowest oxygen gas permeability. Both oxygen and water vapor permeability (WVP) decreased significantly with increasing density, and WVP increased with increasing relative humidity [66].

In an interesting study, Yousefi et al. reported that though microfiber sheet was highly permeable to air (42 ± 7 μm Pa^{-1} S^{-1}), the microfiber-based ACNC became a complete barrier to air (0 μm Pa^{-1} S^{-1}) like other conventional packaging polymers (polyethylene and polypropylene) due to their fully consolidated structure [45].

Ghaderi et al. observed that WVP of bagasse nanofiber–based ACNC films significantly increased with increasing dissolution time. The WVP value reported for nanofiber sheet was 2.46 ± 0.11 gm^{-1} s^{-1} Pa^{-1} × 10^{-11} and increased

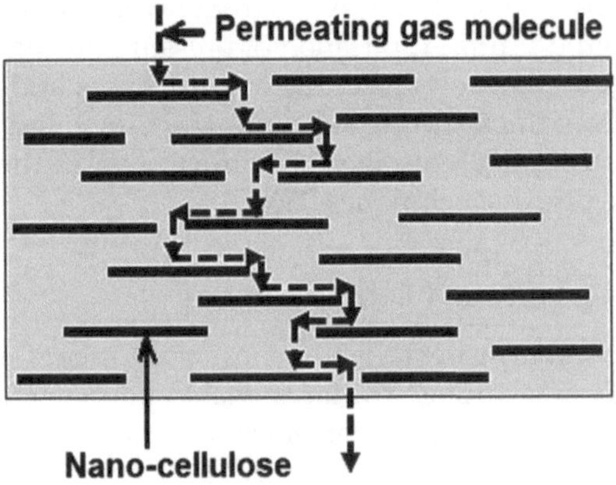

FIGURE 8.29
Schematic representation of increased diffusion path length due to the presence of nanocellulose in cellulose nanocomposite film.

3.5 times when dissolution time was varied from 0 to 120 min. The reasons suggested for the observation at longer dissolution time were: (i) the crystallinity of ACCNCs reduced, while the matrix phase was dominated by less ordered cellulosic chains, resulting in more easy transfer of water vapor through the matrix and (ii) the ACNC sheet showed more shrinkage due to the effect of internal stresses, leading to creation of internal voids [26].

The water retention value (WRV) is primarily a function of capillary spaces within and between fibers. Nilsson et al. studied the WRVs of all-cellulose bio-composites produced by a non-solvent approach—by compression molding of hardwood and softwood pulp (non-beaten or beaten in PFI mill) by varying temperature. The WRV results were clearly related to fibril aggregation. With increasing temperature as fibril aggregation tendency was increased, the surface area accessible to water decreased, resulting in bio-composite plates gradually becoming less water sensitive and showing lower WRV [57]. In this case, there was almost no effect of beating on WRV, as seen from Table 8.6. The swelling capacity of the fibrils were reduced as a result of drying. It was also observed that applied pressure during compression molding had a great role in controlling aggregate's size and the WRV reduced significantly due to application of pressure, compared to only drying at elevated temperature [57].

8.5.4 Hydrophobicity

Being fully cellulose-based material, ACCs are generally highly hydrophilic in nature. As a consequence, in the presence of moisture, ACCs start to

TABLE 8.6
Water Retention Values of the Compression-Molded Hardwood and Softwood Pulps at Different Temperatures

	WRV (g/g)	
Processing Temperature (°C)	**Non-beaten**	**Beaten**
Softwood		
Original pulp	1.80 ± 0.22	–
120	0.72 ± 0.13	0.64 ± 0.01
150	0.56 ± 0.06	0.56 ± 0.03
170	0.46 ± 0.07	0.48 ± 0.02
180	0.48 ± 0.01	0.45 ± 0.03
Hardwood		
Original pulp	2.02 ± 0.10	–
120	0.85 ± 0.07	0.92 ± 0.04
150	0.75 ± 0.01	0.78 ± 0.01
170	0.69 ± 0.05	0.65 ± 0.05
180	0.57 ± 0.01	0.56 ± 0.03

Source: Nilsson, H. et al., *Composi. Sci. Technol.*, 70, 12, 1704, 2010.

FIGURE 8.30
Contact angle measurement with water and the surface of the composites. (From Han, D. and Yan, L. *Carbohydr. Polym.*, 79, 614, 2010. With permission.)

degrade by microbes. Therefore, ACCs have a limitation to be used only in moisture-free zones. However, Han et al. reported that the filter paper–based ACCs not only became transparent but also became hydrophobic to some extent, though filter paper was highly hydrophilic at the onset. The reported contact angle for such type of ACC with dissolution time 12h was about 55° (Figure 8.30) [28]. This property may extend its end use. Similarly, Adak et al. observed that the Lyocell fabric–based ACC laminates produced by treating in ionic liquid for 3 or 4h turned to slightly hydrophobic in nature with sowing contact angle of about 62° and 76°, respectively. Interestingly, the moisture and water absorbency reduced drastically for these ACC laminates. It was a consequence of certain changes in surface and crystallographic structure of cellulose prepared with higher dissolution time [67].

8.5.5 Swelling and Re-swelling of ACC-gel

Wang et al. studied the swelling and re-swelling behavior of CNWs–reinforced ACC gels. The equilibrium swelling ratio (ESR) was gradually decreased from 9.88 to 6.50, with increasing CNWs concentration from 0 to 50wt%. The reason behind that was the increase in cross-linked density with increasing CNWs concentration, which restricted the swelling of RC matrix. This was followed by freeze-drying the gel and subsequently the re-swelling property was checked. The ESR and re-swelling ratio (RSR) were calculated as below:

$$\mathrm{ESR} = \frac{W_s}{W_d} \tag{8.1}$$

$$\mathrm{RSR} = \frac{(W_r - W_d)}{W_s - W_d} \times 100 \tag{8.2}$$

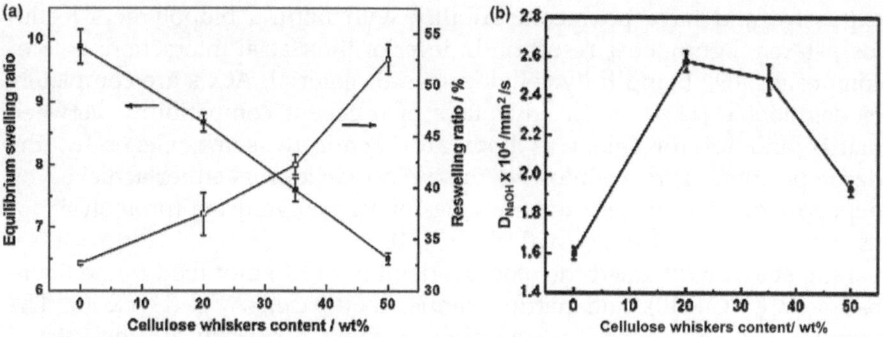

FIGURE 8.31
(a) Swelling and re-swelling properties of regenerated composite gels with different CNWs content. (b) Dependence of NaOH diffusion coefficient through ACNC gels w.r.t CNWs content. The test was conducted in a water bath at 25°C temperature. (From Wang, Y. and Chen, L. *Carbohydr. Polym.*, 83, 2011, 1937. With permission.)

Where, W_s = weight of regenerated gel at 25°C, W_d = weight of the gel at dry state, W_r = weight of re-swelling gel.

It was observed that the extent of re-swelling during immersion of dried gels in water bath was a function of CNWs content. The values for RSRs increased with increasing CNWs content, as shown in Figure 8.31a [68].

8.5.6 Drug-Release Property

Wang et al. studied the in vitro drug (NaOH) release from CNWs–reinforced ACC gels. It was observed that the diffusion behavior of NaOH from the cellulose gel matrices was mainly controlled by "porous membrane." The NaOH diffusion coefficient in water was lowest (1.60×10^{-4} mm²/s) for cellulose gel produced without CNWs, while it increased to a maximum value (2.58×10^{-4} mm²/s) for the cellulose gel produced with 20 wt% CNWs (Figure 8.31b). However, subsequently, with increasing CNWs concentration the diffusion rate of NaOH decreased, due to formation of more jammed hydrogen-bonded network [68].

8.6 Biodegradability of ACCs

Non-biodegradable petroleum-based plastics have become a threat to the environment. Some synthetic polymers, such as polylactic acid (PLA), polyhydroxybutyrate (PHB), and poly (3-hydroxybutyrate-co-3-hydroxyvalerate) (PHBV), are found to be biodegradable. However, these polymers are expensive and have poor compatibility with natural biopolymers (cellulose, starch, lignin, etc.), resulting in inferior interfacial interactions in bio-composites [69]. Being fully cellulose-based material, ACCs are completely biodegradable [14,70], with advantage of excellent compatibility between matrix and reinforcement as both are chemically same. Generally, the decomposition of the cellulose at the molecular level is either achieved by depolymerization by secreted enzymes of microorganisms (proteolysis) or fission by water molecules (hydrolysis) [71].

Yang et al. investigated the biodegradability of RC, modified ramie fiber–based ACC (C-R-10), and natural ramie fiber–based ACC (C-NR-10). The samples (size-15 × 15 cm^2) were buried at about 20 cm beneath the natural soil in an 80-L barrel, enclosing them in a nylon mesh net (mesh size-2 × 2 mm^2). The average temperature, moisture, and pH of soil were maintained as 30°C, 20%, and 6.8, respectively. After 10 days of soil burial, only broken fragments of the films were observed and porous structures were obtained due to decaying of the films caused by the microorganisms in the soil. The RC degraded more quickly than ramie fiber–based ACCs. After 30 days, no fragments of films were found in the soil because of complete biodegradation [41].

In another study, Kalka et al. investigated the end-of-life disposal of rayon fabric–based ACC laminates and compared the result with biodegradation behavior of conventional rayon-PLA bio-composites [72]. Effect of testing conditions (temperature and moisture in soil), time (in days), and presence of fungicide in the composite have also been discussed. Two testing conditions were used: T1 (Set temperature – 22.5°C and moisture content – 51%) and T2 (Set temperature – 36.75°C and moisture content – 28%). It was reported that biodegradability of ACC coupons increased at T2 condition compared with T1 condition, showing strong discoloration and disintegration of tested coupons in T2 condition, as shown in Figure 8.32a [72]. It was because the rate of decomposition of any material under microbial activity is usually dependent on humidity and temperature of the system and the T2 condition was more favorable in this case than T1. In another biodegradability study, Pietikainen et al. reported that fungal and bacterial growth rates were optimum in a temperature range of 25°C–30°C [73]. On the other hand, Tremier et al. reported that the optimal temperature, for composting of organic wastes sludge mixed with pine barks via microorganisms, was about 40°C [74].

Kalka et al. also observed that initially up to 28 days, the mass of the different composites (relative to their dry mass) increased due to water uptake. However, after allowing more days, the mass of the ACC composites started to decrease due to the activity of microorganisms. Figure 8.32b shows the change in dry mass of different composites after 70 days of soil burial. A highest loss in mass (about 73%) was observed for ACC-u-T2, that is, ACC without fungicide in T2 condition [72].

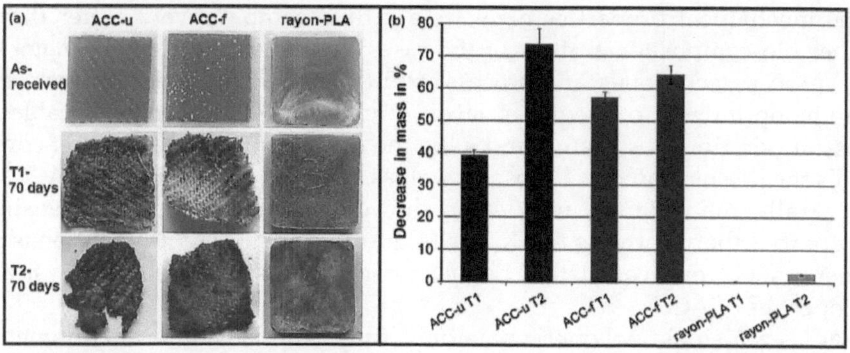

FIGURE 8.32
(a) Photographs of the ACC and rayon-PLA coupons before and after 70 days of soil burial testing at T1 and T2 conditions. Abbreviations: ACC-f = fungicide-treated ACC, ACC-u = untreated ACC (without fungicide); (b) Decrease in dry mass of the various composites after 70 days of soil burial as a function of temperature and fungicidal treatment. (From Kalka, S. et al., *Compos. Part A*, 59, 37, 2014. With permission.)

Interestingly, with T1 condition, the presence of fungicide in the composite (ACC-f-T1) increased the biodegradation compared to ACC-u-T1 (untreated ACC without fungicide). The reasons behind increased biodegradation in the ACC laminates probably are: (i) competitive microbial mechanisms, (ii) temperature- and moisture-dependent decomposition, and (iii) reduced efficacy of the fungicide by the process itself, due to washout of some fungicides in ethanol during the regeneration process. However, in T2 condition, the weight loss and delamination tendency of the individual laminates in ACC-f-T2 were low compared to ACC-u-T2, as presence of fungicide restricted the degradation process [72].

No external degradation was observed in case of rayon-PLA composite, except changing in color to slight dark as a sign of microbial attack (Figure 8.32a) [72]. Although PLA was considered to be biodegradable, the rate of degradation was very slow in the absence of elevated temperature–composting conditions [75], which was the main reason behind non-degradation behavior of rayon-PLA composites here [72].

8.7 Conclusion

In ACCs, the content of fibers/reinforcement is generally higher than other traditional bio-composites. Besides this, due to chemical similarity of the matrix and reinforcing phase, they are highly compatible with each other, which results in effective stress transfer between the two constituents.

The mechanical properties of ACCs are thus comparatively better than other bio-composites in most of the cases. Moreover, by varying materials, process techniques, and process conditions, the mechanical properties can be optimized to a required level. Among different process variables, dissolution time has been found to be the most important factor that controls the dissolution of cellulose as well as strength and stiffness of ACCs. Generally, the unidirectional ACCs showed comparatively better tensile properties than isotropic ACCs, due to the alignment of reinforcing phase. There is a strong correlation between composite structure and physical properties of ACCs.

ACCs are single polymeric materials, fully recyclable and environment friendly. The storage modulus of most of the ACCs showed good values, which is an indication of good dynamic rigidity under cyclic deformation and also good thermomechanical performance. Though the thermal stability of ACCs has not been observed to be very good due to thermal degradation of cellulose at a temperature higher than 200°C, in some cases the ACCs continued to show better thermal stability than other natural fiber–reinforced bio-composites. Besides mechanical properties, the other physical properties such as thermal, viscoelastic, optical, permeability, swelling, water retention property, and biodegradability of ACCs are also important for some special applications. Especially, ACCs performed well in terms of properties such as absorbency, swelling, water retention, biocompatibility, biodegradability, and drug-release, which are important considering a biomedical application, which will be briefly discussed in Chapter 10.

References

1. Qin, C., N. Soykeabkaew, N. Xiuyuan, and T. Peijs. The effect of fibre volume fraction and mercerization on the properties of all-cellulose composites. *Carbohydrate Polymers* 71, no. 3 (2008): 458–467.
2. Soykeabkaew, N., T. Nishino, and T. Peijs. All-cellulose composites of regenerated cellulose fibres by surface selective dissolution. *Composites Part A: Applied Science and Manufacturing* 40, no. 4 (2009): 321–328.
3. Pullawan, T., A. N. Wilkinson, L. N. Zhang, and S. J. Eichhorn. Deformation micromechanics of all-cellulose nanocomposites: comparing matrix and reinforcing components. *Carbohydrate Polymers* 100(2014): 31–39.
4. Garkhail, S. K., R. W. H. Heijenrath, and T. Peijs. Mechanical properties of natural-fibre-mat-reinforced thermoplastics based on flax fibres and polypropylene. *Applied Composite Materials* 7, no. 5–6 (2000): 351–372.
5. Van den Oever, M. J. A., H. L. Bos, and M. J. J. M. Van Kemenade. Influence of the physical structure of flax fibres on the mechanical properties of flax fibre reinforced polypropylene composites. *Applied Composite Materials* 7, no. 5–6 (2000): 387–402.

6. Malik, P. K., *Fibre Reinforced Composites*. 2nd Edition, New York: Marcel Dekker, Inc., 1993.
7. Pullawan, T., A. N. Wilkinson, and S. J. Eichhorn. Discrimination of matrix–fibre interactions in all-cellulose nanocomposites. *Composites Science and Technology* 70, no. 16 (2010): 2325–2330.
8. Adak, B., and S. Mukhopadhyay. Effect of the dissolution time on the structure and properties of lyocell-fabric-based all-cellulose composite laminates. *Journal of Applied Polymer Science* 133, no. 19 (2016).
9. Soykeabkaew, N., N. Arimoto, T. Nishino, and T. Peijs. All-cellulose composites by surface selective dissolution of aligned ligno-cellulosic fibres. *Composites Science and Technology* 68, no. 10–11 (2008): 2201–2207.
10. Askeland, D. R., P. P. Fulay and W. J. Wrigh, *The Science and Engineering of Materials*. 6th Edition, Boston: Cengage Learning, 2010.
11. Nishino, T., I. Matsuda, and K. Hirao. All-cellulose composite. *Macromolecules* 37, no. 20 (2004): 7683–7687.
12. Bledzki, A. K., and J. Gassan. Composites reinforced with cellulose based fibres. *Progress in Polymer Science* 24, no. 2 (1999): 221–274.
13. Ray, D., and B. K. Sarkar. Characterization of alkali-treated jute fibers for physical and mechanical properties. *Journal of Applied Polymer Science* 80, no. 7 (2001): 1013–1020.
14. Huber, T., S. Bickerton, J. Müssig, S. Pang, and M. P. Staiger. Solvent infusion processing of all-cellulose composite materials. *Carbohydrate Polymers* 90, no. 1 (2012): 730–733.
15. Adak, B., and S. Mukhopadhyay. Effect of pressure on structure and properties of lyocell fabric-based all-cellulose composite laminates. *The Journal of the Textile Institute* 108, no. 6 (2017): 1010–1017.
16. Shibata, M., N. Teramoto, T. Nakamura, and Y. Saitoh. All-cellulose and all-wood composites by partial dissolution of cotton fabric and wood in ionic liquid. *Carbohydrate Polymers* 98, no. 2 (2013): 1532–1539.
17. Huber, T., S. Pang, and M. P. Staiger. All-cellulose composite laminates. *Composites Part A: Applied Science and Manufacturing* 43, no. 10 (2012): 1738–1745.
18. Piltonen, P., N. C. Hildebrandt, B. Westerlind, J. P. Valkama, T. Tervahartiala, and M. Illikainen. Green and efficient method for preparing all-cellulose composites with NaOH/urea solvent. *Composites Science and Technology* 135(2016): 153–158.
19. Adak, B., and S. Mukhopadhyay. A comparative study on lyocell-fabric based all-cellulose composite laminates produced by different processes. *Cellulose* 24, no. 2 (2017): 835–849.
20. Ma, H., B. Zhou, H. S. Li, Y. Q. Li, and S. Y. Ou. Green composite films composed of nanocrystalline cellulose and a cellulose matrix regenerated from functionalized ionic liquid solution. *Carbohydrate Polymers* 84, no. 1 (2011): 383–389.
21. Qi, H., J. Cai, L. Zhang, and S. Kuga. Properties of films composed of cellulose nanowhiskers and a cellulose matrix regenerated from alkali/urea solution. *Biomacromolecules* 10, no. 6 (2009): 1597–1602.
22. Zhao, J., X. He, Y. Wang, W. Zhang, X. Zhang, X. Zhang, Y. Deng, and C. Lu. Reinforcement of all-cellulose nanocomposite films using native cellulose nanofibrils. *Carbohydrate Polymers* 104 (2014): 143–150.

23. Duchemin, B. J. C., R. H. Newman, and M. P. Staiger. Structure–property relationship of all-cellulose composites. *Composites Science and Technology* 69, no. 7–8 (2009): 1225–1230.
24. Li, D., X. Sun, and M. A. Khaleel. Materials design of all-cellulose composite using microstructure based finite element analysis. *Journal of Engineering Materials and Technology* 134, no. 1 (2012): 010911.
25. Soykeabkaew, N., C. Sian, S. Gea, T. Nishino, and T. Peijs. All-cellulose nanocomposites by surface selective dissolution of bacterial cellulose. *Cellulose* 16, no. 3 (2009): 435–444.
26. Ghaderi, M., M. Mousavi, H. Yousefi, and M. Labbafi. All-cellulose nanocomposite film made from bagasse cellulose nanofibers for food packaging application. *Carbohydrate Polymers* 104(2014): 59–65.
27. Nishino, T., and N. Arimoto. All-cellulose composite prepared by selective dissolving of fiber surface. *Biomacromolecules* 8, no. 9 (2007): 2712–2716.
28. Han, D., and L. Yan. Preparation of all-cellulose composite by selective dissolving of cellulose surface in PEG/NaOH aqueous solution. *Carbohydrate Polymers* 79, no. 3 (2010): 614–619.
29. Iguchi, M., S. Yamanaka, and A. Budhiono. Bacterial cellulose—a masterpiece of nature's arts. *Journal of Materials Science* 35, no. 2 (2000): 261–270.
30. Yamanaka, S., K. Watanabe, N. Kitamura, M. Iguchi, S. Mitsuhashi, Y. Nishi, and M. Uryu. The structure and mechanical properties of sheets prepared from bacterial cellulose. *Journal of Materials Science* 24, no. 9 (1989): 3141–3145.
31. Gindl, W., K. J. Martinschitz, P. Boesecke, and J. Keckes. Structural changes during tensile testing of an all-cellulose composite by in situ synchrotron X-ray diffraction. *Composites Science and Technology* 66, no. 15 (2006): 2639–2647.
32. Gindl, W., and J. Keckes. Drawing of self-reinforced cellulose films. *Journal of Applied Polymer Science* 103, no. 4 (2007): 2703–2708.
33. Gindl, W., K. J. Martinschitz, P. Boesecke, and J. Keckes. Changes in the molecular orientation and tensile properties of uniaxially drawn cellulose films. *Biomacromolecules* 7, no. 11 (2006): 3146–3150.
34. Pullawan, T., A. N. Wilkinson, and S. J. Eichhorn. Orientation and deformation of wet-stretched all-cellulose nanocomposites. *Journal of Materials Science* 48, no. 22 (2013): 7847–7855.
35. Pang, J. H., X. Liu, M. Wu, Y. Y. Wu, X. M. Zhang, and R. C. Sun. Fabrication and characterization of regenerated cellulose films using different ionic liquids. *Journal of Spectroscopy* 2014 (2014): 1–8.
36. Ding, H., *Handbook of Plastic Industry*. Beijing, China: Chemical Industry Press, 1995.
37. Schuermann, H., T. Huber, and M. P. Staiger. Prepreg style fabrication of all cellulose composites. In *Proceedings of 19th international conference on composite materials, Canada*, pp. 5626–5634. 2013.
38. Adak, B., and M. Samrat. Jute based all-cellulose composite laminates. *Indian Journal of Fibre & Textile Research (IJFTR)* 41, no. 4 (2016): 380–384.
39. Ouajai, S., and R. A. Shanks. Preparation, structure and mechanical properties of all-hemp cellulose biocomposites. *Composites Science and Technology* 69, no. 13 (2009): 2119–2126.
40. Zhao, Q., R. C. M. Yam, B. Zhang, Y. Yang, X. Cheng, and R. K. Y. Li. Novel all-cellulose ecocomposites prepared in ionic liquids. *Cellulose* 16, no. 2 (2009): 217–226.

41. Yang, Q., A. Lue, and L. Zhang. Reinforcement of ramie fibers on regenerated cellulose films. *Composites Science and Technology* 70, no. 16 (2010): 2319–2324.
42. Gindl, W., and J. Keckes. All-cellulose nanocomposite. *Polymer* 46, no. 23 (2005): 10221–10225.
43. Gindl, W., T. Schöberl, and J. Keckes. Structure and properties of a pulp fibre-reinforced composite with regenerated cellulose matrix. *Applied Physics A* 83, no. 1 (2006): 19–22.
44. Duchemin, B. J. C., A. P. Mathew, and K. Oksman. All-cellulose composites by partial dissolution in the ionic liquid 1-butyl-3-methylimidazolium chloride. *Composites Part A: Applied Science and Manufacturing* 40, no. 12 (2009): 2031–2037.
45. Yousefi, H., T. Nishino, M. Faezipour, G. Ebrahimi, and A. Shakeri. Direct fabrication of all-cellulose nanocomposite from cellulose microfibers using ionic liquid-based nanowelding. *Biomacromolecules* 12, no. 11 (2011): 4080–4085.
46. Gindl-Altmutter, W., J. Keckes, J. Plackner, F. Liebner, K. Englund, and M. P. Laborie. All-cellulose composites prepared from flax and lyocell fibres compared to epoxy–matrix composites. *Composites Science and Technology* 72, no. 11 (2012): 1304–1309.
47. Dittenber, D. B., and H. V. S. GangaRao. Critical review of recent publications on use of natural composites in infrastructure. *Composites Part A: Applied Science and Manufacturing* 43, no. 8 (2012): 1419–1429.
48. Persico, P., D. Acierno, C. Carfagna, and F. Cimino. Mechanical and thermal behaviour of ecofriendly composites reinforced by Kenaf and Caroà fibers. *International Journal of Polymer Science* 2011 (2011).
49. Huber, T., S. Bickerton, J. Müssig, S. Pang, and M. P. Staiger. Flexural and impact properties of all-cellulose composite laminates. *Composites Science and Technology* 88(2013): 92–98.
50. Duchemin, B. J. C., M. P. Staiger, N. Tucker, and R. H. Newman. Aerocellulose based on all-cellulose composites. *Journal of Applied Polymer Science* 115, no. 1 (2010): 216–221.
51. Goutianos, S., R. Arévalo, B. F. Sørensen, and T. Peijs. Effect of processing conditions on fracture resistance and cohesive laws of binderfree all-cellulose composites. *Applied Composite Materials* 21, no. 6 (2014): 805–825.
52. Bachtiar, D., S. M. Sapuan, and M. M. Hamdan. The influence of alkaline surface fibre treatment on the impact properties of sugar palm fibre-reinforced epoxy composites. *Polymer-Plastics Technology and Engineering* 48, no. 4 (2009): 379–383.
53. Obasi, H. C., N. C. Iheaturu, F. N. Onuoha, C. O. Chike-Onyegbula, M. N. Akanbi, and V. O. Eze. Influence of alkali treatment and fibre content on the properties of oil palm press fibre reinforced epoxy biocomposites. *American Journal of Engineering Research* 3, no. 2 (2014): 117–123.
54. Mantena, P. R., R. Mann, and C. Nori. Low-velocity impact response and dynamic characteristics of glass-resin composites. *Journal of Reinforced Plastics and Composites* 20, no. 6 (2001): 513–534.
55. Bax, B., and J. Müssig. Impact and tensile properties of PLA/Cordenka and PLA/flax composites. *Composites Science and Technology* 68, no. 7–8 (2008): 1601–1607.
56. Rubio-López, A., A. Olmedo, and C. Santiuste. Modelling impact behaviour of all-cellulose composite plates. *Composite Structures* 122(2015): 139–143.

57. Nilsson, H., S. Galland, P. T. Larsson, E. K. Gamstedt, T. Nishino, L. A. Berglund, and T. Iversen. A non-solvent approach for high-stiffness all-cellulose biocomposites based on pure wood cellulose. *Composites Science and Technology* 70, no. 12 (2010): 1704–1712.
58. Li, D., Z. Liu, M. Al-Haik, M. Tehrani, F. Murray, R. Tannenbaum, and H. Garmestani. Magnetic alignment of cellulose nanowhiskers in an all-cellulose composite. *Polymer Bulletin* 65, no. 6 (2010): 635–642.
59. Lourdin, D., J. Peixinho, J. Bréard, B. Cathala, E. Leroy, and B. Duchemin. Concentration driven cocrystallisation and percolation in all-cellulose nanocomposites. *Cellulose* 23, no. 1 (2016): 529–543.
60. Turi, E. A., *Thermal Characterization of Polymeric Materials.* 2nd Edition, New York, London: Elsevier, 1997.
61. Szcześniak, L., A. Rachocki, and J. Tritt-Goc. Glass transition temperature and thermal decomposition of cellulose powder. *Cellulose* 15, no. 3 (2008): 445–451.
62. Albano, C., J. Gonzalez, M. Ichazo, and D. Kaiser. Thermal stability of blends of polyolefins and sisal fiber. *Polymer Degradation and Stability* 66, no. 2 (1999): 179–190.
63. Nada, A. M. A., and M. L. Hassan. Thermal behavior of cellulose and some cellulose derivatives. *Polymer Degradation and Stability* 67, no. 1 (2000): 111–115.
64. Nair, S. S., J. Y. Zhu, Y. Deng, and A. J. Ragauskas. High performance green barriers based on nanocellulose. *Sustainable Chemical Processes* 2, no. 1 (2014): 23.
65. Pinkert, A., K. N. Marsh, S. Pang, and M. P. Staiger. Ionic liquids and their interaction with cellulose. *Chemical Reviews* 109, no. 12 (2009): 6712–6728.
66. Yang, Q., H. Fukuzumi, T. Saito, A. Isogai, and L. Zhang. Transparent cellulose films with high gas barrier properties fabricated from aqueous alkali/urea solutions. *Biomacromolecules* 12, no. 7 (2011): 2766–2771.
67. Adak, B., and S. Mukhopadhyay. All-cellulose composite laminates with low moisture and water sensitivity. *Polymer* 141(2018): 79–85.
68. Wang, Y., and L. Chen. Impacts of nanowhisker on formation kinetics and properties of all-cellulose composite gels. *Carbohydrate Polymers* 83, no. 4 (2011): 1937–1946.
69. Wu, R. L., X. L. Wang, F. Li, H. Z. Li, and Y. Z. Wang. Green composite films prepared from cellulose, starch and lignin in room-temperature ionic liquid. *Bioresource Technology* 100, no. 9 (2009): 2569–2574.
70. Huber, T., J. Müssig, O. Curnow, S. Pang, S. Bickerton, and M. P. Staiger. A critical review of all-cellulose composites. *Journal of Materials Science* 47, no. 3 (2012): 1171–1186.
71. Ha, S. H., N. L. Mai, G. An, and Y. M. Koo. Microwave-assisted pretreatment of cellulose in ionic liquid for accelerated enzymatic hydrolysis. *Bioresource Technology* 102, no. 2 (2011): 1214–1219.
72. Kalka, S., T. Huber, J. Steinberg, K. Baronian, J. Müssig, and M. P. Staiger. Biodegradability of all-cellulose composite laminates. *Composites Part A: Applied Science and Manufacturing* 59(2014): 37–44.
73. Pietikäinen, J., M. Pettersson, and E. Bååth. Comparison of temperature effects on soil respiration and bacterial and fungal growth rates. *FEMS Microbiology Ecology* 52, no. 1 (2005): 49–58.

74. Tremier, A., A. De Guardia, C. Massiani, E. Paul, and J. L. Martel. A respirometric method for characterising the organic composition and biodegradation kinetics and the temperature influence on the biodegradation kinetics, for a mixture of sludge and bulking agent to be co-composted. *Bioresource Technology* 96, no. 2 (2005): 169–180.
75. Mathew, A. P., K. Oksman, and M. Sain. Mechanical properties of biodegradable composites from poly lactic acid (PLA) and microcrystalline cellulose (MCC). *Journal of Applied Polymer Science* 97, no. 5 (2005): 2014–2025.

9

Derivatized All-Cellulose Composites

9.1 Introduction

Recent advances in material science and engineering as well as environmental awareness all over the world are enforcing development of new and better natural fiber–reinforced bio-composites, which are strong, biodegradable, and recyclable [1–2]. Many studies have reported natural fiber–reinforced bio-composites using biodegradable polymers such as polylactic acid, polycaprolactone (PCL), polyglycolic acid (PGA), and poly(3-hydroxybutyrate-co-3-hydroxyvalerate) (PHBV) [3–4]. The fiber–matrix compatibility issues of conventional bio-composites were tried to be resolved by modifying (physically or chemically) the surface of matrix or reinforcement and thereby increasing interaction between fiber and matrix [5–7]. Subsequently, the concept of "single polymer composite" (SPC) or "self-reinforced composite" (SRC) was developed to overcome the issues related to fiber–matrix incompatibility [8–10]. All-cellulose composite (ACC) is one of the most emerging classes of bio-composites in the field of SRCs. The concept of synthesis, different manufacturing techniques, microstructures, and properties of non-derivatized ACCs have been discussed in Chapters 7 and 8. The main limitations of non-derivatized ACCs are that most of these composites are very much sensitive to moisture/water and as the composite is based on pure cellulose, melt processing is not possible. Therefore, many researches were focused on derivatizing of cellulose, which can be further processed like conventional polymers by hot pressing, for manufacturing of ACCs [8–10]. These types of composites are referred to as "derivatized all-cellulose composites" (DACCs) or "all-plant fiber composites," because natural plant fibers and wood are generally used for making them.

In the manufacturing of DACCs, cellulosic fibers are first derivatized by chemical treatment. One of the most popular processes in the field of derivatized cellulose is "viscose technology." In this process, sodium hydroxide (NaOH) with carbon disulfide (CS_2) are used as solvents, where sodium cellulose xanthate is formed as a cellulose derivative [11].

$$\text{Cellulose} - \text{OH} + \text{NaOH} = \text{Cellulose} - \text{ONa} + H_2O \tag{9.1}$$

$$\text{Cellulose} - \text{ONa} + \text{CS}_2 = \text{Cell} - \text{OCS}_2 - \text{Na}+ \quad (9.2)$$

The main drawback of this process is that it is not eco-friendly, as viscose process requires handling of toxic gaseous by-products, namely, CS_2 and H_2S, during cellulose derivatization [11]. In the last few decades, many solvents or chemical systems have been developed for derivatization of cellulose, which have been used successfully in the synthesis of DACCs.

The plant fibers are composed of cellulose, lignin, hemicellulose, pectin, and some amount of other organic constituents such as waxes and ash. In spite of this heterogeneity, all-plant fiber composites are favorable for interfacial interactions after surface modification or derivatization [12]. The process of derivatization helps to improve thermoplasticity of cellulose, which is useful in making bio-composites.

9.2 Derivatizing Solvents

As the name suggests, "derivatizing solvents" consist of all systems where dissolution takes place in combination with the formation of unstable ester, ether, or acetal derivatives by substituting hydroxyl group of cellulose [13–14]. Different protic acids (H_3PO_4, H_2SO_4, and HNO_3), NaOH/CS_2, DMF/N_2O_4, and so on are examples of derivatizing solvents. Viscose process with NaOH/CS_2-based solvent system is the most common and renowned derivatizing solvent system used so far [11]. Cellulose carbamate, cellulose xanthate, cellulose acetate (CA), cellulose formate, and so on are examples of soluble intermediates that result from reasonably complex derivatizing solvents [13]. The cellulose derivatives subsequently can be regenerated to pure cellulose by additional treatments, or may be further processed by "hot-melt processing," as per requirement for preparation of composites. The list of derivatizing solvents is vast, but these are rarely used for synthesis of ACCs compared to their other counterpart, that is, non-derivatizing solvents.

9.3 Philosophy of Making DACCs

More recently, different approaches have been taken for synthesis of derivatized ACCs, mainly based on benzylation, esterification, and oxypropylation [15–17]. Different derivatized solvents or other chemicals are used for surface modification of cellulosic materials. In this process, the skin part of the cellulosic fibers are converted into thermoplastic material by making a cellulose derivative, while the core part remains unchanged. Subsequently, "hot-melt processing" is used for making of composites, in which the plasticized part forms the matrix and the unplasticized core part acts as reinforcement. Few researchers [18] have reported the conversion of cellulose to thermoset matrix, which was melt processed for one time only, forming cross-links during preparation of composite. Although these types of composites are called as ACCs by the researchers, these are not SPCs in the true sense. Though these composites are primarily based on cellulose, with derivatization of cellulose, the polarity of the matrix alters relative to the fibers, resulting in a different chemical nature of the matrix and reinforcing phase.

9.4 Different Types of DACC

In the last two decades, different approaches have been taken by researchers for making derivatized ACCs. The DACCs, which are also called as all-plant fiber composites can be produced from different cellulose derivatives. The most recent approaches for the production of thermoplastic/thermoset all-plant fiber composites are based on partial transformation of cellulose, mainly via

i. Benzylation
ii. Esterification
iii. Oxypropylation
iv. Carbamation
v. 2,2,6,6-tetramethylpiperidine-1-oxyl radical (TEMPO)–mediated oxidation
vi. Acetylation
vii. Periodate oxidation

9.5 Benzylated Cellulose-Based DACCs

9.5.1 Synthesis

The first approach takes into account the production of all-wood or all-plant fiber composites by a benzylation treatment of the cellulosic source [19–20]. The benzylation of wood is a typical Williamson synthesis reaction, which involves nucleophilic substitution of an alkoxide or a phenoxide ion for a halide ion.

$$\text{Wood-OH} + \text{NaOH} = \text{Wood-O- Na}^+ + H_2O \quad (9.3)$$

$$\text{Wood-O-Na}^+ + C_6H_6\text{-}CH_2Cl = \text{Wood-O-}CH_2 - C_6H_6 + \text{NaCl} \quad (9.4)$$

During benzylation treatment, the cellulosic material (Wood-OH) is first swollen in NaOH and then transferred to benzyl chloride (C_6H_6-CH_2Cl) [15]. Subsequently, the solution is stirred at a temperature higher than 100°C and then washed properly for removal of residual benzyl chloride, inorganic salts, and different by-products, getting a liquid mixture of cellulosic materials. The rate of benzylation is mainly dependent on processing temperature, concentration of NaOH, and also amount of benzyl chloride, but hardly depends on the source of wood [15,21–23].

Due to complex structure and composition of wood and its macromolecules, an indirect technique is used for measuring the extent of benzylation, that is, by measuring the weight gain after chemical treatment. After benzylation, a sheath-core type structure is obtained in the fibers where the core of the fiber is surrounded by a thermoplastic layer. These benzylated fibers can be utilized for making composites by hot pressing, while the fiber volume fraction can reach values up to 40 wt% [15,21–22,24].

Using this technique, Lu et al. produced SRCs from partially benzylated sisal fibers, which were thermally consolidated into a matrix reinforced with the remaining (unsubstituted) cellulose [22]. In another study, unidirectional sisal fibers were used separately as reinforcement where plasticized (benzylated) China fir sawdust was used as matrix phase, producing unidirectional DACC laminates [15]. A similar study was reported by Zhang et al., where they additionally manufactured short sisal fiber–reinforced DACCs [24].

9.5.2 Structure and Properties

Mechanical properties of benzylated ACCs increase with increasing fiber volume fraction and percentage weight gain up to a certain extent. However, subsequently, reduction in mechanical properties were observed. Lu et al. [22] reported that with increasing degree of benzylation of wood (cellulosic materials), crystallinity of cellulose reduces, due to breakage of inter and intramolecular hydrogen bonds of cellulose. It also causes decrease in mechanical properties of benzylated ACCs, after a certain degree of benzylation.

With incorporation of large benzyl molecules in cellulose, larger free volume is introduced, resulting in a supramolecular structure of cellulose. Moreover, with increasing number of benzyl groups, viscosity of the material increases, which leads to insufficient wetting of cellulosic reinforcement, at higher fiber volume fraction (≥40 wt%) [15,22,24]. As a consequence, a weak interface is formed between matrix and reinforcement, causing reduction of tensile strength and Young's modulus (Figure 9.1) for higher extent of benzylation. Almost a similar trend was observed in flexural properties (strength and modulus) of benzylated ACCs [15,22]. An overview of the tensile properties of benzylated ACCs is given in Table 9.1.

The viscoelastic and thermomechanical properties of benzylated ACCs were analyzed by dynamic mechanical analysis and thermomechanical

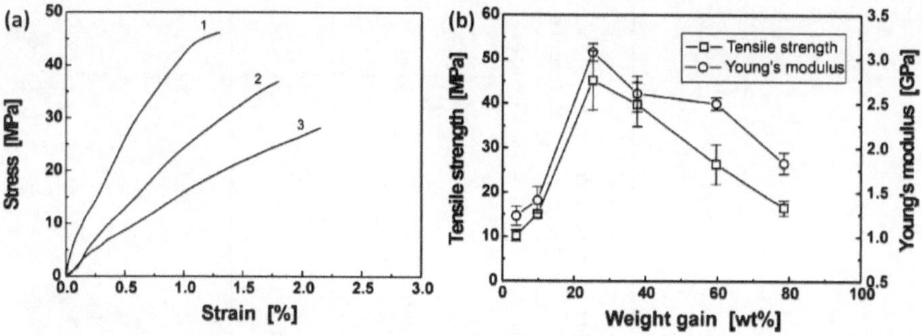

FIGURE 9.1
(a) Typical stress–strain curves of sisal fiber–reinforced DACCs with different weight gains of benzylated sisal: (1) 25.8 wt%, (2) 43 wt%, and (3) 59.8 wt%; (b) Tensile properties of DACCs with respect to weight gain of benzylated sisal. (From Lu, X. et al., *Compos. Sci. Technol.*, 63, 177, 2003. With permission.)

analysis [22,24]. Due to presence of benzylated cellulose at the surface, the benzylated ACCs show a thermoplastic behavior. As a result, it exhibits a softening temperature at about 90°C–120°C, depending on cellulosic source and the extent of benzylation. However, the core fibers remain unchanged, showing no thermoplastic behavior [22].

Lu et al. studied biodegradability of benzylated sisal fiber–based SRCs under the action of enzymes and analyzed the effect of reaction time and temperature, dosage of enzymes, and pH of enzyme solution. The enzymolysis of the composites is a diffusion-controlled process, and due to the impediment of lignin to cellulase (enzyme) solution, the rate of cellulose degradation gradually reduces with increasing time [25]. In another study, Lu et al. showed the effect of enzyme solution as well as aging in water and soil for degradation of benzylated sisal fiber–based SRCs. It was found that the inherent biodegradability of plant fibers was also present in benzylated sisal fiber–based SRCs, resulting in degradation with time with the help of cellulase and fungi [26].

9.6 Esterified Cellulose-Based DACCs

9.6.1 Synthesis

DACCs can also be produced by hot pressing of esterified cellulosic materials. Matsumura et al. produced self-reinforced cellulose-based composites from partially derivatized cellulose esters, which were prepared from dissolving-grade wood pulp fibers. The pulp fibers were treated with a p-toluene sulfonic/hexanoic anhydride–based system and in a cyclohexane based non-swelling reaction medium. The pulp fibers were exposed to heterogeneous

TABLE 9.1

Tensile Properties of DACCs Produced from Benzylated Cellulose

Cellulose source for matrix	Reinforcement	Fiber fraction (vol% or wt%)	Tensile strength (MPa)	Young's modulus (GPa)	Strain at break (%)	Flexural strength (MPa)	Flexural modulus (GPa)	Impact strength (KJ/m^2)	Reference
Wood sawdust	Sisal fiber (Isotropic)	14.3 vol%	32	2.35	–	54	2.85	9	Zhang et al. [24]
Wood sawdust	Sisal fiber (UD)	30.4 vol%	95	16	4.8	160	13	65	Zhang et al. [24]
Wood sawdust	Sisal fiber (UD)	30 vol%	92	15	4.8	160	12.5	–	Lu et al. [15]
Wood sawdust	Sisal fiber (UD)	40 vol%	68	20	4.2	110	24	–	Lu et al. [15]
Sisal fiber	Sisal fiber (UD)	–	4	2.6	–	88	3.25		Lu et al. [22]

hexanoylation reaction in presence of a titanium (IV) isopropoxide–catalyzed system, forming hexanoylated pulp fibers. It was assumed that hexanoylation started from the surface (disordered regions) of each microfibrils and proceeds toward interior (ordered regions) of the microfibrils, thereby creating well-oriented cellulose hexanoates on the surface of cellulosic fibers, under a non-swelling reaction condition. The modified pulp fiber was dispersed in water or methanol and the suspension was filtered to receive a fiber mat of uniform thickness. Subsequently, the fiber mat was hot pressed under a compression molding at 155°C–170°C and at room temperature, producing a self-reinforced cellulose composite [17]. In a similar study, Matsumura et al. also produced self-reinforced cellulose composite from partially modified (esterified) lyocell fibers, having average diameter 12 μm, 1 μm thick surface layer, and an overall degree of substitution (DS) of 0.6 [27].

9.6.2 Structure and Properties

Esterified ACCs were transparent or semitransparent (based on DS) due to a combined effect of the apparent thermoplasticity and molecular consolidation [17,27]. During esterification reaction with increasing DS, the parts of unmodified cellulose I decrease and the parts of cellulose ester increase. The X-ray diffraction study shows that the heterogeneously hexanoylated pulp fibers contained a comingled mixture of crystalline cellulose I and ordered cellulose hexanoate. The crystal size of residual cellulose I in the heterogeneously hexanoylated pulp fibers reduced gradually with increasing DS [17]. The DS of the esterified ACCs is the deciding factor for their mechanical properties. The breaking extension of the esterified ACCs increase gradually with increasing DS. Both the tensile strength and Young's modulus of esterified ACCs also increased up to a certain of DS, but reduced at highest tested DS, that is, 2 [17]. Table 9.2 gives an overview of mechanical properties of different types of DACCs.

Matsumura and Glasser analyzed the thermomechanical performance of lyocell fiber–based esterified ACCs. Due to presence of thermoplastic matrix on the surface of lyocell fibers, an apparent T_g was found at 75°C, which was corresponding to cellulose hexanoate. However, no significant thermal transition was observed for unmodified fiber core [27].

9.7 Oxypropylated Cellulose-Based DACCs

9.7.1 Synthesis

Gandini et al. first reported the oxypropylation treatment of cellulose, for making SRCs [29]. Many researchers have been following this technique

TABLE 9.2

Tensile Properties of Different Types of DACCs

Cellulose source	Treatment	Reinforcement type	Fiber fraction (vol% or wt%)	Tensile strength (MPa)	Young's modulus (GPa)	Strain at break (%)	Reference
Wood pulp	Esterification	Isotropic	–	25	0.8	6	Matsumura et al. [17]
Wood pulp	Esterification	Isotropic	–	20	1.3	5	Matsumura et al. [17]
Filter paper	Oxypropylation (catalyst- DABCO)	Isotropic	–	25.7	1.31	4.91	de Menezes et al. [16]
Filter paper	Oxypropylation (catalyst- KOH)	Isotropic	–	18.7	1.18	2.7	de Menezes et al. [16]
Hardwood kraft pulp	TEMPO-mediated oxidation	NFC	12 wt%	53.2	5.95	1.5	Alcala' et al. [28]

Note: DABCO = 1,4-diazabicyclo [2.2.2] octane.

for preparation of derivatized all-plant fiber composites [16,30]. The oxypropylation of cellulose is done by immersing in a Brønsted base (KOH) or a Lewis base (1,4-diazabicyclo [2.2.2] octane) (DABCO) for activation of hydroxyl groups of cellulose. This step is followed by evaporation of alcohol and an anionic polymerization of propylene oxide (PO) on the surface of cellulosic fibers, which is done in an autoclave under nitrogen atmosphere at temperatures of 130–150 °C, like a grafting process. The PO-homopolymers generated by chain transfer reaction are removed by a Soxhlet extraction using hexane. The degree of oxypropylation, which depends on the amount of PO used, is calculated by measuring the weight gain due to grafting of thermoplastic polymer in the outer surface of cellulosic fibers. This reaction converts cellulose I to an amorphous oxypropylated derivative [16,29–30].

An in-depth oxypropylation treatment of different cellulosic fibers (Rayon, Kraft, Avicel, and filter paper) was done by de Menezes et al. The mono-component and biphasic materials were hot pressed to obtain DACC films where unmodified cellulosic fibers were dispersed in a thermoplastic matrix [30]. In a similar study, the effect of two different catalysts (KOH and DABCO) on activating the OH groups of polysaccharides for polymerization of PO and also on the properties of prepared DACCs were investigated. In oxypropylated cellulose-based DACCs, the fiber volume fraction is generally varied in the range of 10%–40% [16,29–30].

9.7.2 Structure and Properties

The best morphology and properties of oxypropylated ACCs are obtained at optimal molar ratio of PO and cellulose OH groups. With increase in this ratio, the DS increases, which results in increase of percentage weight gain [16]. The oxypropylation of cellulose produces amorphous oxypropylated cellulose, resulting in decrease in crystallinity of cellulose with increasing DS, as shown in Table 9.3.

The differential scanning calorimetry (DSC) analysis (Figure 9.2) shows no T_g for pristine filter paper (cellulose), while a T_g was found at –60°C for isolated PO-homopolymer (PPO) [16,29,31]. In case of oxypropylated cellulose, a T_g was found between –55°C and –45°C. Moreover, with increasing DS or the molar ratio of PO and cellulose OH, the T_g of oxypropylated cellulose was shifted to lower temperature [16].

During oxypropylation treatment, the amorphous parts are more prone to modification compared to the crystalline parts of cellulose. However, with an increase in the amount of PO, the crystalline parts get modified and the cellulose structure is degraded, resulting in reduction in mechanical properties of composites [30]. The tensile properties of oxypropylated ACCs depend on the amount of PO used or the DS. Similar to esterified ACCs, the breaking extension of the oxypropylated ACCs increase with increasing DS. The tensile strength of the oxypropylated ACCs also increase up to a certain DS, while Young's modulus gradually reduces with increasing DS [16].

TABLE 9.3

Crystallinity Index and Weight Gain of FP Pretreated with KOH and DABCO using Different Molar Ratios of [PO]/[Cellulose OH] During Oxypropylation Treatment

Sample Pretreatment	[PO]/[Cellulose OH] Molar Ratio	Weight Gain (%)	Crystallinity Index (%)
KOH	Unmodified	–	85
	PO1	56.8	85
	PO3	58.7	80
	PO5	58.6	25
DABCO	Unmodified	–	85
	PO1	54.1	79
	PO3	55.8	76
	PO5	61.4	70

Source: de Menezes, A. et al., Cellulose, 16, 239, 2009.

9.8 Carbamated Cellulose-Based DACCs

9.8.1 Synthesis

Vo et al. introduced a new method for synthesizing of ACCs from woven fabrics by carbamation treatment of cellulose. When the fabric was initially treated with a mixture of polyethylene glycol, urea, and salt, it underwent carbamation. As cellulose carbamate is alkali soluble, when the carbamated fabrics were immersed in alkali solutions and hot passed through pressure rolls, there was a little dissolution of cellulose. On subsequent neutralization, the dissolved cellulose was regenerated into matrix to bind the fibrous elements [32].

9.8.2 Structure and Properties

X-ray diffraction analysis showed that the crystallinity of cellulose reduced by carbamation treatment and increased slightly by further alkali treatment by applying a higher nip pressure. Both the tensile strength and breaking extension of the carbamated viscose fabric–based composites reduced; a significant reduction in breaking extension was observed due to the remarkable increase in stiffness. With carbamation treatment, adhesion strength between two fabric layers increased significantly, compared to non-carbamated fabric [32].

The air permeability of woven fabric–based single layer laminas followed this sequence: carbamated and alkali-treated pieces < only alkali-treated

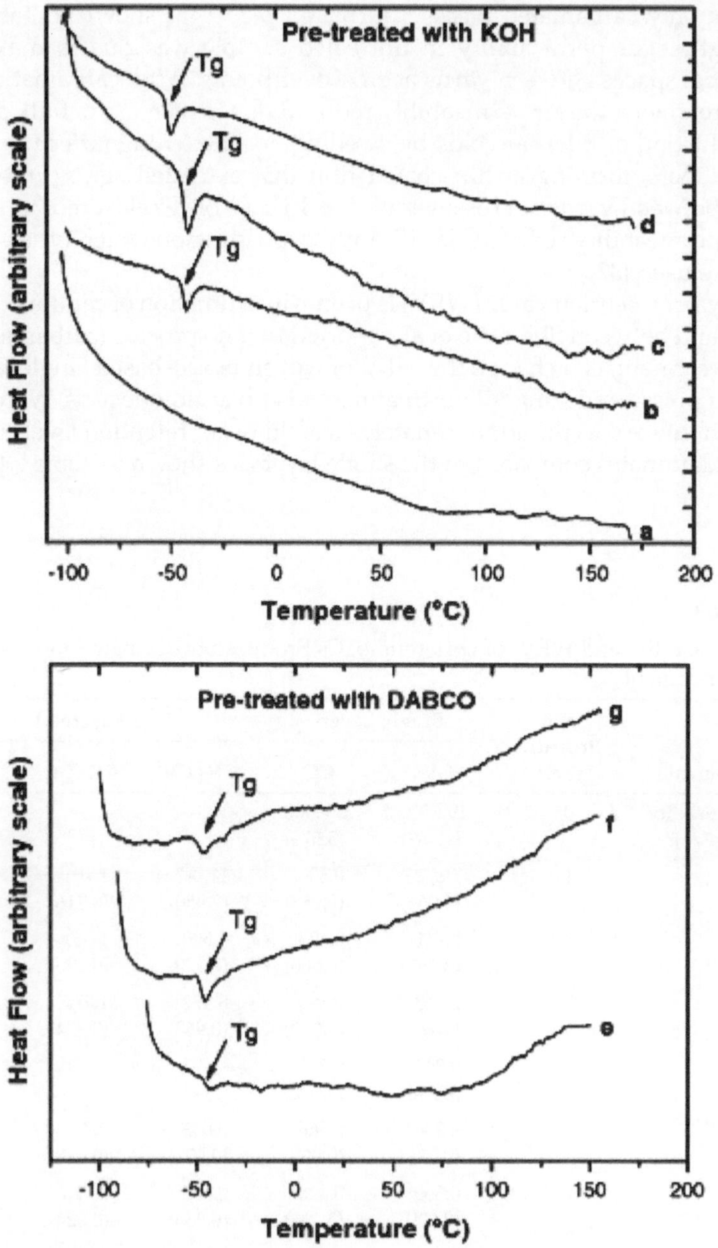

FIGURE 9.2
DSC thermograms of filter paper [(a) FP], oxypropylated filter paper in the presence of KOH and increasing the molar ratio of PO/cellulose OH [(b) FPPO1KOH, (c) FPPO3KOH, (d) FPPO5KOH]; and oxypropylated filter paper in the presence of DABCO and increasing the molar ratio of PO/cellulose OH [(e) FPPO1DABCO, (f) FPPO3DABCO, (g) FPPO5DABCO]. (From de Menezes, A. et al., *Cellulose*, 16, 239, 2009. With permission.)

pieces < only carbamated pieces <untreated pieces (as shown in Table 9.4). The highest air permeability in untreated sample was due to maximum interstitial spaces between yarns at crossover points. After carbamation and alkali treatment, the air permeability reduced due to increase of bulk density and reduction of internal voids by swelling and flattening (after nip pressing) of fibers, forming a thin coated film that extended across interstitial spaces between yarns at crossover points. In case of two-layered laminates, the air permeability reduced significantly depending on which type of laminas were used [32].

The water retention value (WRV) is primarily a function of capillary spaces within and between fibers. Vo et al. reported that no process (carbamation or alkali treatment) can change the WRV of woven fabric–based single lamina. The nip pressure during alkali treatment also had no effect on WRVs. The only difference was the approximately twofold water retention levels in two-layered laminates compared to the single layers, as shown in Table 9.4 [32].

TABLE 9.4

Air Permeability and WRVs of Different ACCs Produced by Carbamation and Alkali Treatment

Measurement	Nip Pressure (bar)	Single Layers		Two-Layered Laminates		
		CV[a]	CC	CV+CV	CC+CC	CV+CC
Air permeability ($cm^3/cm^2/s$); n = 3	0	192.500 (9.383)	177.500 (5.579)	–	–	–
	1	20.722 (4.064)	0.590 (0.053)	11.042 (2.450)	1.701 (0.216)	1.090 (0.133)
	2	26.111 (3.799)	0.473 (0.060)	8.500 (2.170)	1.361 (0.196)	1.069 (0.161)
	3	24.639 (4.187)	0.599 (0.090)	8.472 (0.963)	1.375 (0.224)	1.099 (0.267)
Liquid water retention (ml/g); n = 3	0	0.314 (0.054)	0.315 (0.022)	–	–	–
	1	0.336 (0.036)	0.366 (0.020)	0.648 (0.098)	0.738 (0.085)	0.822 (0.045)
	2	0.350 (0.029)	0.428 (0.052)	0.635 (0.088)	0.710 (0.024)	0.695 (0.043)
	3	0.370 (0.042)	0.389 (0.014)	0.661 (0.079)	0.656 (0.012)	0.727 (0.013)

Source: Vo, L.T.T. et al., *Compos. Sci. Technol.*, 78, 30, 2013.)

[a] CV = untreated (Non-carbamated) viscose fabric piece; CC = carbamated viscose fabric piece; CV+CV, CC+CC, CV+CC indicate the combination in laminates.

"n" signifies the number of repetations.

A nip pressure "0" signifies no alkali treatment.

9.9 Synthesis and Properties of DACC Produced by TEMPO-Mediated Oxidation

Alcala´ et al. prepared nanofibrillated cellulose (NFC) using bleached eucalyptus pulp by TEMPO-mediated oxidation of cellulosic fibers (bleached eucalyptus pulp) at neutral pH conditions and fibrillation was done in a high pressure homogenizer at 60°C–70°C [28,33]. The dried unbleached eucalyptus pulp was disintegrated in water in a very high-speed pulp disintegrator and NFC was added in to the slurry. The resultant slurry was used to fabricate the bio-composites in the form of paper sheets, where the NFC concentration varied from 0–12 wt%. Both the tensile strength and Young's modulus of the bio-composites increased with increasing the NFC concentration up to 12 wt% (Figure 9.3). However, the breaking extension reduced slightly after a certain concentration (9 wt%). The tensile properties were also predicted by mathematical models, which showed higher values than the experimental ones. NFC showed a strong potential in improving mechanical properties of bio-composites and the dispersion of NFC in matrix was one of the key

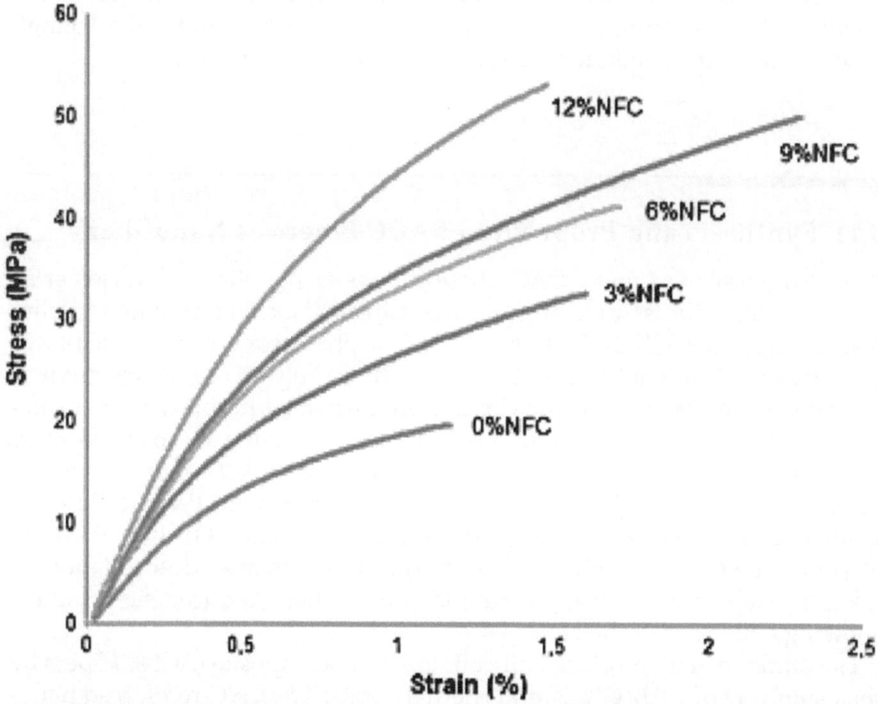

FIGURE 9.3
The stress–strain curve of unbleached eucalyptus pulp/NFC composites produced with varying NFC content. (From Alcala, M. et al., *Cellulose*, 20, 2909, 2013. With permission.)

factors behind the reinforcing effect of ACCs. The increase in revolution of disintegrators helped to disperse NFCs in a better way, resulting in higher tensile strength and modulus [28].

9.10 Comparison of Mechanical Properties: Non-Derivatized ACC vs DACC

The tensile properties of non-derivatized ACCs have been discussed in detail in Chapter 8. Although the DACCs are primarily based on cellulose, derivatization process alters the polarity of the matrix relative to fibers. As a consequence, the mechanical performance of derivatized ACCs is not as good as non-derivatized ACCs. For example, Gindl and Keckes reported that the tensile strength of a derivatized ACC produced from cellulose acetate butyrate (CAB) and reinforced with bacterial cellulose was about 129 MPa [34]. This tensile strength was quite lower in comparison to a bacterial cellulose–based non-derivatized ACC, having tensile strength 411 MPa [35]. In a review on ACCs, Huber et al. reported that the strongest non-derivatized ACCs are about 5–10 times stronger than the derivatized ACCs, whereas the Young's modulus and strain to failure are almost in similar range [36].

9.11 Synthesis and Properties of ACC Fibers or Nanofibers

It is also possible to make DACC-based fibers or nanofibers. Vallejos et al. produced ACC fibers by electrospinning using a dispersion containing cellulose nanocrystals (CNCs, 5 wt%) and CA. The precursor polymer matrix was produced by different DS (1.85, 2.45, and 2.8) in a solvent (2:1 binary mixture of acetone and DMAc) and the CNCs were dispersed properly in that solution. The CA part of electrospun fiber webs was regenerated to cellulose by a deacetylation treatment or alkaline hydrolysis to get ACC fibers that maintained the original morphology of the precursor system. Prior to deacetylation of electrospun webs, a glass transition temperature (T_g) was observed at 145°C, due to the presence of CA matrix. However, after deacetylation by alkali hydrolysis, no T_g was found and water contact reduced significantly, showing a hydrophilic behavior [37].

Hooshmand et al. produced all-cellulose nanocomposite (ACNC) fibers by melt spinning of CAB, CNCs, and triethyl citrate. The CNC-reinforced nanocomposites (2 and 10 wt%) were prepared by sol–gel technique and solution mixing (as described in Figure 9.4). The prepared materials were melt spun in a micro twin-screw extruder and drawn with a draw ratio of 1.5. With the

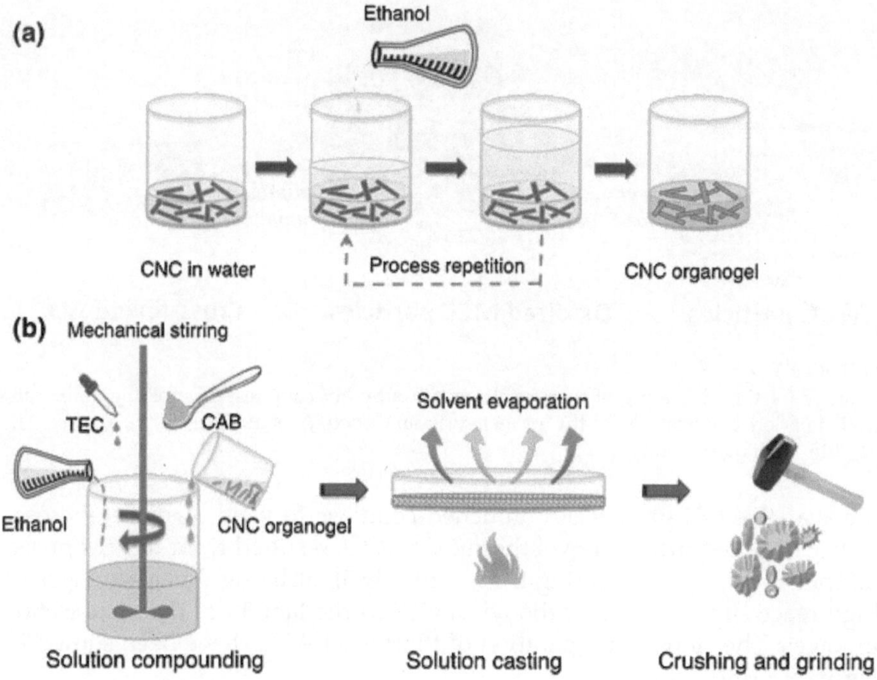

FIGURE 9.4
(a) Schematic view of preparation of CNC organogel by sol–gel technique; (b) Preparation of nanocomposite by solution mixing. (From Hooshmand, S. et al., *Cellulose*, 21, 2665, 2014. With permission.)

addition of CNCs in CAB, smooth and defect-free nanocomposite fibers were produced with significant enhancement of mechanical properties. In case of drawn fiber, with 10 wt% loading of CNC, 59% and 45% improvements were observed in tensile strength and modulus, respectively, in comparison to as-spun ones [38].

9.12 Thermoset DACC

9.12.1 Synthesis

Similar to thermoplastic DACC, the thermoset DACC can be produced by converting the surface of cellulosic precursor to a thermoset polymer, which can be cross-linked during hot pressing at elevated temperatures. Codou et al. prepared ACCs by partial periodate oxidation of microcrystalline cellulose (MCC), which led to oxidation of the specimens at C2 and C3 of the anhydroglucose moieties of cellulose with varying degree of oxidation (DO). The produced carbonyl moieties recombined immediately with

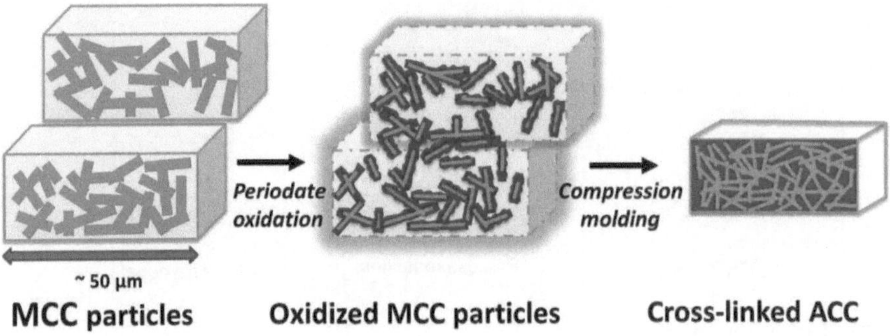

FIGURE 9.5
Strategy for the processing of thermosetting ACCs by hot-compression molding, after controlled periodate oxidation of MCC particles. (From Codou, A. et al., *Compos. Sci. Technol.*, 117, 54, 2015. With permission.)

the available OH groups of untouched cellulose to yield hemiacetal cross-linkages. Subsequently, novel thermoset ACCs resulted from the hot pressing of samples (DOs ranging from 0.2 to 0.85), utilizing the advantages of high reactivity of the neocarbonyl as well as the lability of the hemiacetals linkages. The steps of preparation of thermoset ACCs have been shown in Figure 9.5 [18].

9.12.2 Structure and Properties

The physical properties of the prepared thermoset ACCs were directly dependent on the DO. Interestingly, with increasing DO, the thermoset ACCs gradually became transparent from the white opaque appearance. The increase in transparency was a result of the reduction of optical diffusivity, indicating a better homogeneity in the composite structure. The melting point of the thermoset ACC was well above its degradation temperature and hence not processable further by heat treatment. At highest DO, higher cross-linking led to less ductility in the specimen. As a consequence, optimum flexural properties were obtained at a DO in the range of 0.4–0.65, due to the proper cross-linking between surface-modified cellulose nanocrystals and the oxidized or unoxidized neighbors, resulting in flexural modules and flexural strength as high as 7 GPa and 177 MPa, respectively (Figure 9.6) [18].

9.13 Conclusion

Till date, in comparison to DACC, more extensive research has been done on non-derivatized ACCs [39–44], which have been discussed in Chapters 7 and 8. This chapter focuses on details about the concept and research progress in the

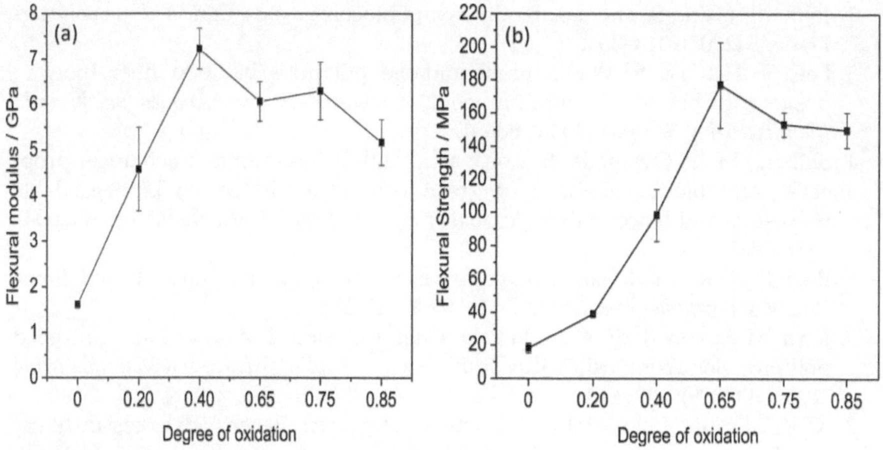

FIGURE 9.6
Flexural modulus and flexural strength of thermoset ACCs as a function of DO. (From Codou, A. et al., *Compos. Sci. Technol.*, 117, 54, 2015. With permission.)

field of DACC. By means of different techniques of derivatization such as benzylation, esterification, oxypropylation, and carbamation, cellulose can be converted to thermoplastic or only one-time melt-processable thermoset polymer. The derivatized cellulosic fibers show a sheath-core type morphology, where the core remains unchanged and the surface become thermoplastic, showing a glass transition temperature. Derivatization of cellulose causes disruption of hydroxyl bonds of cellulose, resulting in decrease in crystallinity of cellulose due to formation of amorphous cellulose derivative. These derivatized cellulose fibers can be used for making bio-composites simply by hot pressing. The structure and morphology of DACCs is strongly dependent on the degree or extent of derivatization of cellulose, which finally control the properties of the composites. Therefore, optimum structure properties of DACCs can be obtained by tailoring the extent of derivatization via controlling the process time, temperature, and the amount of chemical with respect to number of OH groups present in cellulose. As per literature, the mechanical properties of DACCs or all-plant fiber composites are not good as non-derivatized ACCs. Though, after derivatization of cellulose, it becomes slightly hydrophobic, the inherent biodegradability of cellulose is still retained in most of the composites.

References

1. Hassan, M. M., and M. H. Wagner. Surface modification of natural fibers for reinforced polymer composites: a critical review. *Reviews of Adhesion and Adhesives* 4, no. 1 (2016): 1–46.

2. John, M. J., and S. Thomas. Biofibres and biocomposites. *Carbohydrate polymers* 71, no. 3 (2008): 343–364.
3. Lee, S. H., and S. Wang. Biodegradable polymers/bamboo fiber biocomposite with bio-based coupling agent. *Composites Part A: Applied Science and Manufacturing* 37, no. 1 (2006): 80–91.
4. Shibata, M., S. Oyamada, S. Kobayashi, and D. Yaginuma. Mechanical properties and biodegradability of green composites based on biodegradable polyesters and lyocell fabric. *Journal of Applied Polymer Science* 92, no. 6 (2004): 3857–3863.
5. Bledzki, A. K., and J. Gassan. Composites reinforced with cellulose based fibres. *Progress in polymer science* 24, no. 2 (1999): 221–274.
6. John, M. J., and R. D. Anandjiwala. Chemical modification of flax reinforced polypropylene composites. *Composites Part A: Applied Science and Manufacturing* 40, no. 4 (2009): 442–448.
7. Karger-Kocsis, J., H. Mahmood, and A. Pegoretti. Recent advances in fiber/matrix interphase engineering for polymer composites. *Progress in Materials Science* 73(2015): 1–43.
8. Törmälä, P. Biodegradable self-reinforced composite materials; manufacturing structure and mechanical properties. *Clinical materials* 10, no. 1–2 (1992): 29–34.
9. Isayev, A. I., and M. Modic. Self-Reinforced melt processible polymer composites: Extrusion, compression, and injection molding. *Polymer composites* 8, no. 3 (1987): 158–175.
10. Matabola, K. P., A. R. De Vries, F. S. Moolman, and A. S. Luyt. Single polymer composites: a review. *Journal of Materials Science* 44, no. 23 (2009): 6213.
11. Lewin, M. *Hand Book of Fibre Chemistry*. 3rd edition. Boca Raton, London, New York: CRC Press, Taylor & francis group, 2007.
12. Zhang, M. Q., M. Z. Rong, and X. Lu. All-Plant Fiber Composites. In Ye, Lin, Y. W. Mai, and Zhongqing Su, eds. Composites Technologies for 2020: Proceedings of the Fourth Asian-Australasian Conference on Composite Materials (ACCM-4): University of Sydney, Australia, 6–9 July 2004. Woodhead Publishing, 2004.
13. Heinze, T., and A. Koschella. Solvents applied in the field of cellulose chemistry: A mini review. *Polímeros* 15, no. 2 (2005): 84–90.
14. Olsson, C., and G. Westman. Direct dissolution of cellulose: background, means and applications. In *Cellulose-Fundamental Aspects*. Chapter 6, eds T. G. M. Van De Ven. InTech, 2013. DOI: 10.5772/52144
15. Lu, X., M. Q. Zhang, M. Z. Rong, G. Shi, and G. C. Yang. All-Plant fiber composites. I: Unidirectional sisal fiber reinforced benzylated wood. *Polymer composites* 23, no. 4 (2002): 624–633.
16. de Menezes, A. Jr., D. Pasquini, A. A. da Silva Curvelo, and A. Gandini. Self-reinforced composites obtained by the partial oxypropylation of cellulose fibers. 2. Effect of catalyst on the mechanical and dynamic mechanical properties. *Cellulose* 16, no. 2 (2009): 239–246.
17. Matsumura, H., J. Sugiyama, and W. G. Glasser. Cellulosic nanocomposites. I. Thermally deformable cellulose hexanoates from heterogeneous reaction. *Journal of Applied Polymer Science* 78, no. 13 (2000): 2242–2253.
18. Codou, A., N. Guigo, L. Heux, and N. Sbirrazzuoli. Partial periodate oxidation and thermal cross-linking for the processing of thermoset all-cellulose composites. *Composites Science and Technology* 117(2015): 54–61.

19. Wolfrom, M. L., and M. A. El-Taraboulsi. Benzylation and xanthation of cellulose monoalkoxidel. *Journal of the American Chemical Society* 76, no. 8 (1954): 2216–2218.
20. Lorand, E. J., and E. A. Georgi. The mechanism of cellulose benzylation. *Journal of the American Chemical Society* 59, no. 7 (1937): 1166–1170.
21. Lu, X., M. Q. Zhang, M. Z. Rong, G. Shi, and G. C. Yang. All-plant fibre composites: Self reinforced composites based on sisal. *Advanced Composites Letters(UK)* 10, no. 2 (2001): 73–79.
22. Lu, X., M. Q. Zhang, M. Z. Rong, G. Shi, and G. C. Yang. Self-reinforced melt processable composites of sisal. *Composites science and technology* 63, no. 2 (2003): 177–186.
23. Zhang, M. Q., X. Lu, M. Z. Rong, G. Shi, G. C. Yang, and H. M. Zeng. Natural vegetable fibre/plasticised natural vegetable fibre: A candidate for low cost and fully biodegradable composite. *Advanced Composites Letters(UK)* 8, no. 5 (1999): 231–236.
24. Zhang, M. Q., M. Z. Rong, and X. Lu. Fully biodegradable natural fiber composites from renewable resources: All-plant fiber composites. *Composites Science and Technology* 65, no. 15–16 (2005): 2514–2525.
25. Lu, X., M. Q. Zhang, M. Z. Rong, and G. C. Yang. Enzyme degradability of benzylated sisal and its self-reinforced composites. *Polymers for Advanced Technologies* 14, no. 10 (2003): 676–685.
26. Lu, X., M. Q. Zhang, M. Z. Rong, and G. C. Yang. Environmental degradability of self-reinforced composites made from sisal. *Composites Science and Technology* 64, no. 9 (2004): 1301–1310.
27. Matsumura, H., and W. G. Glasser. Cellulosic nanocomposites. II. Studies by atomic force microscopy. *Journal of Applied Polymer Science* 78, no. 13 (2000): 2254–2261.
28. Alcalá, M., I. González, S. Boufi, F. Vilaseca, and P. Mutjé. All-cellulose composites from unbleached hardwood kraft pulp reinforced with nanofibrillated cellulose. *Cellulose* 20, no. 6 (2013): 2909–2921.
29. Gandini, A., A. A. da Silva Curvelo, D. Pasquini, and A. J. de Menezes. Direct transformation of cellulose fibres into self-reinforced composites by partial oxypropylation. *Polymer* 46, no. 24 (2005): 10611–10613.
30. de Menezes, A. Jr, D. Pasquini, A. A. da Silva Curvelo, and A. Gandini. Self-reinforced composites obtained by the partial oxypropylation of cellulose fibers. 1. Characterization of the materials obtained with different types of fibers. *Carbohydrate Polymers* 76, no. 3 (2009): 437–442.
31. Schoenenberger, C., J. F. Le Nest, and A. Gandini. Polymer electrolytes based on modified polysaccharides. 2. Polyether-modified cellulosics. *Electrochimica Acta* 40, no. 13–14 (1995): 2281–2284.
32. Vo, L. T. T., B. Široká, A. P. Manian, H. Duelli, B. MacNaughtan, M. F. Noisternig, U. J. Griesser, and T. Bechtold. All-cellulose composites from woven fabrics. *Composites Science and Technology* 78(2013): 30–40.
33. Besbes, I., M. R. Vilar, and S. Boufi. Nanofibrillated cellulose from alfa, eucalyptus and pine fibres: preparation, characteristics and reinforcing potential. *Carbohydrate Polymers* 86, no. 3 (2011): 1198–1206.
34. Gindl, W., and J. Keckes. Tensile properties of cellulose acetate butyrate composites reinforced with bacterial cellulose. *Composites Science and Technology* 64, no. 15 (2004): 2407–2413.

35. Soykeabkaew, N., C. Sian, S. Gea, T. Nishino, and T. Peijs. All-cellulose nanocomposites by surface selective dissolution of bacterial cellulose. *Cellulose* 16, no. 3 (2009): 435–444.
36. Huber, T., J. Müssig, O. Curnow, S. Pang, S. Bickerton, and M. P. Staiger. A critical review of all-cellulose composites. *Journal of Materials Science* 47, no. 3 (2012): 1171–1186.
37. Vallejos, M. E., M. S. Peresin, and O. J. Rojas. All-cellulose composite fibers obtained by electrospinning dispersions of cellulose acetate and cellulose nanocrystals. *Journal of Polymers and the Environment* 20, no. 4 (2012): 1075–1083.
38. Hooshmand, S., Y. Aitomäki, M. Skrifvars, A. P. Mathew, and K. Oksman. All-cellulose nanocomposite fibers produced by melt spinning cellulose acetate butyrate and cellulose nanocrystals. *Cellulose* 21, no. 4 (2014): 2665–2678.
39. Adak, B., and S. Mukhopadhyay. All-cellulose composite laminates with low moisture and water sensitivity. *Polymer* 141(2018): 79–85.
40. Adak, B., and S. Mukhopadhyay. A comparative study on lyocell-fabric based all-cellulose composite laminates produced by different processes. *Cellulose* 24, no. 2 (2017): 835–849.
41. Huber, T., J. Müssig, O. Curnow, S. Pang, S. Bickerton, and M. P. Staiger. A critical review of all-cellulose composites. *Journal of Materials Science* 47, no. 3 (2012): 1171–1186.
42. Adak, B., and S. Mukhopadhyay. Effect of the dissolution time on the structure and properties of lyocell-fabric-based all-cellulose composite laminates. *Journal of Applied Polymer Science* 133, no. 19 (2016).
43. Bapan, A., and M. Samrat. Jute based all-cellulose composite laminates. *Indian Journal of Fibre & Textile Research (IJFTR)* 41, no. 4 (2016): 380–384.
44. Adak, B., and S. Mukhopadhyay. Effect of pressure on structure and properties of lyocell fabric-based all-cellulose composite laminates. *The Journal of the Textile Institute* 108, no. 6 (2017): 1010–1017.

10

Applications, Current Difficulties, and Future Scope of Single-Polymer Composites

10.1 Introduction

The final properties of any composite material mainly depend on type of matrix, type of reinforcing fibers, and also on process techniques and process parameters [1–6]. Till date, different process routes have been explored and process conditions have been optimized by analyzing their effects on structure and properties of single-polymer composites (SPCs) [7–9]. On the basis of extensive studies on SPCs, the researchers have become successful in developing a wider processing window for SPCs to obtain desired properties [3,7]. However, further research is required in the future in this field to explore their versatile properties and suitable applications.

On the basis of the discussions in previous chapters, SPCs can be classified in two categories in broad sense, these are (i) synthetic polymer (polyolefin, polyester, polyamide, polylactic acid, etc.) based SPCs and (ii) all-cellulose composites (ACCs). In this chapter, the possible applications, the challenges and future scope for these two types of SPCs have been discussed critically.

10.2 Probable Applications of SPCs

10.2.1 Synthetic Polymer–Based SPCs

With energy considerations getting prior importance, development of considerable lightweight, easy reprocessable composites would gain further importance. Synthetic polymer–based SPCs are the key class of materials with the above benefits. The other important advantage of this class of composites is its superior impact properties. Lighter, smaller electrical driven cars are the next big thing in the automotive industry. This drive to lessen the weight of standard cars is forcing materials scientists to think and develop new materials that combine low density with a combination of good mechanical properties. Therefore, synthetic polymer–based SPCs have a great potential in automotive industries in near future.

The application areas of SPCs also include lightweight packaging structures, insulation purposes, and uses involving low-temperature applications. In one of the articles, Jones et al. [10] showed the feasibility of SPCs to be used in automotive applications. High resistance to both impact and abrasion, even at low temperatures makes them useful as automotive exterior components. In a project sponsored by GAIL [11], one of the authors found very low water absorption and excellent damping properties of such composites. The report of high peel strength of such composites would also mean that they can be used for applications involving high flexure.

Another area that could result in a boost in application would be multicomponent single reinforced polymer composites. Development can address the need for hollow containers with improved barrier properties. Among the multistep production methods of multicomponent self-reinforced polymeric materials (SRPMs), those with loose textile assemblies that can be consolidated and shaped simultaneously are in a favored position. There is ongoing research to develop injection-moldable multicomponent SRPMs [12].

Some of the applications that can be perceived based on the properties and information from established companies are industrial cladding, personal protective equipment, sports goods, reusable packaging, travel suitcases, shoe inserts and safety shoe inserts, filter sheets, localized reinforcement, injection-molded parts, compression-molded parts, industrial pallets, and a range of automotive interior and exterior components. Some of the high-end applications of SPCs are in the field of medicine and military.

10.2.2 Companies Using Synthetic Polymer–Based SPCs in Commercial Applications

10.2.2.1 Curv

One of the first commercial websites of SPCs from polypropylene (PP) has been Curv by Propex [13]. Polypropylene single-polymer composites have properties like higher thermoformability, better impact resistance, higher mechanical properties than the homopolymer and for even majority of the PP-glass composite combinations. The notched impact properties have been higher for all classes of composites compared. Best results are achieved using matched tooling at pressures from five bars upwards depending on the complexity of the part. Curv® sheets, as the website claims, can be heated to the required processing temperature (c. 150°C–160°C) by various heating techniques including infrared heating, contact plate heating, or air convection without shrinking or otherwise deforming. Heated sheets are easy to handle either manually or by robot and may be either clamped or allowed for a restricted flow into mold cavity. The choice would be depending on the complexity of the part being produced.

Due to the flexibility of state-of-the-art production equipment, the company was able to produce a range of customized products for specific applications. Examples include Curv in combination with

1. aluminium,
2. aramid,
3. carbon,
4. glass.

Curv has been officially used in leading brands like Samsonite [14] in packaging applications. The detailed technical datasheet of Curv is available on their website [15]. The details of the Samsonite packaging are also available on their website [16]. Propex Fabrics Curv are known to have received the prestigious 2004 Frost & Sullivan Innovation Award for Curv.

Another very interesting application claimed by Curv is in the field of sound. Produced using matched aluminum tooling in a pressure-forming machine (effectively a vacuum forming matching fitted with top and bottom tools). They manufactured it using a cycle time of 80 seconds [17]. Advantages of using Curv: (a) improved acoustic performance (b) self-damping performance (c) lightweight and (d) unaffected by moisture.

10.2.2.2 Armordon

Thrace Group is another established company with their product Armordon [18]. Armordon® has high impact strength and low density, making it the preferred choice in cases where strength and weight saving are key parameters. The major advantages claimed by the company are in line with what has been discussed in Chapter 3 on polyolefins:

1. High impact strength and abrasion resistance
2. Nontoxic and inert
3. Easily thermoformed with a broad thermoprocessing window
4. Unique price/performance benefit
5. Free of glass fiber
6. 100% recyclable

Armordon has used the developed composites in luggage, sports goods, orthotics, packaging, armor applications, oil and gas pipelines, and so on [19]. The company has predicted its use in medical applications as well. These composites are stiffer, stronger, lighter, and more impact resistant than commonly used thermoplastics. Armordon is an ideal material for volume-produced orthotics devices. In contrast to the fiber and resin-based composites presently used for high-end devices, In contrast to the fiber and resin-based composites presently used for high-end devices, this class of composites are easier to handle during the manufacturing process. In addition, the behavior of Armordon is predictably at the limits of its performance envelope allowing for safe, planned replacement.

In the automotive industry, like other SPCs, Armordon has also been predicted to be of great use. The company claims the availability of this class of composites with a range of decorative and functional finishes. Armordon components support the drive to produce lighter, more fuel-efficient vehicles. Working with foremost automotive materials partners "Don & Low" have developed Armordon/expanded foam composites that suggest substantial strength and weight benefits over existing materials.

The other interesting application is military applications. Armordon offers excellent cost-effective performance in areas of ballistic and blast protection, helping armoring specialists to encounter increasing need for protective solutions. Armordon has also been combined with other hard and soft armoring materials, reflecting increased trend and requirement for composite armoring solutions in today's environment. The company has produced a series of hybrid weaves [20], pairing Armordon with a range of other ballistic materials and expanding the options available for armoring technicians. Another very important perspective from the military point of view is the need to make it invisible. Armordon combines lightweight and ruggedness with excellent signal transparency, making it an ideal material for transmitter or receiver equipment housings. Other possible applications would include automotive interiors, sporting goods, personal equipment, packaging, luggage, and use in cold temperature.

10.2.2.3 PURE

The third company that exists in this direction is DIT Weaving with their developed product as PURE [21]. PURE® is a patented 100% PP composite material; thus, a mono-material concept that is fully recyclable. The PURE tapes are coextruded and consist of a highly oriented, high strength and modulus core and a specially formulated skin on both sides for joining the tapes together in a compaction process using a hot press or continuous belt press.

PURE materials have shown excellent properties with respect to impact resistance, even at very low temperatures. This has been a common feature with all the SPCs discussed so far. The material shows a "soft" or ductile crash behavior. PURE, as the company claims, does not splinter but fails in a more ductile manner. Such a "soft" crash behavior is of noteworthy attention for various applications, since it will lead to safer products. Also, in blast protection, PURE has shown an excellent behavior.

The company gives an overview of their pilot plant [22]. In the plant, the company has capabilities to test several materials and processes. This includes extrusion, winding, sheet consolidation, and thermoforming. These facilities enable them to develop new products and offer excellent customer support. The pilot plant is supported by our laboratory for testing and characterization. The company has good facilities for thermoforming and filament winding. A lab-scale press with heating and cooling system is available for consolidating sheets for research purposes. A 250 tons press is available for prototyping, testing different molds, and molding technologies. The company has explored various molding technologies and served customers with thermoforming knowhow, support, and expertise before going into serial production. The properties of PURE sheet are available on the website [23] and they also produce fabric from the starting material [24].

Other than the applications claimed by the first two companies, PURE also claims applications in the construction sector. Standard PURE panels show very good impact behavior, even at low (–40°C) temperatures. PURE can be easily transformed into thick (>3 mm) panels. There are no technical limits toward thickness of the panels. This will also allow customers with sheet-pressing facilities to produce PURE panels. Fabrics and panels from the company can be supplied with UV protection. PURE on PU, PVC, or honeycomb are also possible.

The application possibilities of this class of composites are immense. There are a few other companies [25] divulging the details of their processing sequence [26].

10.2.3 All-Cellulose Composites

All-cellulose composites are one of the latest classes of SPCs. The versatile properties of non-derivatized ACCs have been discussed in Chapter 8. Among these, one of the most important and interesting aspects of ACCs is its positive environmental impact. On the basis of this property, green ACCs are at present showing potential for applications almost in all fields ranging from structural to biomedical applications. However, applications of ACCs are still limited because of lack of durability. Figure 10.1 shows some potential and futuristic applications of different ACCs.

A fully bio-based nanocomposite film with high mechanical properties, transparency, and barrier property, together with the potential to be made from very low value cellulose resources is very demanding. Ghaderi et al. developed a cheap and fast processing route to produce ACC films from low-cost bagasse fiber and checked their suitability for food packaging applications. The ACC films produced with very good tensile properties, toughness, biodegradability, and acceptable level of low water vapor permeability, were very suitable for food packaging applications [27].

In another study, Pang et al. showed that tensile strength of ACC films produced from cotton linter is better than conventional PP and polyethylene (PE) films. It indicates that these films have a potentiality to replace nonbiodegradable PP and PE films for food packaging and agricultural purposes [28].

Duchemin et al. prepared cellulose-based "aerogel," which was named as "aerocellulose." This was a highly porous all-cellulose nanocomposite

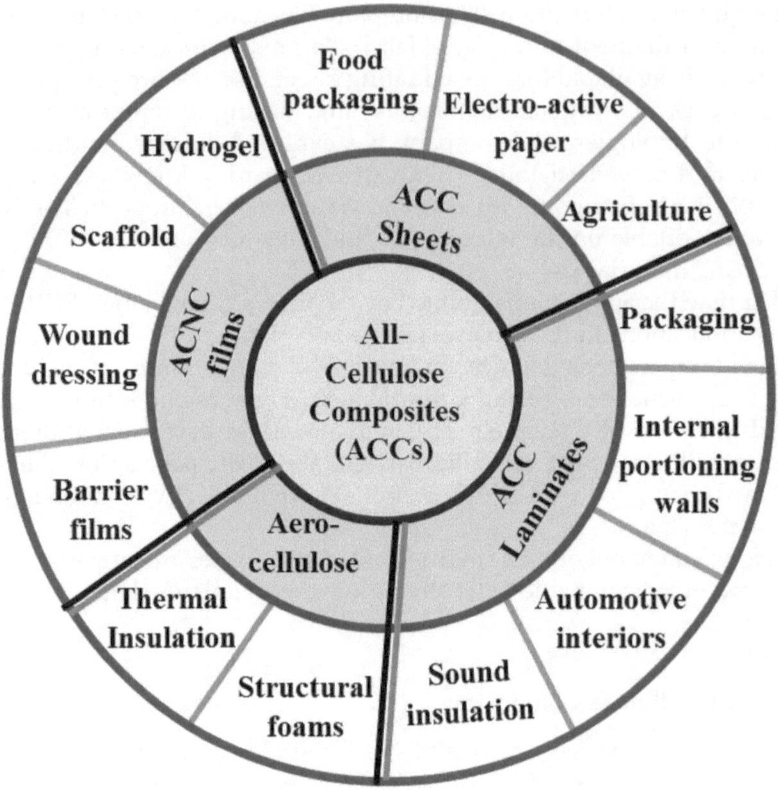

FIGURE 10.1
Some potential and futuristic applications of ACCs.

(ACNC) prepared via freeze drying or supercritical drying [29]. Due to interconnected porous network and high surface-to-volume ratio, aerogel may be applicable in various fields. The lightweight, bio-based, environmentally friendly, and biodegradable aerocellulose is an alternative to synthetic foams, with a wide range of applications from structural foams to packaging [29–30].

Yousefi et al. prepared ACNCs by solvent-based nanowelding technique using cellulose microfibers as starting material. It was suggested as a fully environment friendly and economical product having high mechanical properties, barrier properties, lightweight, and transparency. These properties are potentially useful for wide range of applications like biomedical engineering, sports equipment, aerospace, highly flexible electronic magnetic devices, and batteries [31].

A serious problem of non-derivatized ACCs is its tendency of decomposition when in contact with water during outdoor application. However, thermoplastic or thermoset matrix–based derivatized ACCs (DACCs) can perform better in this respect, showing a potential to be used for many structural applications [32]. Yousefi et al. produced an ACNC film having water-repellent property by using a silane coupling agent (dodecyltriethoxysilane). With increasing silane concentration, the water contact angle of treated ACNC films increased and reached a maximum value of 93°. It was also reported that mechanical performance of ACNCs also increased with increasing silane concentration. This finding may rid of the restriction of outdoor application of ACNCs [33]. Very recently, Tervahartiala et al. reported production of ACC-based corrugated board from different chemical pulp raw materials [34].

There is a large application of softwood-based paper and board products in the packaging industry [34]. Nilsson et al. produced a high-density ACC material from low-cost wood pulp by beating, pressing, and applying temperature. Depending on its environment friendly processing and very good tensile properties, it has a potential to replace conventional softwood-based paper board products for packaging [35].

Nishino et al. showed that mechanical performance of ramie fiber–based ACC is comparable or better than conventional glass fiber–reinforced composites (GFRC). In addition to this, ramie fiber–based ACC showed a very low linear thermal expansion coefficient (about 10^{-7} K^{-1}), which was much lower than metals like iron, silicon, and so on, and even quite lower than the cellulose matrix used [36]. Nowadays, natural fiber–reinforced composites are replacing GFRCs as a viable alternative in various applications, particularly in the automotive industry [37–38]. On the basis of this concept, natural fiber–reinforced thick ACC or laminates may also be used inside different interior parts of automotives such as door panels, seat backs, headliners, and dash boards, providing good strength, light weight, flexibility, dimensional stability, as well as end-of-life disposal.

Due to very unique properties of bacterial cellulose (BC), that is, high purity, high crystallinity, high mechanical strength, high water-holding capacity, high porosity, and good biocompatibility, the ACCs based on BC may be applied in paper, textile, and food industries, to acoustic diaphragm for loudspeakers, and in different biomedical applications [38–39]. Bacterial cellulose has a potential for making scaffold in tissue engineering of cartilage [40]. Bacterial cellulose–based composite films have been successfully applied in many biomedical applications such as topical wound dressings, artificial skin, artificial blood vessels, drug delivery, barrier films, and specialty membranes [39]. Cross-linked cellulose nanowhiskers–reinforced composite films can be used as hydrogel [41].

All-cellulose nanocomposites based on regenerated cellulose and cellophane can also be used as "smart material." Cellulose nanopaper, termed electro-active paper, can be used as sensors and actuators, having many advantages such as an electric stimulus, ease of processing, good mechanical properties, low density, low cost, biodegradability, low actuation voltage, and low power consumption [42].

10.3 Current Difficulties, Major Challenges, and Future Scope of SRCs

10.3.1 Synthetic Polymer–Based SPCs

10.3.1.1 Current Difficulties and Major Challenges

The main challenges for synthetic polymer–based SPCs are

1. Finding and enhancing the temperature window:

   In spite of the advantages of SPCs over traditionally reinforced composites in terms of chemical compatibility and recyclability, the minor difference in melting temperature between reinforcing phase and matrix poses a big challenge during fabrication for polymers that can be melt processed. This has been discussed in Chapters 3 and 4 covering polyolefins and polyamides. The major challenge has been to find a processing window that is large enough to preserve integrity of fibers after consolidation. There have been many routes that have been explored. The efforts by the research group of Bárány are worth a mention [43]. They effectively used polymorphism of PP as a possibility for increasing the processing window in all-PP composites. A temperature difference of up to 25°C between the α-PP and the β-PP has been reported with the α-PP having the higher melting point. These composites were primarily prepared by the film-stacking method. However, more efforts can be directed toward enhancing the temperature of processing, so that the safe zone of manufacturing gets enhanced.

2. Research on recycling:

   SPCs have been very often referred to as the "ultimate" materials for recycling. Interestingly, the aspect of recyclability has not been extensively studied. Recycling via re-melting of SPCs was mostly studied in repeated injection molding. On the other hand, the works performed on this field [44] clearly show that the number of reprocessing/re-melting cycles, which are not accompanied with property deterioration, agrees fairly with the related value derived for injection-molded neat polymers. This result was received in works using PP-based systems. In another research [45] to use recycled material, study was done on recycled polyamide 6 (PA6) cloth that was used as the only raw material to prepare SPC by partially melting the PA6 fibers. During the process, a part of PA6 fibers was only melted to fill the gap between the remainder fraction of fibers. The control of temperature in the melting temperature range of fibers was critical. The matrix was formed in situ by recrystallization of molten part and bonded the remainder of unmolten fibers together with cooling.

iii. Quality assurance of SPCs:

The quality assurance of SPC in manufactured parts is a critical issue that needs to be addressed. Quality of the products has been assessed by various destructive methods (different kind of mechanical loading) coupled with further techniques such as acoustic emission [46]. Inspections subsequent to failure using light and scanning electron microscopy have also been explored. Nondestructive techniques were rarely used for quality management though ultrasonic testing [47] and X-ray microcomputed tomography showed to be one of the promising tools [48]. This is still an issue of research, especially if in line quality inspection is the target. Nondestructive testing would also hint at partial interlaminar damage for SPCs.

10.3.1.2 Future Scope

With respect to matrix/reinforcement combinations, amorphous/semicrystalline and semicrystalline/semicrystalline combinations are the winners. Attempts can be further made to modify the reinforcement in SPCs by the use of nanofillers, especially the use of those reinforcements with high aspect ratios. The incorporation of nanofillers can be directed toward increase of stiffness, strength, and thermal stability. The fire resistance of SPCs need to be improved. This is a high priority issue for further potential applications. Pioneering activity in this direction yielded some interesting results [49]. Polymer systems with single polymer PP composites, when exposed to heat, form a special, compact-charred surface layer, which hinders the release of pyrolysis gases and where the flame retarding action of P- and N- containing compounds are effective at low volume fractions, and thus at higher concentration, resulting in effective fire extinction. The observed new synergism is capable of being efficiently utilized in the formation of cost-effective, fire retardant, self-reinforced composites (SRCs). However, this area needs to be further investigated. SPCs of more complex structure, such as panels containing honeycomb or foam cores, may help to acquire new applications fields [50]. One of the authors found honeycomb structures to be very useful in designing practical SPCs using PPs [11].

Building bigger structures need joining of manufactured SPCs. Joining of SPCs via various methods has been a great challenge and needs further exploration. The group of Kiss [48] researched on the joining process of this class of composites. In their research, composite sheets were produced by film-stacking method and compression molded with different thicknesses (1 mm, 2 mm) with different contents at varying processing temperatures, maintaining a constant holding time and pressure. The self-reinforced polypropylene composite sheets were welded by ultrasonic welding machine with different welding parameters. The welds were characterized by mechanical and microscopic tests and results showed that thermoplastic reinforcement did not melt; therefore, the reinforcement was kept the strength-increasing effect.

Another interesting development can be the production of (multi)functional SPCs. In this respect, attention may be focused on the shape memory performance of SPCs [51]. Further works are required to check the potential of electrospun reinforcements in SPCs. Incorporation of suitable (nano) particles in the matrix or matrix-giving component acting as hot spots (heat sources) by external triggering (electromagnetic field, microwave, etc.) seems to be a sound strategy. To solve the quality assurance in line during production is also imperative.

Describing the nature of infiltration of the molten resin into the reinforcing structure and modeling of the structure–property relationships in SPCs are further important tasks that need to be carried out [52]. Searching the patent database with keywords SPC or "self-reinforced composite" results only in one relevant result [53]. One of the authors has filed a joint patent with GAIL [54] on the development of SRCs from PP. This shows that there is tremendous possibility of practically using this class of composites.

10.3.2 All-cellulose Composites

Although many literatures have been published emphasizing microstructure and different properties of ACCs, till date ACCs have not been commercialized in the industrial level. Production of ACCs is still limited to laboratory scale. Therefore, more effective researches are required to find more suitable industrial applications as well as for evolution of ACCs from laboratory to industrial scale.

So far, most of the researches were confined to manufacturing of nano/micro level ACC films and less work has been explored producing thicker type of ACC laminates. Therefore, applications of ACCs are also confined in small areas. Thus, more initiatives can be taken in preparation of thicker type of ACC laminates, with finding proper applications. Furthermore, very few studies [5,27] covered the use of low-cost biofibers (bagasse, jute, banana, etc.), cellulosic wastes, or agricultural by-products. The reason can be attributed to the presence of higher lignin/hemicellulose content in lingo-cellulosic natural fibers, which is an obstacle in dissolving cellulose to a greater extent. These non-cellulose contents dissolve very poorly in solvents under normal process conditions of ACCs [5,55]. However, use of these low-cost natural fibers or cellulosic wastes as raw materials can reduce the production cost-effectively. Moreover, till date no research has been reported on the mechanism of interaction of solvents with other non-cellulosic components (lignin, hemicellulose, pectin) of natural biofibers.

Over the past decades, different solvent systems have been tried for dissolution of cellulose, which has been discussed in detail in Chapter 7. However, conventional solvent systems have many disadvantages like toxicity, high cost, limited dissolution capability, uncontrollable side reactions, solvent recovery issue, instability during cellulose processing, derivatization, formation of by-product, and so on [28]. Therefore, selection of proper solvent is very important for preparation of ACCs.

Until now, in the preparation of ACCs, most of the researchers used LiCl/DMAc as cellulose-solvent. However, due to the toxicity of DMAc, it inhibits the use of prepared ACCs for several applications, especially in biomedical applications [56]. Additionally, the disposal of nonrecyclable and toxic solvents also adds extra cost to the process. So, studies should be more concentrated on the use of nontoxic solvents for cellulose to get completely "greener" alternative of existing materials. Use of nonhazardous and recycled solvents such as ionic liquids and N-methylmorpholine-N-oxide (NMMO) has a potential to develop cost-effective eco-friendly process for ACCs. The recovery of NMMO can be as high as 99% [57] and for ionic liquid may be up to 93% [58]. However, ionic liquids are generally very costly and some ionic liquids are also reported as toxic [59]. Therefore, there remains a question whether it is really a "greener" and economical alternative to the existing materials? Hence, selection of proper ionic liquids is also vital for ACC preparation considering all factors, that is, cost, toxicity, and also effectiveness to cellulose dissolution.

Many studies discuss the phase transformation of cellulose during processing of ACCs. However, no study enclosed the correlation of how properties of ACCs change with phase transformation of cellulose.

For proper dissolution of cellulose in ionic liquids, about 90–110°C temperature is used. The processing temperature with NMMO is also in the same or slightly higher range. During synthesis of ACCs at these higher temperatures, there is always a chance of degradation of cellulose. Nilsson et al. observed that after processing of ACCs at elevated temperature in compression molding the molar mass of cellulose reduced due to oxidative and hydrolytic degradation [35]. The molar mass distribution data of bio-composites obtained from size exclusion chromatography (SEC) analysis was plotted against processing temperature, showing a significant reduction of molar mass above 150°C (Figure 10.2). Although in few papers [60–62] the matter of cellulose degradation has been highlighted while prolong treatm ent was done at higher temperatures (about 100°C), till now no detailed study has been published, studying the degradation of cellulose and correlating it with properties of ACCs. Moreover, research should be driven toward a low-temperature processing of ACCs to avoid cellulose degradation as well as for cost-effectiveness. Therefore, many recent researchers reported the use of NaOH/Urea as solvent, where the effective temperature for dissolution of cellulose is about –12°C [63–66].

The solvents can effectively dissolve only those cellulosic materials that have a low degree of polymerization (DP) and lower lignin/hemicellulose content. For example, having higher DP, cotton shows comparatively low solubility compared to regenerated viscose fibers. Therefore, right selection of cellulosic raw material is also important. In case of natural fiber–based

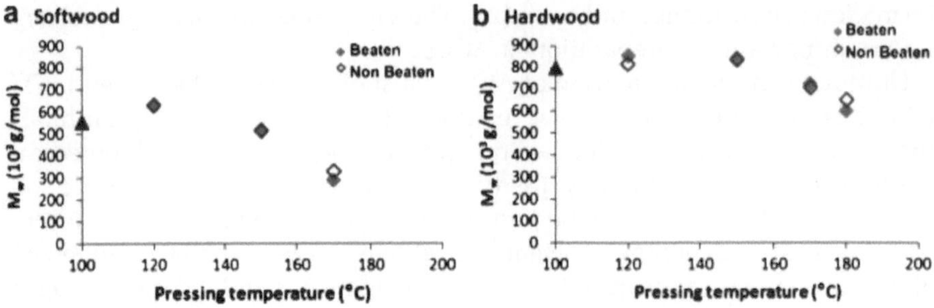

FIGURE 10.2
Weight average molar mass (M_w) determined by SEC with respect to the pressing temperature for bio-composite plates made of (a) softwood pulp and (b) hardwood pulp. The beaten, non-beaten, and reference pulps have been shown by filled rhombus, unfilled rhombus, and filled triangle, respectively. (From Nilsson, H. et al., *Compos. Sci. Technol.* 70, 1704, 2010. With permission.)

ACCs, the quality of cellulose source can play a major role in controlling mechanical properties of ACCs.

Duchemin et al. observed almost a complete dissolution of 10% microcrystalline cellulose in LiCl/DMAc, although a trace of residual crystalline cellulose was detected at this concentration [67]. Therefore, more work is required to analyze in which concentration complete transition occurs and whether the degree of transition is sufficient to provide satisfactory mechanical properties or not.

Fiber volume fraction plays a key role in controlling morphology and properties of ACCs. However, calculation of fiber volume fraction is very difficult in case of nano-derivatized ACCs, especially those produced by "surface-selective dissolution" method. This can be attributed to the difficulty in measuring the amount of cellulose that has dissolved partially, which finally acts as matrix after regeneration. Though a few references [68] have mentioned fiber volume fraction of ACCs, the technique of measurement has not been divulged. Therefore, development of a proper measurement technique for determination of fiber volume of ACC is very essential.

Though non-derivatized ACCs are supposed to be completely biodegradable, there is no effective documentation on it, except only a work by Kalka et al. [69]. Therefore, further research work may broadly cover this area. Biodegradability of cellulosic materials sometimes are also considered as its disadvantage. It is because of poor longevity of cellulose in moist or wet conditions. Due to highly hydrophilic nature of cellulose, any cellulosic material performs best only in dry conditions. Therefore, to use these materials outdoors, their environmental behavior and susceptibility to moisture needs to be investigated in detail. Hygroscopic nature of cellulose is also a reason for shrinkage of ACCs [9], especially in thicker type ACC laminates, during regeneration and drying. Although in some of the studies [4] the shrinkage in ACCs has been reported during regeneration and drying, measurements have not been documented. Therefore, a detailed study calculating exact shrinkage in ACCs and correlating it with process parameters and properties may be appreciable. Moreover, though the shrinkages in ACCs were controllable in lab scale [4], they might be problematic in industrial scale.

Another very important thing that has not been discussed by any researcher is the detailed studies on interface between matrix and reinforcement of ACCs. In most of the works, assessment for fiber–matrix adhesion has been done on the basis of SEM micrographs. Therefore, some research on this topic includes single fiber pull-out test or interlaminar adhesion strength test. Nevertheless, in most of the literatures, mainly tensile properties of non-derivatized ACCs have been explored and other mechanical properties such as flexural properties, impact properties, and compressibility have not been covered. Therefore, a great opportunity lies in these areas.

10.4 Conclusion

SPCs have a huge potential to develop next generation's most sustainable composites leading on traditional natural fiber–based bio-composites, based on its excellent tensile properties, good fiber–matrix adhesion, and many other interesting properties. Many companies have started commercial production of synthetic polymer–based SPCs due to their excellent properties and suggested their materials for different applications. However, a lots of challenges lie in process optimization, recycling, and quality control of synthetic fiber–based SPCs. All-cellulose composites are another category of SPCs. Most of the processing techniques of ACCs are simple. On the other hand, production of ACCs is very economical as the raw materials for ACCs are very cheap and the processing cost is low. The wide variety of cellulosic sources, different solvent systems, and various processing possibilities promise to produce ACCs for various applications. However, a lot of research can be directed to bring slight automation in the processing as well as to scale it up to industrial scale from laboratory trials.

References

1. Huber, T., S. Bickerton, J. Müssig, S. Pang, and M. P. Staiger. Solvent infusion processing of all-cellulose composite materials. *Carbohydrate Polymers* 90, no. 1 (2012): 730–733.
2. Zafeiropoulos, N. E., C. A. Baillie, and F. L. Matthews. A study of transcrystallinity and its effect on the interface in flax fibre reinforced composite materials. *Composites Part A: Applied Science and Manufacturing* 32, no. 3–4 (2001): 525–543.
3. Fakirov, S. Nano-and microfibrillar single-polymer composites: A review. *Macromolecular Materials and Engineering* 298, no. 1 (2013): 9–32.

4. Adak, B., and S. Mukhopadhyay. Effect of pressure on structure and properties of lyocell fabric-based all-cellulose composite laminates. *The Journal of the Textile Institute* 108, no. 6 (2017): 1010–1017.
5. Adak, B., and S. Mukhopadhyay. Jute based all-cellulose composite laminates. *Indian Journal of Fibre & Textile Research (IJFTR)* 41, no. 4 (2016): 380–384.
6. Adak, B., and S. Mukhopadhyay. A comparative study on lyocell-fabric based all-cellulose composite laminates produced by different processes. *Cellulose* 24, no. 2 (2017): 835–849.
7. Matabola, K. P., A. R. De Vries, F. S. Moolman, and A. S. Luyt. Single polymer composites: A review. *Journal of Materials Science* 44, no. 23 (2009): 6213.
8. Duchemin, B. J. C., R. H. Newman, and M. P. Staiger. Structure–property relationship of all-cellulose composites. *Composites Science and Technology* 69, no. 7–8 (2009): 1225–1230.
9. Gindl, W., T. Schöberl, and J. Keckes. Structure and properties of a pulp fibre-reinforced composite with regenerated cellulose matrix. *Applied Physics A* 83, no. 1 (2006): 19–22.
10. http://www.temp.speautomotive.com/SPEA_CD/SPEA2002/pdf/a01.pdf (accessed February 07, 2018).
11. Mukhopadhyay, S., and B. L. Deopura, Single polymer composites from Polyethylene, Sponsored project GAIL India Ltd., Unpublished work.
12. Kmetty, Á., T. Bárány, and J. Karger-Kocsis. Self-reinforced polymeric materials: A review. *Progress in Polymer Science* 35, no. 10 (2010): 1288–1310.
13. http://www.curvonline.com/ (accessed February 07, 2018).
14. https://www.cnbc.com/video/2012/02/23/samsonite-ceo-on-innovation.html (accessed February 07, 2018).
15. http://www.curvonline.com/pdf/datasheet_long.pdf (accessed February 07, 2018).
16. http://www.curvonline.com/pdf/FireLite20120316.pdf (accessed February 07, 2018).
17. http://www.curvonline.com/info/audio.html (accessed February 07, 2018).
18. http://www.thracegroup.com/es/en/technical-fabrics/advanced-fabrics-composites/armordon/ (accessed February 07, 2018).
19. http://www.thracegroup.com/uploads_file/2014/03/17/p18j7inn081vs51d-hq1nt0179kmfm5.pdf (accessed February 07, 2018).
20. http://www.armordon.com/gb/en/applications/amourmilitary/ (accessed February 07, 2018).
21. http://www.ditweaving.com/about_pure.php?page=pure_technology (accessed February 07, 2018).
22. http://www.ditweaving.com/pilot_plant.php?page=facilities_pilot_plant (accessed February 07, 2018).
23. http://www.ditweaving.com/pdf/datasheet%20pure%20sheet.pdf (accessed February 07, 2018).
24. http://www.ditweaving.com/pdf/datasheet%20pure%20tape-fabric.pdf (accessed February 07, 2018).
25. https://omnexus.specialchem.com/tech-library/brochure/self-reinforced-composite-fabric-tds (accessed February 07, 2018).
26. https://omnexus.specialchem.com/centers/self-reinforced-composite-fabrics/processing (accessed February 07, 2018).
27. Ghaderi, M., M. Mousavi, H. Yousefi, and M. Labbafi. All-cellulose nanocomposite film made from bagasse cellulose nanofibers for food packaging application. *Carbohydrate Polymers* 104(2014): 59–65.

28. Pang, J. H., X. Liu, M. Wu, Y. Y. Wu, X. M. Zhang, and R. C. Sun. Fabrication and characterization of regenerated cellulose films using different ionic liquids. *Journal of Spectroscopy* 2014(2014): 1–8.
29. Duchemin, B. J. C., M. P. Staiger, N. Tucker, and R. H. Newman. Aerocellulose based on all-cellulose composites. *Journal of Applied Polymer Science* 115, no. 1 (2010): 216–221.
30. Kistler, S. S. Coherent expanded aerogels and jellies. *Nature* 127, no. 3211 (1931): 741.
31. Yousefi, H., T. Nishino, M. Faezipour, G. Ebrahimi, and A. Shakeri. Direct fabrication of all-cellulose nanocomposite from cellulose microfibers using ionic liquid-based nanowelding. *Biomacromolecules* 12, no. 11 (2011): 4080–4085.
32. Zhang, M. Q., M. Z. Rong, and X. Lu. Fully biodegradable natural fiber composites from renewable resources: All-plant fiber composites. *Composites Science and Technology* 65, no. 15–16 (2005): 2514–2525.
33. Yousefi, H., T. Nishino, A. Shakeri, M. Faezipour, G. Ebrahimi, and M. Kotera. Water-repellent all-cellulose nanocomposite using silane coupling treatment. *Journal of Adhesion Science and Technology* 27, no. 12 (2013): 1324–1334.
34. Tervahartiala, T., N. C. Hildebrandt, P. Piltonen, S. Schabel, and J. P. Valkama. Potential of all-cellulose composites in corrugated board applications: Comparison of chemical pulp raw materials. *Packaging Technology and Science* 31, no. 4 (2018): 173–183.
35. Nilsson, H., S. Galland, P. T. Larsson, E. K. Gamstedt, T. Nishino, L. A. Berglund, and T. Iversen. A non-solvent approach for high-stiffness all-cellulose biocomposites based on pure wood cellulose. *Composites Science and Technology* 70, no. 12 (2010): 1704–1712.
36. Nishino, T., I. Matsuda, and K. Hirao. All-cellulose composite. *Macromolecules* 37, no. 20 (2004): 7683–7687.
37. John, M. J., and S. Thomas. Biofibres and biocomposites. *Carbohydrate Polymers* 71, no. 3 (2008): 343–364.
38. Soykeabkaew, N., C. Sian, S. Gea, T. Nishino, and T. Peijs. All-cellulose nanocomposites by surface selective dissolution of bacterial cellulose. *Cellulose* 16, no. 3 (2009): 435–444.
39. Wan, Y. Z., Y. Huang, C. D. Yuan, S. Raman, Y. Zhu, H. J. Jiang, F. He, and C. Gao. Biomimetic synthesis of hydroxyapatite/bacterial cellulose nanocomposites for biomedical applications. *Materials Science and Engineering: C* 27, no. 4 (2007): 855–864.
40. Svensson, A., E. Nicklasson, T. Harrah, B. Panilaitis, D. L. Kaplan, M. Brittberg, and P. Gatenholm. Bacterial cellulose as a potential scaffold for tissue engineering of cartilage. *Biomaterials* 26, no. 4 (2005): 419–431.
41. Goetz, L., A. Mathew, K. Oksman, P. Gatenholm, and A. J. Ragauskas. A novel nanocomposite film prepared from crosslinked cellulosic whiskers. *Carbohydrate Polymers* 75, no. 1 (2009): 85–89.
42. Qiu, X., and S. Hu. "Smart" materials based on cellulose: A review of the preparations, properties, and applications. *Materials* 6, no. 3 (2013): 738–781.
43. Bárány, T., A. Izer, and J. Karger-Kocsis. Impact resistance of all-polypropylene composites composed of alpha and beta modifications. *Polymer Testing* 28, no. 2 (2009): 176–182.
44. Bárány, T., A. Izer, and A. Menyhárd. Reprocessability and melting behaviour of self-reinforced composites based on PP homo and copolymers. *Journal of Thermal Analysis and Calorimetry* 101, no. 1 (2010): 255–263.

45. Gong, Y., and G. Yang. Single polymer composites by partially melting recycled polyamide 6 fibers: Preparation and characterization. *Journal of Applied Polymer Science* 118, no. 6 (2010): 3357–3363.
46. Romhany, G., T. Barany, T. Czigany, and J. Karger-Kocsis. Fracture and failure behavior of fabric-reinforced all-poly (propylene) composite (Curv®). *Polymers for Advanced Technologies* 18, no. 2 (2007): 90–96.
47. Heim, H. P., W. Tillmann, A. Ries, N. Sievers, B. Rohde, and R. Zielke. Visualisation of the degress of compaction of self-reinforced polypropylene composites by means of ultrasonic testing. *Journal Plastic Technology* 9, no. 6 (2013): 275–94.
48. Kiss, Z., Á. Kmetty, and T. Bárány. Investigation of the weldability of the self-reinforced polypropylene composites. In *Materials Science Forum*, vol. 659, pp. 25–30, eds P. J. Szabó and T. Berecz. Trans Tech Publications, Switzerland, 2010.
49. Bocz, K., T. Bárány, A. Toldy, B. Bodzay, I. Csontos, K. Madi, and G. Marosi. Self-extinguishing polypropylene with a mass fraction of 9% intumescent additive-A new physical way for enhancing the fire retardant efficiency. *Polymer Degradation and Stability* 98, no. 1 (2013): 79–86.
50. Cabrera, N. O., B. Alcock, and T. Peijs. Design and manufacture of all-PP sandwich panels based on co-extruded polypropylene tapes. *Composites Part B: Engineering* 39, no. 7–8 (2008): 1183–1195.
51. Kolesov, I. S., and H. J. Radusch. Multiple shape-memory behavior and thermal-mechanical properties of peroxide cross-linked blends of linear and short-chain branched polyethylenes. *Express Polymer Letters* 2, no. 7 (2008): 461–473.
52. Houshyar, S., R. A. Shanks, and A. Hodzic. Modelling of polypropylene fibre-matrix composites using finite element analysis. *Express Polymer Letters* 3, no. 1 (2009): 2–12.
53. Isayev, A. Self reinforced thermoplastic composite laminate. U.S. Patent 5,275,877, issued January 4, 1994.
54. Mukhopadhyay, S., and B. L. Deopura, GAIL India Ltd, Indian patent pending
55. Huber, T., S. Pang, and M. P. Staiger. All-cellulose composite laminates. *Composites Part A: Applied Science and Manufacturing* 43, no. 10 (2012): 1738–1745.
56. Sharma, M., C. Mukesh, D. Mondal, and K. Prasad. Dissolution of α-chitin in deep eutectic solvents. *Rsc Advances* 3, no. 39 (2013): 18149–18155.
57. Huber, T., J. Müssig, O. Curnow, S. Pang, S. Bickerton, and M. P. Staiger. A critical review of all-cellulose composites. *Journal of Materials Science* 47, no. 3 (2012): 1171–1186.
58. Duchemin, B. J. C., A. P. Mathew, and K. Oksman. All-cellulose composites by partial dissolution in the ionic liquid 1-butyl-3-methylimidazolium chloride. *Composites Part A: Applied Science and Manufacturing* 40, no. 12 (2009): 2031–2037.
59. Swatloski, R. P., J. D. Holbrey, and R. D. Rogers. Ionic liquids are not always green: Hydrolysis of 1-butyl-3-methylimidazolium hexafluorophosphate. *Green Chemistry* 5, no. 4 (2003): 361–363.
60. Adak, B., and S. Mukhopadhyay. Effect of the dissolution time on the structure and properties of lyocell-fabric-based all-cellulose composite laminates. *Journal of Applied Polymer Science* 133, no. 19 (2016).
61. Zhao, Q., R. C. M. Yam, B. Zhang, Y. Yang, X. Cheng, and R. K. Y. Li. Novel all-cellulose ecocomposites prepared in ionic liquids. *Cellulose* 16, no. 2 (2009): 217–226.

62. Adak, B., and S. Mukhopadhyay. All-cellulose composite laminates with low moisture and water sensitivity. *Polymer* 141(2018): 79–85.
63. Piltonen, P., N. C. Hildebrandt, B. Westerlind, J. P. Valkama, T. Tervahartiala, and M. Illikainen. Green and efficient method for preparing all-cellulose composites with NaOH/urea solvent. *Composites Science and Technology* 135(2016): 153–158.
64. Wang, Y., and L. Chen. Impacts of nanowhisker on formation kinetics and properties of all-cellulose composite gels. *Carbohydrate Polymers* 83, no. 4 (2011): 1937–1946.
65. Pullawan, T., A. N. Wilkinson, L. N. Zhang, and S. J. Eichhorn. Deformation micromechanics of all-cellulose nanocomposites: Comparing matrix and reinforcing components. *Carbohydrate Polymers* 100(2014): 31–39.
66. Dormanns, J. W., J. Schuermann, J. Müssig, B. J. C. Duchemin, and M. P. Staiger. Solvent infusion processing of all-cellulose composite laminates using an aqueous NaOH/urea solvent system. *Composites Part A: Applied Science and Manufacturing* 82(2016): 130–140.
67. Benoît, J. D., R. H. Newman, and M. P. Staiger. Phase transformations in microcrystalline cellulose due to partial dissolution. *Cellulose* 14, no. 4 (2007): 311–320.
68. Qin, C., N. Soykeabkaew, N. Xiuyuan, and T. Peijs. The effect of fibre volume fraction and mercerization on the properties of all-cellulose composites. *Carbohydrate Polymers* 71, no. 3 (2008): 458–467.
69. Kalka, S., T. Huber, J. Steinberg, K. Baronian, J. Müssig, and M. P. Staiger. Biodegradability of all-cellulose composite laminates. *Composites Part A: Applied Science and Manufacturing* 59(2014): 37–44.

Index

A

AAROP. *see* Activated anionic ring-opening polymerization (AAROP)
ACCs. *see* All-cellulose composites (ACCs)
Acetobacter xylinum, 125
ACNCs. *see* All-cellulose nanocomposites (ACNCs)
Activated anionic ring-opening polymerization (AAROP), 73
AFM. *see* Atomic force microscopy (AFM)
AGU. *see* D-anhydroglucose units (AGU)
Aliphatic polyamides, 63
All-cellulose composites (ACCs), 231–233
  cellulose
    materials, 124–126
    physical and chemical properties of, 118–119
    polymorphism of, 117–118
    pros and cons of, 120–121
    solid-state structures of, 116–117
    sources of, 119–120
  classification
    matrix phase, 123
    reinforcement type, 124
  derivatized (*see* Derivatized All-Cellulose Composites (DACCs))
  flowchart of, 122
  FTIR spectra analysis, 145–146
  impregnation technique, 133–135
  microstructural analysis of, 148–151
  non-derivatized (*see* Non-derivatized all-cellulose composites)
  non-solvent approach, 138–139
  partial dissolution technique, 135–137
  Raman spectroscopy, 147
  wide-angle x-ray diffraction analysis, 140–143
All-cellulose nanocomposites (ACNCs), 135, 149, 150, 220
  bagasse nanofiber–based, 169
  dynamic mechanical properties of, 186
  films, 188–190, 194
  nanofiber–based, 169
  tensile properties of, 168
All-polyamide (all-PA)
  composites, 66–67, 69
  good recyclability of, 71, 74
D-anhydroglucose units (AGU), 116, 118
Anionic polymerization, 66, 67, 215
Armordon®, 229–230
Aromatic polyamide, 63, 65
Arrhenius equation, 52
Atomic force microscopy (AFM), 148, 149
Attenuated total reflection Fourier-transform infrared (ATR-FTIR), 24

B

Bacteria cellulose (BC), 124, 125, 169
BD laminates. *see* Bidirectional (BD) laminates
Benzylation
  cellulose, 209–211
  of wood, 209, 210
Bicomponent PET multifilament yarns, 90–91
Bidirectional (BD) laminates, 99
Bio-based polyesters, 81–82
Biodegradability, 197–199
Bocell fibers, 164–165, 185–186
Brittle fracture, 179, 183
1-butyl-3-methylimidazolium chloride, 131
1-butyl-3-methylimidazolium hexafluorophosphate, 129

C

CAB. *see* Cellulose acetate butyrate (CAB)
ε-caprolactam (CL), 66, 72
Carbamated cellulose, 216–218
Carbon disulfide (CS_2), 207
Cellulose
  based natural fibers, 115
  benzylated, 209–211
  carbamated, 216–218
  esterified, 211–213
  oxypropylated, 213–216
Cellulose acetate butyrate (CAB), 220
Cellulose nanocrystals (CNCs), 134, 193, 220–221
Cellulose nanowhiskers (CNWs), 124, 168, 173, 186, 191, 197
  in cellulosic matrix, 171
  negative diamagnetic anisotropy of, 139
  reinforced nanocomposite films, 134, 138
Cellulosic material, 133, 135, 148, 210, 238
CL. *see* ε-caprolactam (CL)
CNCs. *see* Cellulose nanocrystals (CNCs)
CNWs. *see* Cellulose nanowhiskers (CNWs)
Compaction pressure, 9–10
Co-PET films, 85, 86
CrI value. *see* Crystallinity index (CrI) value
Crystallinity index (CrI) value, 141
Curv® sheets, 228–229

D

3D. *see* Three-dimensional (3D)
DCB. *see* Double cantilever beam (DCB)
Degree of oxidation (DO), 221, 222
Degree of polymerization (DP), 118
Derivatized All-Cellulose Composites (DACCs), 207–208, 223
  benzylated cellulose, 209–211
  carbamated cellulose, 216–218
  esterified cellulose, 211–213
  fibers or nanofibers, 220–221
  making, 208–209
  non-derivatized ACC *vs.*, 220
  oxypropylated cellulose, 213–216
  solvent system, 208
  techniques of, 223
  TEMPO-mediated oxidation, 219–220
  thermoset, 221–222
  types of, 209
Dielectric spectroscopy, 24
Differential scanning calorimetry (DSC), 52, 82, 100, 105–106, 215
DMTA. *see* Dynamic mechanical thermal analysis (DMTA)
DO. *see* Degree of oxidation (DO)
Double cantilever beam (DCB), 178
Ductile fracture, 179–180
Dynamic mechanical analysis (DMA), 67, 88, 102, 184–187
Dynamic mechanical thermal analysis (DMTA), 50, 52

E

Encyclopaedia Britannica
  polyamide, 63
  polyester, 81
  polyolefins, 33
Equilibrium swelling ratio (ESR), 196
Esterified cellulose, 211–213

F

FEM. *see* Finite element model (FEM)
Fiber/polymer interfaces, 28, 29
Fiber-reinforced composite (FRC), 1, 29, 159, 179
Filament winding, 107–109
Film-casting technique, 69–70
Film-stacking
  method, 49–51, 77
  technique, 55, 67–69, 88, 107
  undercooling melt, 44–46
Finite element model (FEM), 179, 184
Fire-retardant (FR), 107, 108
Flax fiber, 163
Flexural strength, 176–177
FR. *see* Fire-retardant (FR)

Fracture behavior, of all-cellulose composites, 179–184
FRC. *see* Fiber-reinforced composite (FRC)
FTIR spectrum analysis, 145–146

G

Gel-spun PE fibers, 6, 35–36
Gibbs free energy equation, 3

H

HDPE. *see* High-density polyethylene (HDPE)
Heat deflection temperature (HDT), 97
High-density polyethylene (HDPE), 3, 24, 28, 38, 39
High-velocity impact testing, 178–179
Hot-compacted material, 3
Hot-melt processing, 208–209
Hybrid yarn, 86
Hydrogen (H)-bonds network, 125, 139, 169, 194, 197
Hydrophobicity, 195–196

I

In situ polymerization, 72–74
Isotactic polypropylene (iPP), 3, 5, 11–12
  composites cross-section, 54
  fiber/matrix composites, 22–24
Isotactic PP Homopolymer
  α and β crystal forms of, 50–51
  α- *and* β- polymorphs of, 50
Isothermal crystallization, 23

L

Lactic acid. *see* Poly(lactic acid) (PLA)
Linear low-density polyethylene (LLDPE), 38
Lithium chloride/N, N-dimethylacetamide (LiCl/DMAc), 127, 130, 131, 135
LLDPE. *see* Linear low-density polyethylene (LLDPE)
Low-density polyethylene (LDPE), 38–40
Low-velocity impact testing, 178–179
Lyocell fibers, 163, 164, 175, 185, 187

M

Mandrel, 108
Matrix–fiber adhesion, 49
MCC. *see* Microcrystalline cellulose (MCC)
Mercerization, 117, 163
Microcellular injection-molding, 40–43
Microcellular PP SPCs (MPPSPCs), 41–43
Microcrystalline cellulose (MCC), 124, 134, 142, 147, 176, 182, 221
Microencapsulation, 70–72
Microstructural analysis, of all-cellulose composites, 148–151
Montmorillonite (MMT) nanoclays, 71
mPET. *see* Multifilaments matrices (mPET)
MPPSPCs. *see* Microcellular PP SPCs (MPPSPCs)
Multifilaments matrices (mPET), 89

N

Nanocomposite films
  all-cellulose, 124
  self-reinforced poly(lactic acid), 105–106
Nanocrystalline cellulose (NCC), 134, 167–168
Nanofibrillated cellulose (NFC), 219
Natural fibers, 12, 97
  cellulose-based, 115
  classification of, 119
  low-cost, 236
  reinforced bio-composites, 120, 200, 239
  reinforced PLA composites, 97
NCC. *see* Nanocrystalline cellulose (NCC)
NFC. *see* Nanofibrillated cellulose (NFC)
N-methylmorpholine-N-oxide (NMMO), 127, 128, 130, 134, 237
NMR. *see* Nuclear magnetic resonance (NMR)

Non-derivatized all-cellulose composites, 159–160
  biodegradation of, 197–199
  cellulose regeneration, 132–133
  cellulose–solvent systems for, 126–129
  density, 192–193
  dissolution mechanism of cellulose, 129–132
  drug-release property, 197
  drying, 133
  dynamic mechanical analysis, 184–187
  flexural properties, 176–178
  fracture behavior, 179–184
  gas barrier property, 193–195
  hydrophobicity, 195–196
  influencing factors, 160
  optical transparency, 189–192
  permeability test, 193–195
  rule of mixtures, 161
  swelling and re-swelling, 196–197
  tensile properties (*see* Tensile properties of ACCs)
  thermal expansion coefficient, 188, 189
  thermogravimetric analysis, 187–188
  thickness, 193
Nuclear magnetic resonance (NMR), 36
Nylons
  film-casting technique, 69–70
  film-stacking technique, 67–69
  microencapsulation, 70–72
  resin transfer molding, 66–67
  in situ polymerization, 72–74

O

Optical transparency of ACCs, 189–192
Oxypropylated cellulose, 213–216

P

PA. *see* Polyamide (PA)
PBS. *see* Poly (butylene succinate) (PBS)
PDLLA. *see* Poly(D, l-lactic acid) (PDLLA)
PE. *see* Polyethylene (PE)
Peel strength, 179
PE/PE. *see* Polyethylene fibers/polyethylene matrix (PE/PE)
PET. *see* Polyethylene terephthalate (PET)
PLA. *see* Poly(lactic acid) (PLA)
Plant fibers, 119, 120, 208, 211
Plastic deformation, 42
PLLA fibers. *see* Poly (l-lactic acid) (PLLA) fibers
PO. *see* Propylene oxide (PO)
PO-homopolymer (PPO), 215
Polarized optical microscopy (POM), 21
Poly (butylene succinate) (PBS), 103
Poly(D, l-lactic acid) (PDLLA), 100, 106
Poly(lactic acid) (PLA)
  application of, 106–108
  comparative mechanical properties of, 110–111
  fibers, 100–102
  pellets, 105
  self-reinforced composites, 97–102
    importance of temperature, 103–105
    importance of time, 105–106
  SRC/PBS, comparison of, 104
  unidirectional and bidirectional laminates, 99
Polyamide (PA), 63–66
  mechanical properties of, 75–76
  nylons
    film-casting technique, 69–70
    film-stacking technique, 67–69
    microencapsulation, 70–72
    resin transfer molding, 66–67
    in situ polymerization, 72–74
Polyester, 81–82
  mechanical properties of, 93
  starting materials
    bicomponent multifilament yarns, 90–91
    fibers, 82–85
    tapes, 85–86
    uncommingled yarn, 86–89
    woven sheets, 91–93
Polyethylene (PE)
  grades of, 38
    HDPE/LDPE, 39
    UHMPE/LDPE, 40

UHMWPE/HDPE, 39
matrix, 33
molecular modulus, 54–55
oriented fibers and tapes, 36–38
starting materials
different grades of, 38–40
oriented fibers and tapes, 36–38
TENFOR, 34–36
structure of, 38
Polyethylene fibers/polyethylene matrix (PE/PE), 33
Polyethylene terephthalate (PET), 81, 93
fibers, 82, 84
multifilament yarns, 90
self-reinforced, 88–89
sheets of woven, 91
tapes, 85–86
Poly (l-lactic acid) (PLLA) fibers, 100–103
Polymer
fibers testing, 53–54
film casting, 69
matrix, 97
semicrystalline, 63
Polyolefins
mechanical properties of, 56–57
overview, 33–34
Polypropylene (PP), 5–9
different approaches
film stacking, 49–50
microcellular injection-molding, 40–43
undercooling melt film-stacking method, 44–46
woven materials, hot compaction of, 46–49
molecular modulus, 54–55
starting material, importance of, 50–53
structure of, 40
tape, 51–53
testing polymer fibers, 53–54
yarns, 50–51
POM. *see* Polarized optical microscopy (POM)
Porous membrane, 197
PP. *see* Polypropylene (PP)
PPO. *see* PO-homopolymer (PPO)
Propylene oxide (PO), 215
Pulp fibers, 211, 213
PURE®, 230–231

R

Raman spectroscopy, 139, 147
Ramie fiber–based ACCs, 162, 165, 167, 182, 188, 191, 198
Rayon fibers, 164, 175, 176
Recyclability, 2
Resin transfer molding, 66–67
Re-swelling ratio (RSR), 196, 197
Rice husk (RH), 182, 186
Ring-opening polymerization, 63, 97
Room temperature ionic liquids, 129
RSR. *see* Re-swelling ratio (RSR)
RTM
schematic of, 66
structure of, 73
Rule of mixtures, 161

S

Scanning electron microscopy (SEM), 36, 67, 148–150
SCF. *see* Supercritical fluid (SCF)
Self-reinforced composites (SRCs), 49, 100, 207, 211, 235–236
Self-reinforced PET (srPET), 88–89
Self-reinforced poly(lactic acid) (SR-PLA), 105, 106
composites, 97–102
importance of temperature, 103–105
importance of time, 105–106
fibers, 100–102
nanocomposite films, 106, 109
unidirectional and bidirectional laminates, 99
Self-reinforced polymer (SRP), 97
Self-reinforced PP composites (SRPPCs), 49–50
SEM. *see* Scanning electron microscopy (SEM)
Semicrystalline polymers, 19, 63
Shish–kebab crystallization, 6
Single polymer composites (SPCs)
advantages of, 2, 234
applications of, 227–228

Single polymer composites (SPCs) (*cont.*)
  approaches, 1
  commercial applications, 228–233
  compaction pressure, 9–10
  elevation of melting point, 3–6
  fibers/tapes (research), 2–3
  hot-compacted material, 12
  isotactic polypropylene, 21, 23
  melting temperature, 6–7
  morphology, 6
  parameters affecting production of, 12
  quality assurance of, 235
  recycling research, 234
  synthetic polymer, 227–228
  thermal expansion, 7–9
  transcrystallization, 10–11
Solvent infusion processing (SIP), 136, 137, 176
SPCs. *see* Single polymer composites (SPCs)
SPECTRA 1000 fiber, 6, 35
SRP. *see* Self-reinforced polymer (SRP)
srPET. *see* Self-reinforced PET (srPET)
SR-PLA. *see* Self-reinforced poly(lactic acid) (SR-PLA)
Starting materials
  polyester
    bicomponent multifilament yarns, 90–91
    fibers, 82–85
    tapes, 85–86
    uncommingled yarn, 86–89
    woven sheets, 91–93
  polyethylene
    grades of polyethylene, 38–40
    oriented fibers and tapes, 36–38
    TENFOR, 34–36
Supercritical fluid (SCF), 41
Surface-selective dissolution, 135
Swelling of ACC-gel, 196–197
Synthetic polymer-based SPCs
  difficulties and major challenges, 234–235
  future scope, 235–236

T

Tanδ factor, 186
Tape-winding process, schematic of, 51–52. *see also* Polypropylene (PP)
TEM. *see* Transmission electron microscopy (TEM)
TEMPO-mediated oxidation, of cellulosic fibers, 219–220
TENFOR, 34–35
Tensile properties, of ACCs
  isotropic, 167–176
  unidirectional, 160–167, 171–174
Tensile strength, 45
Textile technology, 125
TGA. *see* Thermogravimetric analysis (TGA)
Thermal expansion, 7–9, 188–189
Thermogravimetric analysis (TGA), 187–188
Thermoplastic coating, 12
Thermoset, derivatized all-cellulose composites, 221–222
Three-dimensional (3D), 104–105
T-peel test, fracture surfaces of, 48
Transcrystalline layers (TCL)
  fiber introduction temperature on, 21–22
  function, 19–20
  growth, 22–24, 29
  matrix morphology, 27–29
  surface change, effect of, 24–27
Transmission electron microscopy (TEM), 148, 149

U

UD laminates. *see* Unidirectional (UD) laminates
Ultra-high-molecular-weight polyethylene (UHMWPE), 9–10, 24, 28, 37, 38, 40
Unidirectional (UD) laminates, 99

V

Viscoelasticity, 184

W

Water retention value (WRV), 195, 218
Water vapor permeability (WVP), 194
Wide-angle X-ray diffraction (WAXD), 21, 24, 38, 72, 73, 84, 140–143
Williams–Landel–Ferry model, 52
Woven materials, hot compaction of, 46–49. *see also* Polypropylene (PP)
WRV. *see* Water retention value (WRV)
WVP. *see* Water vapor permeability (WVP)

Y

Young's modulus, 47, 65, 164, 165, 167, 220
  of novel ACC films, 170
  of PLA-SRC, 109